T0310835

ELEMENTS OF THE RANDOM WALK

An Introduction for Advanced Students and Researchers

Random walks have proven to be a useful model in understanding processes across a wide spectrum of scientific disciplines. *Elements of the Random Walk* is an introduction to some of the most powerful and general techniques used in the application of these ideas.

The mathematical construct that runs through the analysis of each of the topics covered in this book, and which therefore unifies the mathematical treatment, is the generating function. Although the reader is introduced to modern analytical tools, such as path integrals and field-theoretical formalism, the book is self-contained in that basic concepts are developed and relevant fundamental findings fully discussed. The book also provides an excellent introduction to frontier topics such as fractals, scaling and critical exponents, path integrals, application of the GLW Hamiltonian formalism, and renormalization group theory as they relate to the random walk problem. Mathematical background is provided in supplements at the end of each chapter, when appropriate.

This self-contained text will appeal to graduate students across science, engineering, and mathematics who need to understand the application of random walk techniques, as well as to established researchers.

JOSEPH RUDNICK earned his Ph.D. in 1970. He has held faculty positions at Tufts University and the University of California, Santa Cruz, as well as a visiting position at Harvard University. He is currently a Professor in the Department of Physics and Astronomy at the University of California, Los Angeles.

GEORGE GASPARI is currently Emeritus Professor at the University of California, Santa Cruz. He has held visiting positions at the University of Bristol, UK; Stanford University; and the University of California Los Angeles. He has been a Sloan Foundation Fellow.

ELEMENTS OF THE RANDOM WALK

An Introduction for Advanced Students and Researchers

JOSEPH RUDNICK
Department of Physics and Astronomy University of California, Los Angeles

GEORGE GASPARI
Department of Physics University of California, Santa Cruz

CAMBRIDGE
UNIVERSITY PRESS

University Printing House, Cambridge CB2 8BS, United Kingdom

Cambridge University Press is part of the University of Cambridge.

It furthers the University's mission by disseminating knowledge in the pursuit of education, learning and research at the highest international levels of excellence.

www.cambridge.org
Information on this title: www.cambridge.org/9780521828918

First published 2004

A catalogue record for this publication is available from the British Library

ISBN 978-0-521-82891-8 Hardback
ISBN 978-0-521-53583-0 Paperback

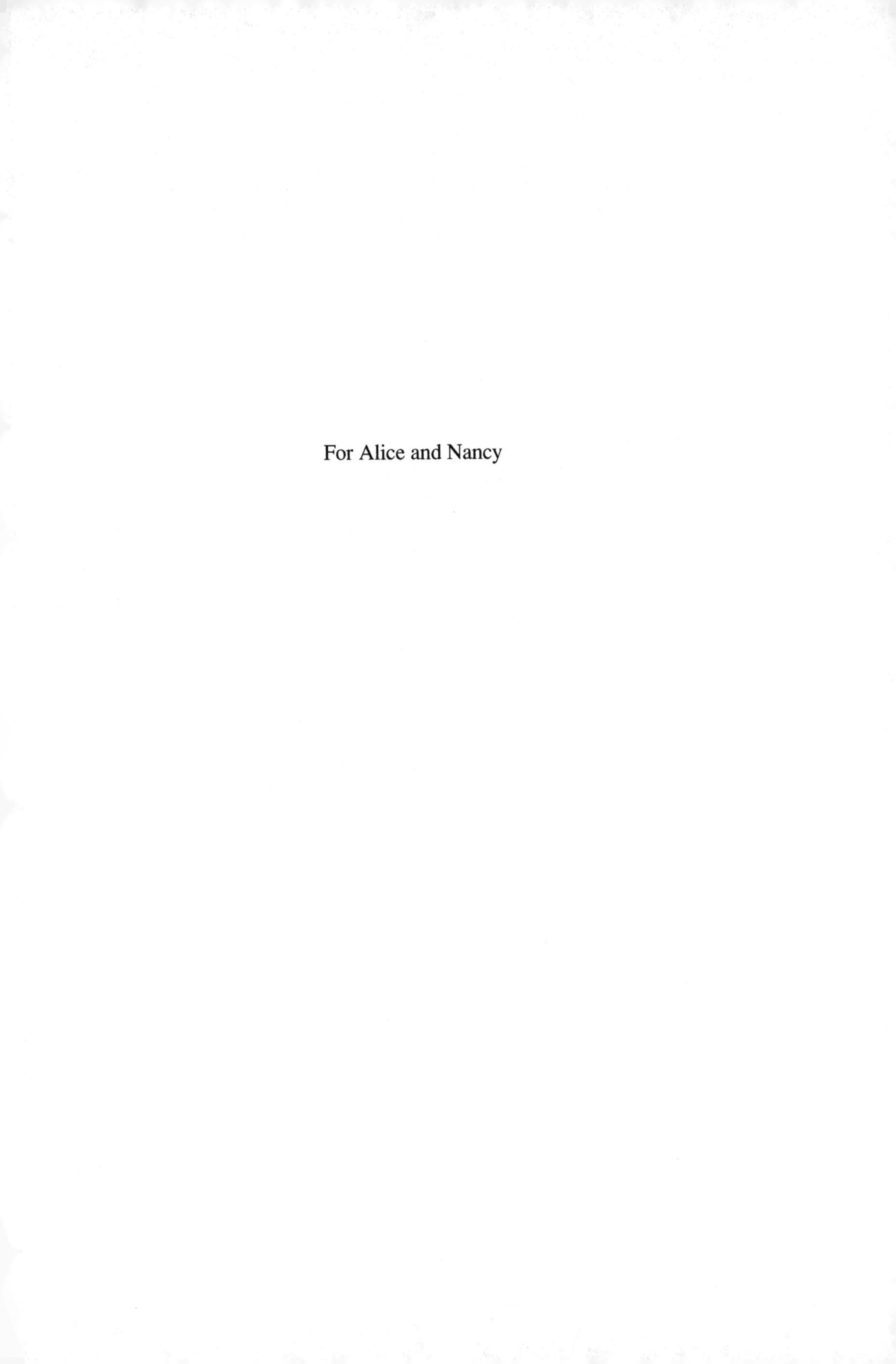

For Alice and Nancy

Contents

Preface

We begin this preface by reporting the results of an experiment. On April 23, 2003, we logged onto INSPEC – the physical science and engineering online literature service – and entered the phrase "random walk." In response to this query, INSPEC delivered a list of 5010 articles, published between 1967 and that date. We then tried the plural phrase, "random walks," and were informed of 1966 more papers. Some redundancy no doubt reduces the total number of references we received to a quantity less than the sum of those two figures. Nevertheless, the point has been made. Random walkers pervade science and technology.

Why is this so? Think of a system – by which we mean just about anything – that undergoes a series of relatively small changes and that does so at random. It is more likely than not that important aspects of this system's behavior can be understood in terms of the random walk. The canonical manifestation of the random walk is Brownian motion, the jittering of a small particle as it is knocked about by the molecules in a liquid or a gas. Chitons meandering on a sandy beach in search of food leave a random walker's trail, and the bacteria *E. coli* execute a random walk as they alternate between purposeful swimming and tumbling. Go to a casino, sit at the roulette wheel and see what kind of luck you have. The height of your pile of chips will follow the rules governing a random walk, although in this case the walk is biased (see Chapter 5), in that, statistically speaking, your collection of chips will inevitably shrink.

We could go on. Random walks play a role in the analysis of the movements of stock prices. *A Random Walk down Wall Street*, by Burton Malkiel has just been published completely revised, following eight previous editions. *Random Walks, Critical Phenomena, and Triviality in Quantum Field Theory*, by Roberto Fernandez *et al.* focuses on the behavior of quantum field theory in higher dimensions. There are also *Random Walks and Other Essays: Ruminations of a so-so Manager*, by René Azurin; *Random Walk: A Novel for a New Age*, by Lawrence Block, and the record *Random Walks Piano Music*, by David Kraehenbuehl and Martha

Braden. Which is to say, the idea of the random walk has seeped into our collective unconscious.

In this book, we hope to acquaint the reader with powerful techniques for the analysis of random walks. The book is intended for the interested student or researcher in physics, chemistry, engineering, or mathematics. It is our hope that the level, style, and content of the book will be appealing and useful to advanced undergraduate students, graduate students, and research scientists in all disciplines. The mathematical techniques used in developing the theory are either explained in the text proper or relegated to supplements at the end of each chapter when it was thought that their inclusion would interrupt the flow of the discussion. We are hopeful that a student with a good understanding of calculus ought to be able to follow much of the analytical manipulations. However, there are instances where more advanced mathematical familiarity would be helpful.

The first five chapters of this book focus on features of a variety of unrestricted walks – that is to say, the trails left by walkers that retain no memory of where they have visited previously – including biased walks, persistent walks, continuous time walks, continuous flow walks, and walks confined to restricted regions of space. The treatment is standard for the most part. However, we attempt to introduce a language and a point of view based on generating functions, which is consistent with a more modern field-theoretical approach to the subject. This method will be fully developed in the later chapters, when we must confront the complications introduced by requiring the walk to be self-avoiding, meaning that the walker's path can never intersect itself. The generating function not only provides for a field-theoretical representation of the walker's statistical behavior, but also allows for the connection to a statistical mechanical model of magnetism. The identification of random walks and magnetism has led to a quantum jump in our understanding of the effects of self-avoidance; it makes available to the theorist the full arsenal of analytical techniques that proved so successful in unraveling the complex properties of systems that undergo continuous phase transitions.

A brief overview of the subjects covered in this book is as follows. Chapter 1 begins with a discussion of the properties of a one-dimensional walk. The chapter is intended as a sort of overture, in that points of view and tricks are introduced that we develop more fully in later chapters. Chapter 2 contains a serious discussion of the meaning, nature, and implementation of the generating function in the context of the random walk. In Chapter 3, we utilize the generating function to investigate various aspects of unrestricted walks, including recurrence, mean number of sites visited, and first passage times. Chapter 4 – which relies heavily on the wonderful little book *Random Walks in Biology*, by Berg, and contains discussions of the effects of boundary conditions on walks – introduces the electrostatic analogy for the analysis of a walk in the steady state. Biased and persistent walks make their appearance in Chapter 5. We generalize the method of treating persistent walks in one dimension to higher dimensional walks and present complete solutions for persistent walks in two and three spatial dimensions. Chapter 6 is devoted entirely to the problem of characterizing the average shape of the trail left by a random walker. We focus on a particular quantitative measure of the shape of an object that is unusually well suited to the kind of analytical tools that now exist for the characterization of the properties of a random walk.

It should be mentioned that, in each of these chapters, we attempt to point out the usefulness of the concepts of models of actual physical and biological processes as the subject is developed. No attempt is made at a comprehensive comparison between predictions of the model and experiments on particular systems. We direct the reader to Weiss' book *Aspects and Applications of the Random Walk* for such detailed comparisons.

The random walk is one of the most important and intuitively appealing examples of a statistical field theory. It is a useful pedagogical model with which to introduce someone to the latest techniques of such a theory, such as Ginzburg–Landau–Wilson effective Hamiltonians, renormalization group theory, and graphical techniques. Finding and understanding the original literature, particularly when one is branching out beyond his or her field of specialty, can be a daunting task. We have tried to reorganize and synthesize the most recent advances in the subject, which in many cases are quite formidable in formulation. In so doing, we intended to make these theories of random walks accessible to those who will find the model useful but are not well versed in the mathematical techniques upon which many recent theoretical developments are based. We set out to accomplish this task in Chapters 7 through 12.

A reading of the table of contents clearly indicates what each of these chapters entails. Here we only point out a few of the features which we found to be particularly interesting. In Chapter 7 we embark on a field theory formulation of the random walk problem *à la* S.F. Edwards, by establishing a path integral expression for the generating function. Once this is accomplished, it is straightforward to generate a perturbation expansion in a quantity which measures "self-avoidance." Doing this allows for a gentle introduction to Feynman-like graphs and an exposition of the associated graphical algebraic techniques. Using rather simple scaling arguments, we clearly demonstrate the crucial role played by dimensionality in determining the behavior of the walker, a feature that is stressed throughout the book. Finally, a mean field theory of self-avoidance is identified which then permits the infinite perturbation series to be summed. The mean field generating function yields an expression for statistical properties of the walk which shows it is exactly equivalent to Flory's treatment of self-avoidance. Chapter 8 contains a brief review of general scaling notions as they apply to the random walk.

In Chapter 9, we establish the connection between the generating function and the correlation function of a fictional magnetic system, the $O(n)$ model. This is an extremely important result, for it brings to the theorist a new set of mathematical tools developed over years by statistical physicists in their study of critical phenomena, which can now be applied to the random walk problem. Critical point scaling, critical exponents, universality, effective Hamiltonians, and renormalization group theory are now at our disposal. These topics are covered in the remaining chapters.

Once the connection between magnetism and random walks has been established, mean field theory and its extensions can be studied in well-known ways. This is done in Chapters 9 and 10. The mean field result is shown to be identical to that found previously, thereby independently demonstrating the correctness of the $O(n)$ representation of random walks. Fluctuations are incorporated in a spin wave approximation, leading to a reasonable physical rendering of the condensed state of the magnetic system as it relates to the random walker. We outline the conceptual underpinnings of the renormalization group approach and present some simple realizations of the method in Chapter 11. Chapter 12 contains a full treatment of the renormalization group as it applies to self-avoiding random walks.

We have interspersed problems throughout each chapter. These are intended to be an aid in understanding the material and to provide a way for the reader to participate in the exploration of the subject. They were not designed to be excessively long or difficult. We suggest students attempt their solution as they work their way through the chapter as a way of gauging their understanding of the material. This book is intended to be a textbook, appropriate for a stand-alone course on random walks or as a supplemental text in a field in which an understanding of random walks is required. For example, this text might prove useful in a course on polymers, or one on advanced topics in statistical mechanics, or even quantum field theory. Since our purpose here is to create a textbook, we have decided not to encumber the presentation with a plethora of footnotes and an associated comprehensive bibliography. It is our hope that the references we have included can be used to track down the original articles dealing with the various aspects of the book. We apologize to all those researchers who have made major contributions to the field and whose work is not cited herein.

This book took shape over several years, and the authors have benefited from the contributions of a number of people. We would like to express our gratitude to Professor Fereydoon Family, who stimulated our initial interest in the subject of random walks, and to Arezki Beldjenna whose contributions to the joint research that underlies much of our chapter on shapes were especially important. We are grateful to the students who sat in on the graduate seminar on random walks at

UCLA for their enthusiasm and useful comments. Special thanks go to Maria R. D'Orsogna for her careful reading of the notes that eventually became the text of this book. The problem of the shape of a random walk was brought to our attention by Professor Vladimir Privman, and for this we extend our heartfelt thanks.

The possibility of our writing a book on random walks was initially raised by Professor Lui Lam. Our decision to publish with Cambridge University Press arose from discussions with Rufus Neal. We thank him for brokering what has turned out to be an enjoyable relationship with CUP, and for introducing us to Simon Capelin, who has proven to be everything we could want in an editor. We thank Fiona Chapman for her careful, and most cheerful, efforts as copy editor. We are also grateful to Professor Warren Esty for permission to reproduce the images used in Figures 1.1 and 1.2.

Finally, we are especially indebted to Professor Peter Young, who carefully read the next-to-final version of this manuscript. His queries, comments, and suggestions resulted in a greatly improved final version.

One of the authors (GG) expresses his appreciation to Tara and Bami Das for making the early years among the best. He is also indebted to Nancy for her loving support from the beginning to the end of this project. The other author (JR) thanks his wife Alice for support, love, advice, and forbearance.

1
Introduction to techniques

This entire book is, in one way or another, devoted to a single process: the random walk. As we will see, the rules that control the random walk are simple, even when we add elaborations that turn out to have considerable significance. However, as often occurs in mathematics and the physical sciences, the consequences of simple rules are far from elementary. We will also discover that random walks, as interesting as they are in themselves, provide a basis for the understanding of a wide range of phenomena. This is true in part because random walk processes are relevant to so many processes in such a wide range of contexts. It also follows from the fact that the solution of the random walk problem requires the use of so many of the mathematical techniques that have been developed and applied in contemporary twentieth-century physics. We'll start out simply, but it won't be long before we enounter aspects to the problem that invite – indeed require – intense scrutiny.

We begin our investigations by looking at the random walk in its most elementary manifestation. The reader may find that most of what follows in this chapter is familiar material. It is, nevertheless, useful to read through it. For one thing, review is always helpful. More importantly, connections that are hinted at in the early portions of this book will play an important role in later discussion.

1.1 The simplest walk

In the simplest example of a random walk the walker is confined to a straight line. This kind of walk is called, appropriately enough, a one-dimensional walk. In this case, steps take the walker in one direction or the other. We will call those two directions "right" and "left." This makes everything easy, as we can now describe the location of the walker by drawing a horizontal line on the page and showing where on the line the walker happens to be. Let's imagine that the walker decides where its next step takes it by flipping a coin. If the coin falls heads up the walker takes a step to the right; if the coin falls tails up the walker takes a step to the left.

The outcome of a flip of the coin is equally likely to be heads or tails, so the walk is clearly unbiased, in that there is no preference for progress to the left or the right.

Suppose the walker has taken N steps. It will have flipped the coin N times. If there were n heads and $N - n$ tails, the walker will have taken n steps to the right and $N - n$ steps to the left. Suppose that each step is l meters long. Then the walker will have moved a distance

$$\begin{aligned} d &= nl - (N - n)l \\ &= l(2n - N) \end{aligned} \tag{1.1}$$

to the right. The walker will thus end up Nl meters to the left of where it started, Nl meters to the right, or somewhere in between.

Before proceeding with the analysis of the behavior of the one-dimensional walker, it is useful to inquire as to the relevance of the notion of such a walker to the real world. As it turns out, the one-dimensional walk models a number of interesting physical and mathematical processes. There is, for example, the diffusive spreading, in one dimension, of a group of molecules or small particles as the result of thermal motion. The one-dimensional walk also represents an idealization of a chain-like polymer whose monomeric units can take on one of two possible conformations. The outcome of a simple game of chance – for instance, one governed by the flip of a coin – can also be described in terms of the eventual location of a one-dimensional random walker. In this last context, one of the first applications of notions eventually associated with the random walk is due to the mathematician de Moivre in the solution of the "gambler's ruin" problem (Montroll and Shlesinger, 1983).

An immediate and fairly obvious question about the walker is the sort one generally asks about the outcome of a random process, and that is with what probability the walker ends up at a given location. That question is equivalent to asking with what probability the walker throws a certain number of heads and tails in N tosses of the coin. Another way to visualize this problem is to consider the act of flipping a coin a "trial" and to call all flips that lead to heads a success. Then, clearly, the above probability is the same as the probability of obtaining n successes in N trials. Note that this interpretation applies to trials with more than two outcomes.

Back to the random walker. Suppose we want to know the probability that the walker has gone a distance d to the right of its original position. In terms of the net distance traveled, $d = l(2n - N)$, the number of heads that were thrown is given by

$$n = \frac{1}{2}\left(\frac{d}{l} + N\right) \tag{1.2}$$

Fig. 1.1. A particular outcome of three flips of a Roman coin displaying an image of Emperor Septimius Severus (AD 193–211). Shown, left to right, is a head, then a tail, then a head.

and the number of tails is

$$N - n = \frac{1}{2}\left(N - \frac{d}{l}\right) \tag{1.3}$$

Now, the probability of throwing a specific sequence that consists of n heads and $N - n$ tails in N coin tosses is equal to $(1/2)^N$. See Figure 1.1. We arrive at the result $(1/2)^N$ for this probability by noting that the probability of either result is one half. Specifying the exact sequence of heads and tails is the same as specifying the sequence of outcomes in a set of N trials, each of which has two possible results. To obtain the probability of this sequence of outcomes, we multiply together the probabilities of each outcome in the sequence. We obtain the probability in this way because each toss of the coin is statistically independent of all other coin flips. That is, the probability of a given flip yielding a head is 1/2, regardless of how all previous tosses turned out.

The probability of throwing n heads and $N - n$ tails in *any* order is $(1/2)^N$ multiplied by the number of sequences of n heads and $N - n$ tails. See, for example, Figure 1.2. This number is simply the binomial coefficient:

$$\binom{N}{n} = \frac{N!}{n!(N-n)!}. \tag{1.4}$$

To derive the combinatorial factor in (1.4) in the case of the coin flips depicted in Figures 1.1 and 1.2, imagine the sequence of flips in Figure 1.1 as an array of coins. Then shuffle the coins in all possible ways. There are $3 \times 2 \times 1 = 3!$ ways of doing this (three possibilities for the leftmost coin, two for the next in line and only one left to place at the far right). However, in shuffling the coins, you have overcounted the number of ways in which heads and tails can turn out. Switching the first and third coins in Figure 1.1 does not change the sequence of heads and tails as both are heads. To compensate for this overcounting, we divide 3! by 2!, the number of ways of shuffling, or permuting, the two heads. This leaves us with three

Fig. 1.2. All outcomes of three flips of the Roman coin in Figure 1.1 in which one of the flips turns up tails and the other two turn up heads.

distinct ways of having two heads and a tail turning up. In general, one computes the number ways in which one can end up with n heads and $N - n$ tails in N flips of a coin by imagining the results of the flip being lined up as in Figure 1.1. Then one shuffles the coins in all possible ways, leading to the factor $N!$, which one divides by the number of ways of shuffling the n heads among themselves and the number of ways of shuffling the $N - n$ tails among themselves (Boas, 1983).

The factor in (1.4) is clearly the one that accounts for all distinct walks. It is not hard to see that the combinatorial factor $N!/n!(N - n)!$ is also equal to the number of different ways that the walker can take n steps to the right and $N - n$ steps to the left. Put another way, the factor $N!/n!(N - n)!$ *is equal to the number of walks that consist of n steps to the right and N − n steps to the left.*

All this leads to the result that the likelihood that the one-dimensional walker will take n steps to the right and $N - n$ steps to the left is

$$\frac{1}{2^N}\frac{N!}{n!(N-n)!} \tag{1.5}$$

Exercise 1.1

How does the result (1.5) change when the coin is "biased" and the probability of a heads at each toss is $p \neq 1/2$? Assume that p does not change from one coin toss to the next.

We can recast our expressions in terms of the location of the walker. Using (1.2) and (1.3), we have for the number of N-step walks that take the walker a distance

d to the right of its original location

$$C(N, d) = N! \Big/ \left(\frac{N + d/l}{2}\right)! \left(\frac{N - d/l}{2}\right)! \tag{1.6}$$

and for the probability that the walker ends up a distance d to the right of its starting point

$$\begin{aligned} P(N, d) &= \frac{1}{2^N} C(N, d) \\ &= \frac{1}{2^N} N! \Big/ \left(\frac{N + d/l}{2}\right)! \left(\frac{N - d/l}{2}\right)! \end{aligned} \tag{1.7}$$

The quantity $P(N, d)$ in (1.7) is called the binomial probability distribution.

The results above allow one to calculate a good deal about the one-dimensional random walk. However, there is much that can be found out without direct recourse to them. In the next few sections we will see how much information can be extracted from fairly simple and general arguments.

1.2 Some very elementary calculations on the simplest walk

The first question that we will answer about the walker is where, on the average, it ends up. Now, the answer to that question is one that you can come up with without having to do an actual calculation. On the average, the walker will take as many steps to the right as it does to the left. The mean distance to the right from the point of departure is equal to zero.

We can do a little better than the above argument. We imagine an ensemble of walkers, performing their walks in lockstep. We note the location of each of them, and we calculate the average position by adding up the locations of all the walkers and dividing by the number of walkers in the ensemble. If the position of the ith walker is x_i, then the mean position of a set of M walkers is

$$\overline{x} = \frac{1}{M} \sum_{i=1}^{M} x_i \tag{1.8}$$

If we denote by $w(x)$ the number of walkers who have ended up at x, then $\overline{x}$ as given by the above equation is also equal to

$$\overline{x} = \frac{1}{M} \sum_{x} x w(x) \tag{1.9}$$

Given that it is equally likely that a walker will take a step to the right as to the left, we know that there will be as many walkers at $-x$ as at x, at least on the average. This means that the two terms $xw(x)$ and $-xw(-x)$ will cancel each other out in the sum in (1.9).

Of course, the cancellation will not be perfect in an actual ensemble of walkers. However, if we consider an enormous number of such ensembles, and take a sort of "super" average, then such cancellation is, indeed, achieved.

This doesn't mean that a given walker inevitably ends up where it started, or even that it ends up near its starting point. To refine our picture of the random walk, let's calculate the mean square displacement from the point of departure. This quantity, x^2, is given for a particular walk by

$$x^2 = \left(\sum_{j=1}^{N} \Delta_j \right)^2 \tag{1.10}$$

Here, Δ_j is the displacement at the jth step.[1] That is to say that at the jth step the walker moves a distance Δ_j to the right. Using

$$x = \sum_{j=1}^{N} \Delta_j \tag{1.11}$$

one can argue that $\overline{x} = 0$ by pointing out that $\overline{\Delta_j} = 0$. This is an alternative derivation of the result immediately above. In the case of $\overline{x^2}$, we expand the right hand side of (1.10) and then average.

$$\begin{aligned}
\overline{x^2} &= \overline{\left(\sum_{j=1}^{N} \Delta_j \right)^2} \\
&= \sum_j \overline{\Delta_j^2} + \sum_{j \neq k} \overline{\Delta_j \Delta_k} \\
&= \sum_j \overline{\Delta_j^2} + \sum_{j \neq k} \overline{\Delta_j} \times \overline{\Delta_k}
\end{aligned} \tag{1.12}$$

The last line in Equation (1.12) expresses the fact that each decision to take a step to the right or left is independent of every other decision. Because $\overline{\Delta_j} = 0$ for all Δ_j, the contribution of the cross terms is equal to zero. We are left with $\sum_j \overline{\Delta_j^2}$. We suppose that the length of each step is the same, so the square of the displacement at each step is equal to a fixed number, which we will call l. This means that

$$\overline{x^2} = Nl^2. \tag{1.13}$$

The root mean square displacement, which measures how far away from its starting point the walker has gotten, on the average, is, then given by

$$\sqrt{\overline{x^2}} = l\sqrt{N} \tag{1.14}$$

[1] If the walker takes a step to the left, then Δ_j is negative

The distance that the one-dimensional random walker has wandered away from its starting point goes as the square root of the number of steps it has taken.

Note that (1.14) implies that the net displacement of a random walker from the origin scales as a fractional power of the number of steps that the walker has taken. We will see that power laws pervade any quantitative discussion of the average behavior of a random walker.

Worked-out example

Generate a formula for $\overline{x^n}$ for arbitrary values of n.

Solution

The quantities on the left hand sides of (1.8) – (1.14) are known as **moments** of the random walk distribution. The quantity $\overline{x^n}$ is referred to as the nth moment of the distribution. The general form of this quantity is

$$\overline{x^n} = \frac{1}{2^N} \sum_{m=0}^{N} ((2m - N)l)^n \frac{N!}{m!(N-m)!} \tag{1.15}$$

Here, we have made use of (1.1) and (1.5). Suppose we were interested in one of the higher moments of the distribution. For example, suppose we wanted to find x^4. How would we go about doing that? We might expand the sum of the Δ_i's, raised to the fourth power, in a version of the calculation indicated in (1.12). This would lead, in due course, to an answer. In fact, calculations of $\overline{x^3}$ and $\overline{x^4}$ using (1.11) are posed as a problem later on in this chapter. However, there is another approach, based on the notion of a *generating function* (Wilf, 1994), that yields a straightforward algorithm for obtaining all moments of the distribution. What we do is note that the quantity $N!/m!(N-m)!$ is a binomial coefficient. That is, this quantity appears as the coefficient of the term w^m in the expansion of $(1+w)^N$ in powers of w:

$$(1+w)^N = \sum_{m=0}^{N} \frac{N!}{m!(N-m)!} w^m \tag{1.16}$$

Replace w by e^y, and divide by 2^N. We have

$$\frac{1}{2^N}\left(1+e^y\right)^N = \frac{1}{2^N} \sum_{m=0}^{N} \frac{N!}{m!(N-m)!} e^{my} \tag{1.17}$$

Let's call the function on the left hand side of (1.17) $g(y)$. Suppose we set $y = 0$ in (1.17). We generate the equality $g(0) = (1/2^N) \sum_{m=0}^{N} N!/m!(N-m)! = (1+1)^N/2^N = 1$.

To find the moments, we take derivatives. For example,

$$
\begin{aligned}
\left.\frac{\mathrm{d}}{\mathrm{d}y}g(y)\right|_{y=0} &= \left.\frac{1}{2^N}\sum_{m=0}^{N}\frac{N!}{m!(N-m)!}me^{my}\right|_{y=0} \\
&= \frac{1}{2^N}\sum_{m=0}^{N}\frac{N!}{m!(N-m)!}m \\
&\equiv \overline{m} \\
&= \left.\frac{\mathrm{d}}{\mathrm{d}y}\frac{1}{2^N}(1+\mathrm{e}^y)^N\right|_{y=0} \\
&= \left.N\mathrm{e}^y\frac{(1+\mathrm{e}^y)^{N-1}}{2^N}\right|_{y=0} \\
&= N\frac{(1+1)^{N-1}}{2^N} \\
&= \frac{N}{2}
\end{aligned} \tag{1.18}
$$

This tells us both that $\overline{m} = N/2$ and that $\overline{m} = \mathrm{d}g(y)/\mathrm{d}y|_{y=0}$. We can readily generalize this result to

$$
\overline{m^n} = \left.\frac{\mathrm{d}^n}{\mathrm{d}y^n}g(y)\right|_{y=0} \tag{1.19}
$$

Making use of this result – and noting that $x = N - 2m$ – we can rewrite the expression for x^n as follows:

$$
x^n = l^n\left.\left(N - 2\frac{\mathrm{d}}{\mathrm{d}y}\right)^n g(y)\right|_{y=0} \tag{1.20}
$$

We can do a bit more. We rewrite the function $g(y)$ as follows:

$$
\begin{aligned}
g(y) &= \mathrm{e}^{Ny/2}\left(\frac{\mathrm{e}^{y/2}+\mathrm{e}^{-y/2}}{2}\right)^N \\
&= \mathrm{e}^{Ny/2}\cosh(y/2)^N
\end{aligned} \tag{1.21}
$$

Then, we note that

$$
\begin{aligned}
&\left(N - 2\frac{\mathrm{d}}{\mathrm{d}y}\right)\mathrm{e}^{Ny/2}\cosh(y/2)^N \\
&= \cosh(y/2)^N\left(N - 2\frac{\mathrm{d}}{\mathrm{d}y}\right)\mathrm{e}^{Ny/2} - 2\frac{\mathrm{d}}{\mathrm{d}y}\cosh(y/2)^N \\
&= 0 - 2\frac{\mathrm{d}}{\mathrm{d}y}\cosh(y/2)^N
\end{aligned} \tag{1.22}
$$

We can carry this calculation out for the case of higher powers of $N - 2\mathrm{d}/\mathrm{d}y$ as applied to the function $g(y)$, and we find in general that

$$\left(N - 2\frac{\mathrm{d}}{\mathrm{d}y}\right)^n g(y) = (-2)^n \frac{\mathrm{d}^n}{\mathrm{d}y^n} \cosh(y/2)^N \tag{1.23}$$

Exercise 1.2

Prove (1.23) by induction, or any other method you like.

This means that

$$\overline{x^n} = (-2l)^n \frac{\mathrm{d}^n}{\mathrm{d}y^n} \cosh(y/2)^N \Big|_{y=0} \tag{1.24}$$

Then,

$$\begin{aligned}
\overline{x^2} &= 4l^2 \frac{\mathrm{d}^2}{\mathrm{d}y^2} \cosh(y/2)^N \Big|_{y=0} \\
&= 4\left(\frac{N\cosh(y/2)^N}{4} + \frac{(-1+N)N\cosh(y/2)^{-2+N}\sinh(y/2)^2}{4}\right)\Bigg|_{y=0} \\
&= Nl^2
\end{aligned} \tag{1.25}$$

as found earlier (see (1.13)).

The next non-zero moment is $\overline{x^4}$. We find

$$\begin{aligned}
\overline{x^4} &= 16l^4 \frac{\mathrm{d}^4}{\mathrm{d}y^4} \cosh(y/2)^N \Big|_{y=0} \\
&= \frac{l^4}{8} N \cosh\left(\frac{y}{2}\right)^{-4+N} ((-4+N)(8+3(-4+N)N) \\
&\quad - 4(-4+(-4+N)(-2+N)N)\cosh(y) + N^3\cosh(2\,y))\big|_{y=0} \\
&= l^4 N(-2+3N)
\end{aligned} \tag{1.26}$$

Note that the average of the fourth power of the distance of a one-dimensional random walker from its point of origin has a term going as the square of the number of steps, N, and also a term going linearly in N. At large values of N, the term going as N^2 dominates the expression. Comparing (1.26) and (1.25), we see that when N is very large

$$\overline{x^4}/(\overline{x^2})^2 \approx 3 \tag{1.27}$$

Exercise 1.3

Use the relationship $x = \sum_{j=1}^{N} \Delta_j$ for a one-dimensional walk – where Δ_j is the displacement to the right of the walker at the ith step – to find $\overline{x^3}$ and $\overline{x^4}$. Make use of the fact that $\overline{\Delta_i^n}$ is equal to zero when n is odd and also that $\overline{\Delta_i^n} = l^n$ when n is even. You will also make use of the fact that $\overline{\Delta_{j_1}^{n_1} \Delta_{j_2}^{n_2} \cdots \Delta_{j_m}^{n_m}} = \overline{\Delta_{j_1}^{n_1}} \times \overline{\Delta_{j_2}^{n_2}} \times \cdots \times \overline{\Delta_{j_m}^{n_m}}$ when $j_1 \neq j_2 \neq \cdots \neq j_m$.

1.3 Back to the probability distribution

Let's return to the combinatorial factor in (1.4). Although the expression is complete, in that we know perfectly well how to calculate each term in it, it is not of immediate analytical use, especially when N, the number of random walk steps, is large. We will now remedy this shortcoming by making use of Stirling's formula for the factorial to produce an expression more amenable to calculation. Stirling's formula is

$$\ln n! \approx n \ln\left(\frac{n}{\mathrm{e}}\right) + \frac{1}{2}\ln(2\pi n) \tag{1.28}$$

An approximation that holds with greater accuracy as n is increased.

1.3.1 Derivation of Stirling's formula

The approximate form that we will use follows from the well-known expression for the gamma function

$$\Gamma(x) = \int_0^\infty w^{x-1} \mathrm{e}^{-w}\, \mathrm{d}w \tag{1.29}$$

The relationship between the gamma function and the factorial is

$$N! = \Gamma(N+1) \tag{1.30}$$

This means

$$N! = \int_0^\infty w^{N} \mathrm{e}^{-w}\, \mathrm{d}w \tag{1.31}$$

The proof of the equality is readily established by integration by parts.

Figure 1.3 is a plot of the integrand in (1.31) when $N = 10$. Superimposed on that plot is a Gaussian, shown as a dashed curve, which will be used to approximate the integrand in the derivation of Stirling's formula. How do we arrive at the approximation by a Gaussian? First, we notice that the integrand is maximized

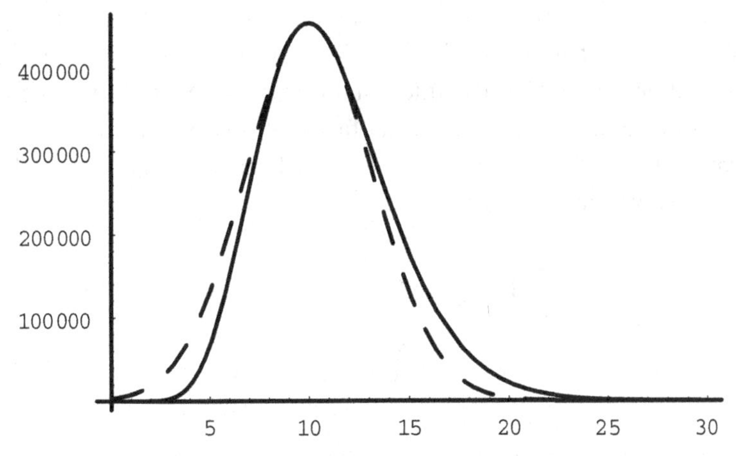

Fig. 1.3. The integrand $w^N e^{-w}$, in (1.31), when $N = 10$, along with the Gaussian curve, shown dashed here, which will be used to approximate that integrand, in the derivation of Stirling's formula.

with $w = N$. We notice this by replacing w in the integrand by $N + \delta$, and then by exponentiating everything in the integrand. This exponentiation yields

$$N! = \int_{-N}^{\infty} \mathrm{d}w \exp\left[N \ln(N + \delta) - N - \delta\right] \tag{1.32}$$

Focusing on the exponent and expanding in powers of δ:

$$\begin{aligned} N \ln(N + \delta) - (N + \delta) &= N \ln N + N \ln(1 + \delta/N) - N - \delta \\ &= (N \ln N - N) + N\left(\frac{\delta}{N} - \frac{1}{2}\frac{\delta^2}{N^2} + \cdots\right) - \delta \\ &= (N \ln N - N) - \frac{\delta^2}{2N} + O\left(\delta^3\right) \\ &\equiv (N \ln N - N) - \frac{(w - N)^2}{2N} + \cdots \end{aligned} \tag{1.33}$$

The Gaussian curve in Figure 1.3 is the function $\exp[10 \ln 10 - 10 - (w - 10)^2/20]$. Suppose we replace the integrand by that Gaussian approximation. We then get the following result for the integration, leading to the factorial

$$N! \approx \sqrt{2\pi N} \exp\left[N(\ln N - N)\right] \tag{1.34}$$

To see how good an approximation it is, let's compare the natural logarithm of 10! with the natural logarithm of the right hand side of (1.34).

$$\ln 10! = 15.1044 \tag{1.35}$$

$$\ln\left(\sqrt{2\pi N} \exp\left[N(\ln N - N)\right]\right) = 15.0961 \tag{1.36}$$

The fractional difference between the right hand side of (1.35) and the right hand side of (1.36) is about five parts in 10^4. The approximation gets even better as N increases. When $N = 100$, the fractional difference between the logarithm of Stirling's formula and the log of the exact factorial is two parts in 10^6.

The formula works if n is large compared to 1. This means that the combinatorial factor in (1.4) is well-approximated by

$$\exp\left[N\ln\left(\frac{N}{\mathrm{e}}\right)+\frac{1}{2}\ln(2\pi N)-n\ln\left(\frac{n}{\mathrm{e}}\right)+\frac{1}{2}\ln(2\pi n)\right.$$
$$\left.-(N-n)\ln\left(\frac{N-n}{\mathrm{e}}\right)+\frac{1}{2}\ln(2\pi(N-n))\right] \quad (1.37)$$

Let $n = \frac{N}{2} + m$ with $m \ll N$. Then the exponent in (1.37) can be expanded as follows. We start with

$$\ln\left(\frac{N}{2}+m\right)=\ln\left(\frac{N}{2}\right)+\frac{2m}{N}+\frac{1}{2}\left(\frac{2m}{N}\right)^2+\cdots \quad (1.38)$$

With the use of this equation, one obtains

$$N\ln 2-\frac{n^2}{2N}-\frac{1}{2}\ln 2\pi N+O\left(\frac{n^3}{N^2},\frac{n}{N}\right) \quad (1.39)$$

so the combinatorial factor has the form

$$\frac{2^N}{\sqrt{2\pi N}}\exp\left[-\frac{n^2}{2N}+O\left(\frac{n^3}{N^2},\frac{n}{N}\right)\right] \quad (1.40)$$

The number m is equal to $n - N/2$, and since by (1.2) $n = d/2l + N/2$, we have $m = d/2l$. This means that the likelihood that a walker will end up a distance d from its point of departure is given by

$$\frac{1}{\sqrt{2\pi N}}\exp\left(-\frac{d^2}{2Nl^2}\right). \quad (1.41)$$

In arriving at (1.41), we have divided by the requisite factor of 2^N to arrive at a probability density that is normalized to one.

The expression in (1.41) is a Gaussian. We will encounter this ubiquitous form repeatedly in the course of our investigation of random walk statistics. It reflects the consequences of the central limit theorem of statistics (Feller, 1968), as it applies to the random walk process.

Exercise 1.4

If N is large, then we can approximate the derivative of the log of $N!$ as follows:

$$\frac{d}{dN} \ln N! \approx \frac{\ln N! - \ln(N-1)!}{N-(N-1)}$$

Use this approximation to derive the leading contribution to Stirling's formula for $N!$ (the first term on the right hand side of (1.28)).

1.4 Recursion relation for the one-dimensional walk

There is another way to investigate the one-dimensional random walk. The number of N-step walks that begin at a given location and end up at another one can be related to the number of $N-1$-step walks that start at the same location and end up nearby. If $C(N;x,y)$ is equal to the number of walks that start at x and end up at y then

$$C(N;x,y) = C(N-1;x,y-l) + C(N-1;x,y+l) \tag{1.42}$$

The formula (1.42) states mathematically that the number of N-step walks starting at x and ending at y is equal to the sum of the number of $N-1$-step walks that start at x and end up at all points adjacent to y. This statement reflects the fact that the last step that a walker takes before ending up at the point y is from a neighboring location. This fact tells us that the sequence of steps taken by our walker, considered as a sequence of random events has the form of a *Markovian* process of the first order (Boas, 1983; Feller, 1968). That is, the probability of occurrence of a given event is independent of the history consisting of all previous events. In future chapters we will encounter higher order Markovian chains, and even some that are non-Markovian.

The recursion relation (1.42) can be utilized to derive a familiar formula for the combinatorial factors. Recall (1.4). This expression is for the number of N-step walks that consist of n steps to the right and $N-n$ steps to the left. Replacing the terms in (1.42) by the equivalent expressions in terms of the combinatorial factors, we have

$$\frac{N!}{n!(N-n)!} = \frac{(N-1)!}{(n-1)!(N-n)!} + \frac{(N-1)!}{n!(N-n-1)!} \tag{1.43}$$

That (1.43) is true can be readily verified. It is also the relation between combinatorial factors that leads to Pascal's triangle.

There is more that can be done with this recursion relation. If one assumes a gentle dependence on the end point y – which is, in fact, the case when N is large – the recursion relation can be approximated by a differential equation. This is accomplished by rewriting (1.42) as follows:

$$\begin{aligned} C(N+1;x,y) &= (C(N;x,y-l)+C(N;x,y+l)-2C(N;x,y)) \\ &\quad +2C(N;x,y) \\ &= l^2\left(\frac{C(N;x,y-l)+C(N;x,y+l)-2C(N;x,y)}{l^2}\right) \\ &\quad +2C(N;x,y) \\ &\approx l^2\frac{\partial^2 C(N;x,y)}{\partial y^2}+2C(N;x,y) \end{aligned} \tag{1.44}$$

Another way to write this equation is

$$C(N+1;x,y)-2C(N;x,y)=l^2\frac{\partial^2 C(N;x,y)}{\partial y^2} \tag{1.45}$$

Suppose we replace $C(N;x,y)$ by $2^N P(N;x,y)$, where $P(N;x,y)$ is the probability that a walker starting out at y ends up at x after N steps. Equation (1.45) becomes

$$P(N+1,x,y)-P(N;x,y)=\frac{l^2}{2}\frac{\partial^2 P(N;x,y)}{\partial y^2} \tag{1.46}$$

Again, imagine that N is large and that $P(N;x,y)$ is a slowly varying function of N. Then, we approximate the right hand side of (1.46) by $\partial P(N;x,y)/\partial N$, and we are left with the equation

$$\frac{\partial P(N;x,y)}{\partial N}=\frac{l^2}{2}\frac{\partial^2 P(N;x,y)}{\partial y^2} \tag{1.47}$$

This equation occupies a place of central importance, not only in the study of the random walk, but also in physical and biological sciences, as well as in engineering. It is the diffusion equation. While there are a variety of ways in which it can be solved, we will write down a solution with the understanding that one can verify that it works. We assume that the reader has encountered the equation and its solution previously, and assert that it is

$$P(N;x,y)=\frac{\alpha}{\sqrt{N}}\exp\left(\frac{(x-y)^2}{2l^2N}\right) \tag{1.48}$$

Exercise 1.5

Verify (1.48) by direct substitution into (1.47).

1.5 Backing into the generating function for a random walk

So far our analysis of the statistics of the one-dimensional random walk problem is the standard introduction to the subject that one finds in any elementary exposition of the process. In the remaining portion of the section we would like to play some games with the solution that we derived in (1.6) as a gentle way of initiating the reader to the more advanced analytical techniques presented in subsequent sections. It may seem at first that the development represents a retreat from the results obtained earlier on, in that the solution is, in a sense, "hidden" in the expressions to be derived. However, what we will have at the end is a set of definitions and relationships that can be generalized into a powerful approach to the properties of more generally defined random walks.

Recall that the expression on the right hand side of that equation is the combinatorial factor in (1.4), with n, the number of steps to the right, expressed in terms of d through (1.2). The factor $N!/n!(N-n)!$ also appears in the expansion of the expression $(1+w)^N$ in powers of w. That is

$$(1+w)^N = \sum_{n=0}^{N} \frac{N!}{n!(N-n)!} w^n \tag{1.49}$$

It is a straightforward exercise to verify that the right hand side of (1.6) is the coefficient of $\mathrm{e}^{\mathrm{i}qd}$ in

$$\left(\mathrm{e}^{\mathrm{i}ql} + \mathrm{e}^{-\mathrm{i}ql}\right)^N \equiv \chi(q)^N \tag{1.50}$$

The above means that the number of N-step walks that take the walker a distance d to the right of its starting point is the coefficient of $\mathrm{e}^{\mathrm{i}qd}$ in the expansion in terms of $\mathrm{e}^{\mathrm{i}q}$ of the expression $(2\cos q)^N$. There is a simple way to obtain that term. One merely performs the integral

$$\frac{l}{2\pi} \int_{-\pi/l}^{\pi/l} \mathrm{e}^{-\mathrm{i}qd} \chi(q)^N \, \mathrm{d}q \tag{1.51}$$

In other words, the number of walks of interest is obtained by taking the inverse Fourier transform of $\chi(q)$ raised to the Nth power. How (1.51) comes about is more fully explained in the next section.

It is possible to regain the Gaussian form for the number of N-step walks by performing the integral above, with the use of a couple of tricks and approximations. First, we rewrite the expression for $\chi(q)$ as follows.

$$\begin{aligned}
\chi(q) &= 2\cos ql \\
&= \mathrm{e}^{\ln(2\cos ql)} \\
&= 2\mathrm{e}^{\ln(\cos ql)} \\
&= 2\mathrm{e}^{\ln(1-q^2l^2/2+\cdots)} \\
&= 2\mathrm{e}^{-q^2l^2/2+O(q^4)}
\end{aligned} \tag{1.52}$$

If we neglect the terms of order q^4 in the exponent,

$$\chi(q)^N \approx 2^N e^{-Nq^2l^2/2} \tag{1.53}$$

Plugging this result into (1.51), we obtain

$$\frac{l}{2\pi}\int_{-\pi/l}^{\pi/l} e^{-iqd} 2^N e^{-Nq^2l^2/2}\, dq \approx \frac{2^N}{\sqrt{2\pi N}} e^{-d^2/2Nl^2} \tag{1.54}$$

This is the same result as is given in (1.41). The integral was evaluated by assuming that the upper and lower limits can be taken to plus and minus infinity. The error that this assumption entails can be shown to be negligible.

There is more. Suppose we perform the geometrical sum

$$g(z,q) = \sum_{N=0}^{\infty} \chi(q)^N z^N = \frac{1}{1 - z\chi(q)} \tag{1.55}$$

The quantity $\chi(q)^N$, entering into (1.51), is the coefficient of z^N in the expansion of the right hand side of (1.55) in powers of the quantity z. The functions defined by expansions of the kind given in (1.55) are called *generating functions*. A large portion of this book is devoted to the exploration of their properties. If we perform a further expansion of the "structure function" $\chi(q)$ in powers of q, we have

$$\chi(q) \approx 2e^{-q^2l^2/2} \approx 2\left(1 - \frac{q^2l^2}{2}\right) \tag{1.56}$$

This expansion is accurate for our purposes as long as the number of steps in the walk, N, is large and the end-to-end distance, d, is not too great. As a practical matter, we require $d \ll N$.

The right hand side of (1.55) is, then, replaced by

$$\frac{1}{1 - 2z + zq^2l^2} \tag{1.57}$$

and the number of N-step walks that displace the walker a distance d from its starting point is equal to the inverse Fourier transform of the coefficient of z^N of the expression in (1.57).

At this point, it may have seemed as if the transformations that have been performed have had the effect of complicating, rather than simplifying, the problem at hand. After all, the inverse Fourier transform is bad enough. The extraction of a coefficient in a power series can be an arbitrarily difficult procedure. In the case at hand, one has the right hand side of (1.55). As it turns out, there are also some general prescriptions and a class of tricks that ease the difficulty of the latter procedure. To find the coefficient of z^N in the function $f(z)$, one simply performs the

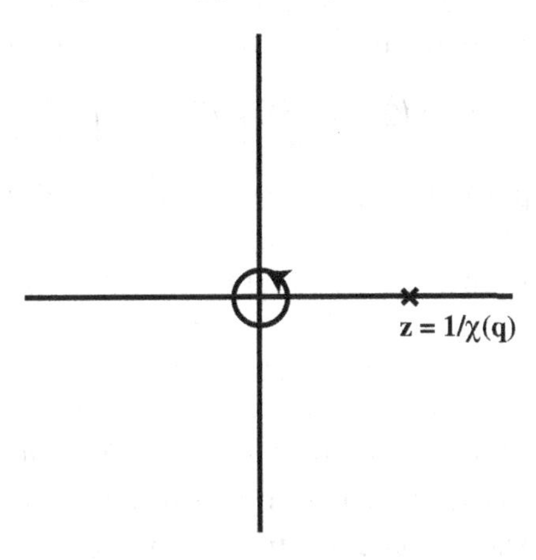

Fig. 1.4. Contour for the integral in (1.58). The pole at $z = 1/\chi(q)$ when $f(z)$ is given by (1.55) is indicated.

following contour integral

$$\frac{1}{2\pi i}\oint_{\mathrm{c}} \frac{f(z)}{z^{N+1}}\,\mathrm{d}z. \tag{1.58}$$

The contour encircles the origin, as shown in Figure 1.4. It is assumed that there is no significant singularity of the function $f(z)$ inside the closed contour. The alert reader will recognize this as Cauchy's formula (Jeffreys, 1972) for $N!\mathrm{d}^N/\mathrm{d}z^N\ f(z)|_{z=0}$. The evaluation of the contour integral when $f(z)$ looks like the right hand side of (1.55) is relatively straightforward. One deforms the contour so that it encloses the pole at $z = 1/\chi(k)$. Applying the formulas that apply to integration around simple poles, one recovers the appropriate coefficient.

There is another way to recover the result for the number of walks from the generating function. This method simplifies the extraction of the desired expression when the number of steps, N, is large. Let the generating function be given by the approximate form in (1.57). The integral over z is accomplished by exponentiating the generating function and the denominator z^{N+1}. The integral is now over the function

$$\exp\left[-(N+1)\ln z - \ln\left(1 - 2z + zq^2l^2\right)\right] \tag{1.59}$$

The integral is evaluated by looking for an extremum in the exponent of the expression above. The equation for the extremum is

$$-\frac{N+1}{z} + \frac{(2 - q^2l^2)}{1 - 2z + zq^2l^2} = 0 \tag{1.60}$$

Because N is large, the denominator of the second term on the left hand side of (1.60) will be small. Writing $z = 1/(2 - q^2l^2) + \delta$, we find δ of order $1/(N+1)$.

Substituting into the expression in (1.59), we find for the integrand

$$\exp\left[(N+1)\ln\left(2-q^2l^2\right)+O(\ln N)\right]$$
$$=\exp\left[(N+1)\ln 2-(N+1)\frac{q^2l^2}{2}+O(\ln N)\right] \quad (1.61)$$

Neglecting the terms of order $\ln N$, and replacing $(N+1)$ with N, we obtain for the integrand

$$2^N \mathrm{e}^{-Nq^2l^2/2} \quad (1.62)$$

This is just the result for $\chi(q)^N$ that is displayed in (1.53). Of course, there is still an integration to do, but the method that is utilized here, known as the *method of steepest descents*, produces the leading order contribution in the asymptotic expansion, i.e. the expansion in powers of $1/N$, of the integral (Jeffreys, 1972). We will comment more extensively on the method of steepest descents in the supplement at the end of this chapter.

1.5.1 More on generating functions: analytical structure

We defined the generating function $g(z,q)$ by (1.55). The Fourier integral of this function defines another generating function

$$\begin{aligned}G(z,d) &= \frac{1}{2\pi}\int_{-\pi/l}^{\pi/l}\mathrm{e}^{\mathrm{i}qd}g(z,q)\,\mathrm{d}q\\ &= \sum_{N=0}^{\infty}\left(\frac{1}{2\pi}\int_{-\pi/l}^{\pi/l}\mathrm{e}^{\mathrm{i}qd}\chi(q)^N\,\mathrm{d}q\right)z^N\\ &= \sum_{N=0}^{\infty}C(N;d)z^N\end{aligned} \quad (1.63)$$

Obviously, $G(z,d)$ "generates" the number of N-step walks that cover a distance d. Now, d ranges from $-N$ to N, so if we sum on d, the left hand side becomes $g(z,0)$, which will then generate the total number of N-step walks:

$$g(z,0)=\sum_{N=0}^{\infty}\Gamma(N)z^N \quad (1.64)$$

where $\Gamma(N)$, the total number of N-step walks, is equal to 2^N in the case of the one-dimensional walker. On summing the series on the right hand side of (1.64), we find

$$g(z,0)=\frac{1}{1-2z} \quad (1.65)$$

As $z\to 1/2$, the function $g(z,0)$ diverges.

One might expect that for more complicated walks this singularity persists, and the analytic behavior of $g(z, 0)$ generalizes to

$$g(z, 0) \sim (z_c - z)^{-\gamma} \tag{1.66}$$

as $z \rightarrow z_c$, hinting at an analogy to the behavior of a system undergoing a phase transition in the vicinity of a thermodynamic critical point. There will be more on this subject later on.

1.5.2 Moments of the random walk

Here we show that the moments of the distribution of random walks follow from the generating function. The key to the calculation of moments lies in the relation of those moments to the Fourier transform of the generating function. From the inverse transform of $g(z, q)$, we can easily derive a general expression for the moments of the random walk. As an example, we carry through the calculation of the second moment, $\overline{d_N^2}$, for an N-step walk. Recall

$$\begin{aligned} g(z, q) &= \sum_{N=0}^{\infty} \chi(q)^N z^N \\ &= \sum_{N=0}^{\infty} \sum_{d=-N}^{N} C(N, d) \mathrm{e}^{\mathrm{i}qd} z^N \end{aligned} \tag{1.67}$$

Taking two derivatives with respect to q:

$$-\frac{\mathrm{d}^2 g}{\mathrm{d}q^2} = \sum_{N=0}^{\infty} \sum_{d=-N}^{N} d^2 C(N, d) \mathrm{e}^{\mathrm{i}qd} z^N \tag{1.68}$$

This implies

$$-\left.\frac{\mathrm{d}^2 g}{\mathrm{d}q^2}\right|_{q=0} = \sum_{N=0}^{\infty} \overline{d^2} z^N \tag{1.69}$$

A general expansion for $\overline{d^2}$ follows from Cauchy's formula

$$\overline{d^2} = -\frac{1}{2\pi \mathrm{i}} \oint \frac{\left(\mathrm{d}^2 g/\mathrm{d}q^2\right)_{q=0}}{z^{N+1}} \, \mathrm{d}z \tag{1.70}$$

Two points should be noted about (1.70). First, the expression is quite general and applies to more complicated paths than are considered here. For instance, the formula is also valid when the walker suffers restrictions, such as the requirement that it not cross its own path. Second, the result for $\overline{d^2}$ is trivially generalized to walks taking place in higher spatial dimensions.

Exercise 1.6
Calculate $\overline{d^2}$ in the limit of large N.

Exercise 1.7
Calculate $\overline{d^4}$ in the limit of large N. Show that $d^4 = 3(d_N^2)^2$.

1.6 Supplement: method of steepest descents

There is a general method for extracting the coefficient of z^N in the power series expansion of a function of z that is free of singularities in the vicinity of $z = 0$. Let's see how it works. If $f(z)$ is the function of interest, and if it admits the power series expansion

$$f(z) = \sum_{N=0}^{\infty} C_N z^N \tag{1:S-1}$$

then a fundamental theorem of complex analysis tell us that

$$C_N = \frac{1}{2\pi \mathrm{i}} \oint \frac{f(z)}{z^{N+1}} \,\mathrm{d}z \tag{1:S-2}$$

where the closed integration contour encircles the origin and does not enclose any singularity of $f(z)$. In the limiting regime $N \gg 1$ we extract the leading contribution to the integral above by rewriting the integral as follows

$$C_N = \frac{1}{2\pi \mathrm{i}} \oint \mathrm{e}^{\ln(f(z))-(N+1)\ln(z)} \,\mathrm{d}z \tag{1:S-3}$$

Then, we imagine that we can deform the contour so that it passes through an extremum of the real part of the exponential, at $z = z^*$ along a path of steepest descent. The actual requirement on the path is that the imaginary part of the function in the exponential in (1:S-3) is fixed and equal to its value at the extremum point (Jeffreys, 1972). If we are smart enough to have chosen the correct extremum, and if N is large enough, the dominant contribution to the integral will be given by the integrand at the extremum, multiplied by the result of an integration in the immediate vicinity. Let's apply the method to an integral of the general form

$$\frac{1}{2\pi \mathrm{i}} \oint h(z) \mathrm{e}^{Np(z)} \,\mathrm{d}z, \quad N \gg 1 \tag{1:S-4}$$

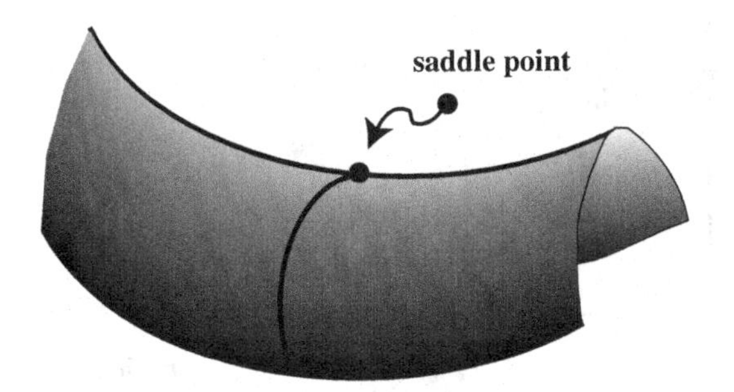

Fig. 1.5. The saddle point.

where $\mathrm{d}p(z)/\mathrm{d}z = 0$ at $z = z^*$, and $h(z)$ varies slowly in the vicinity of the extremum. The quantity z^* is actually a saddle point. Figure 1.5 displays the general behavior of the real part of a function in the neighborhood of its saddle point and a path of steepest descent. It is also assumed that $\mathrm{d}^2 p(z)/\mathrm{d}z^2 \neq 0$ at $z = z^*$. The method applies equally well when the second derivative vanishes at the extremum point, but is a bit more complicated. The major contribution to the integral comes in the vicinity of the extremum when the path can be deformed to pass along a line of steepest descent – that is, valley to valley through z^*. Then the integral is dominated by the factor $\mathrm{e}^{N\,\mathrm{Re}\,p(z_*)}$ in the integral as $N \to \infty$. All other contributions along the path are vanishingly small. In this region we can write

$$p(z) = p(z^*) + \frac{1}{2}\frac{\mathrm{d}^2 p}{\mathrm{d}z_*^2}\left(z - z^*\right)^2 \tag{1:S-5}$$

and

$$h(z) = h(z^*) \tag{1:S-6}$$

Then, the leading term in the asymptotic expansion of the integral is given by

$$\frac{h(z^*)}{2\pi\mathrm{i}}\mathrm{e}^{Np(z^*)}\int_{\mathrm{c}} \mathrm{e}^{\frac{N}{2}p''(z^*)(z-z^*)^2}\,\mathrm{d}z \tag{1:S-7}$$

and the remaining integral is easily evaluated along the steepest descent line. Let's see how it works. In anticipation that the second derivative of $p(z)$ is complex at z^*, we write

$$p''(z^*) = -\left|p''(z^*)\right|\mathrm{e}^{2\mathrm{i}(\beta-\pi/2)} \tag{1:S-8}$$

and the integrand becomes

$$\exp\!\left(N\left|p''(z^*)\right|\mathrm{e}^{2\mathrm{i}(\beta-\pi/2)}(z-z^*)^2\right) \tag{1:S-9}$$

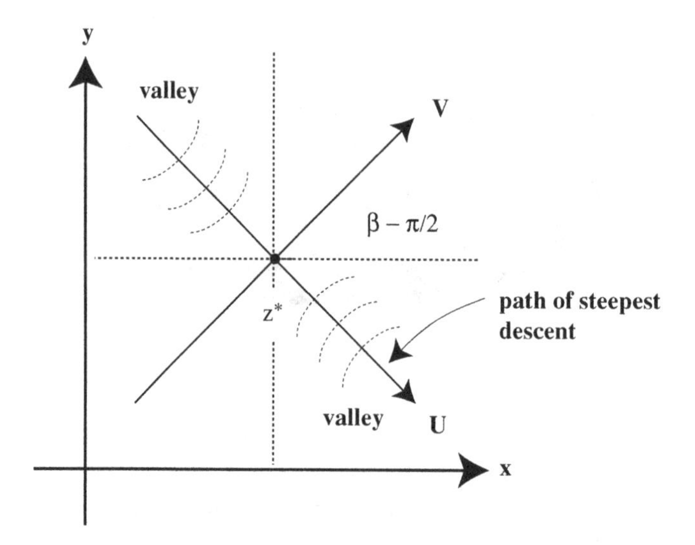

Fig. 1.6. The new path of integration.

This reduces rather nicely by introducing the new variable

$$\chi = (z - z^*)e^{i(\beta - \pi/2)} \\ \equiv U + iV \tag{1:S-10}$$

The relation between the two coordinate systems can be seen in Figure 1.6.

In terms of the new variable χ, our integral looks like

$$e^{-i(\beta-\pi/2)} \int_{-\infty}^{\infty} e^{-N|p''|\chi^2/2}\, d\chi \tag{1:S-11}$$

with the steepest descent path being along the real axis of the new coordinate system. Putting everything together, we obtain for the integral

$$\frac{1}{2\pi i} \oint h(z)e^{Np(z)}\, dz \sim \frac{h(z^*)}{2\pi i} e^{p(z^*)} \sqrt{\frac{2\pi}{N\,|p''|}} e^{i(\pi/2-\beta)} \tag{1:S-12}$$

As an example of how the method works, take $f(z)$ to be given by $f(z) = (z_c - z)^{-\alpha}$. Identifying corresponding terms in (1:S-3) and our general form we see that $h(z) = 1$ and $p(z) = -\alpha \ln(z_c - z) - (N+1)\ln z$. The extremum equation is

$$\frac{dp(z)}{dz} = 0 = \frac{\alpha}{z_c - z} - \frac{N+1}{z} \tag{1:S-13}$$

When N is large, we can write $z^* = z_c - \delta$, where the quantity δ is small. Then, to a good approximation, $\delta = \alpha z_c/(N+1)$, and the integrand at the extremum point is equal to

$$\exp\left[-\alpha \ln\left(\frac{\alpha}{N+1} z_c\right) - (N+1)\ln(z_c) + \alpha\right] + \cdots$$
$$= (N+1)^{\alpha} z_c^{-N-\alpha-1} \alpha^{-\alpha} e^{\alpha} + \cdots \qquad (1\text{:S-}14)$$

where the ellipses refer to corrections that are symptotically smaller than the terms displayed. To complete the identification note that

$$p''(z^*) = \frac{(N+1)^2}{\alpha z_c^2}$$

Combining this result with (1:S-14), we find for the coefficient of z^N in $(z_c - z)^{-\alpha}$

$$\frac{1}{2\pi} \alpha^{-\alpha+1/2} e^{\alpha} (N+1)^{\alpha-1} z_c^{-N-\alpha} \qquad (1\text{:S-}15)$$

This result is close to the exact one, which will be derived in section 2.4.

2

Generating functions I

2.1 General introduction to generating functions

This book makes extensive use of generating functions. In that respect the discussions here are consistent with the approach that condensed matter physicists generally take when calculating properties of the random walk as it relates to problems of contemporary interest. This chapter is devoted to a discussion of the generating function and to an exploration of some of the ways in which the generating function method can be put to use in the study of the random walk. Many of the arguments in later chapters will call upon techniques and results that will be developed in the pages to follow. Thus, the reader is strongly urged to pay close attention to the discussion that follows, as topics and techniques that are introduced here will crop up repeatedly later on.

2.1.1 What is a generating function?

The generating function is a mathematical stratagem that simplifies a number of problems. Its range of applicability extends far beyond the mathematics of the random walk. Readers who have had an introduction to ordinary differential equations will have already seen examples of the use of the method of the generating function in the study of special functions. The generating function also plays a central role in graph theory and in the study of combinatorics, percolation theory, classical and quantum field theory and a myriad of other applications in physics and mathematics.

Briefly, a generating function is a mathematical expression, depending on one or more variables, that admits a power series expansion. The coefficients of the expansion are the members of a family, or sequence, of numbers or functions. The expansion can thus be said to "generate" the sequence of functions or numbers. A particularly simple example of a generating function is the exponential $e^x = 1 + x/1! + x^2/2! + \cdots + x^n/n! \cdots$. The coefficient of x^n is $1/n!$, so the exponential generates the sequence of numbers $1/n!$. Now, unless we do something clever with this result, it is not particularly useful or interesting. A far more powerful relationship is the following:

$$\frac{1}{\sqrt{z^2 - 2z\,x + 1}} = \sum_{n=0}^{\infty} P_n(x) z^n \tag{2.1}$$

where $P_n(x)$ is the nth order Legendre polynomial. The expression on the left hand side of (2.1) is, thus, a generating function for Legendre polynomials. Similar identities allow one to generate other special functions that are solutions of important linear differential equations of mathematical physics. Many of the key properties of special functions are easily demonstrated with the use of generating functions.

Why use generating functions?

Equation (2.1) provides a clear indication of one of the advantages of the generating function, in that it represents a prescription for the construction of the special function that it generates. If you need to know the fifth order Legendre polynomial, $P_5(x)$, you can find out what its form is by expanding the right hand side to fifth order in y. In this sense, the generating function encapsulates all information with regard to the functions that it generates. Furthermore, it contains this information in an exremely compact form.

In addition, the generating function can often be derived with relative ease. This will certainly be true in the case of the generating function of ordinary random walks. Finally, as we will soon see, there are cases in which the generating function is the quantity of ultimate interest.

Exercise 2.1

Use (2.1) to derive the expression for $P_2(x)$.

Exercise 2.2

Use (2.1) to verify that $\int_{-1}^{1} P_n(x)\,dx$ is equal to 2 if $n = 0$ and that the integral is equal to zero if $n \neq 0$.

Exercise 2.3

Making use of the result

$$\int_{-1}^{1} \frac{dx}{\sqrt{\left(y^2 - 2yx + 1\right)\left(z^2 - 2zx + 1\right)}} = \frac{1}{\sqrt{yz}} \ln\left(\frac{1+\sqrt{yz}}{1-\sqrt{yz}}\right)$$

find

$$\int_{-1}^{1} P_n(x) P_m(x)\,dx$$

for arbitrary integers, m and n.

Exercise 2.4

The integral order Bessel functions satisfy

$$e^{(z-1/z)\rho/2} = \sum_{n=-\infty}^{\infty} J_n(\rho) z^n$$

Show that

$$J_n(\rho) = \frac{1}{2\pi} \int_{-\pi}^{\pi} e^{i(\rho \sin\theta - n\theta)} \, d\theta$$

Statistical mechanics: the generating function as the ultimate goal of an investigation

Generating functions play an important role in statistical mechanics, though they are not usually identified as such. They appear in relations connecting partition functions of the various ensembles. For instance, the grand canonical partition function (also called the grand partition function), Q, can be expressed in the following way:

$$Q(\alpha) = \sum_N Z(N) e^{-\alpha N} \tag{2.2}$$

In this equation, the quantity N is the number of particles a system contains, and $Z(N)$ is the partition function of that system in the canonical ensemble: $Z(N) = \sum_r e^{-E_r/kT}$. The quantity α is given in terms of thermodynamic variables by $\alpha = -\mu/kT$, μ being the chemical potential. Defining the fugacity z via $z = e^{-\alpha}$ one is led to the identification of Q as a generating function having the form $Q(z) = \sum_N Z(N) z^N$. This function generates the canonical partition function for a system of N particles.

Notice that in the previous example one's goal might well have been the generating function itself – in this case the grand partition function – and not necessarily the sequence it generates. There are other physical systems whose properties are most naturally characterized in terms of an appropriate generating function, which thus becomes the object of primary focus. For example, consider a collection of long chain polymers in chemical equilibrium with a reservoir of monomers. The monomer bath controls the molecular weights of the polymers by establishing a chemical potential, μ – or, alternatively, a fugacity z – per monomeric unit. The ensemble of interest then consists of a distribution of polymers having a variety of molecular weights, and the Boltzmann factor associated with an n-unit polymer contains the factor z^n. Random walk statistics are appropriate to the statistical mechanics of this ensemble of polymers. These statistics derive directly from the generating function.

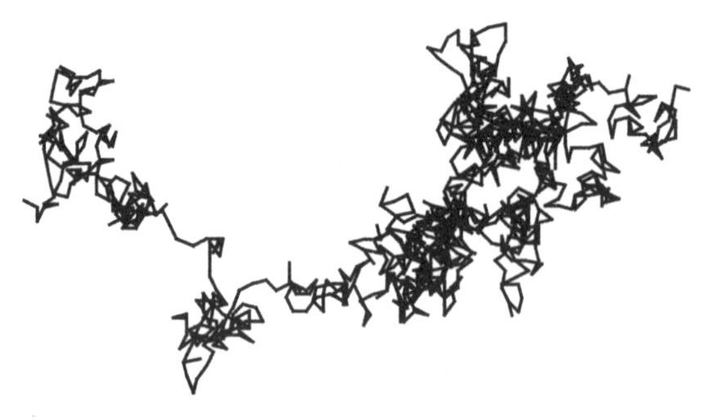

Fig. 2.1. The trail left by a random walker.

One may also have occasion to consider an ensemble of *dynamical* walkers ("Brownian particles") that face the possibility of extinction at each step. If the probability of surviving to take a subsequent step is equal to z, then the probability that a given walker will make exactly n steps before changing or losing its identity is equal to $z^n(1-z)$. The properties of an ensemble of such walkers also follows directly from an appropriate generating function.

Exercise 2.5

The partition function of the monatomic ideal gas is given by

$$Z(N) = \frac{1}{N!}\left(CT^{3/2}V\right)^N$$

where T is the absolute temperature, V is the volume occupied by the gas and C is a collection of physical constants. Find the grand partition function of the monatomic ideal gas.

2.1.2 Generating function for a random walk on a lattice

Let's turn our attention back to the process of primary interest and take a look at a quantity directly relavant to random walks. Suppose $C(N;\vec{x},\vec{y})$ is equal to the number of N-step walks that start out from the point $\vec{x}$ and end up at the point $\vec{y}$. Figure 2.1 illustrates the trail left by such a walker. Then, consider the following function of the variables z, $\vec{x}$ and $\vec{y}$:

$$G(z;\vec{x},\vec{y}) = \sum_{N=0}^{\infty} z^N C(N;\vec{x},\vec{y}) \tag{2.3}$$

By the definition of the function $G(z;\vec{x},\vec{y})$, the coefficient of z^N in its expansion as a function of z is clearly $C(N;\vec{x},\vec{y})$. Thus, $G(z;\vec{x},\vec{y})$ is the generating function for the family of functions $C(N; \vec{x}, \vec{y})$. Note that the probability that an N-step walk begins at $\vec{x}$ and ends at $\vec{y}$ is simply $C(N; \vec{x}, \vec{y})$ divided by the total number of walks. Thus, the family of functions generated by $G(z; \vec{x}, \vec{y})$ is also proportional to the probability, $P_N(\vec{x}, \vec{y})$, that an N-step walk starting at $\vec{x}$ ends up at $\vec{y}$.

If what we have been saying about the advantages of generating functions holds in the case at hand, we ought to be able to derive an explicit expression for the generating function $G(z; \vec{x}, \vec{y})$. As we will see, this task is well within our abilities. Indeed, in the previous chapter we derived explicit expressions for both $G(z; x, y)$ and $C(N; x, y)$ in the case of a walker proceeding along a straight line. There, we found

$$C(N, \vec{x}, \vec{y}) = \frac{1}{2\pi} \int_{-\pi}^{\pi} \mathrm{e}^{-\mathrm{i}q(x-y)} \chi(q)^N \, \mathrm{d}q \tag{2.4}$$

and

$$\begin{aligned} G(z; x, y) &= \sum_{N=0}^{\infty} C(N; x, y) z^N \\ &= \frac{1}{2\pi} \int_{-\pi}^{\pi} \frac{\mathrm{d}q}{1 - \chi(q)z} \mathrm{e}^{-\mathrm{i}q(x-y)} \end{aligned} \tag{2.5}$$

where

$$\chi(k) = 2\cos q \tag{2.6}$$

Here, q is measured in units of the inverse of the step length, l.

Not surprisingly, the results for the one-dimensional walk are easily extended to walks taking place in d spatial dimensions. The trick is to utilize a recursion relation that relates the statistics of an N-step walk to those of a walk consisting of $N + 1$ steps. To keep the discussion simple we will assume that the walker is confined to the vertices of a three-dimensional, simple cubic lattice and that steps are along the bonds joining nearest neighbor sites (see Figure 2.2 for a picture of a walker on such a lattice (Pólya, 1919)). Such walkers are called "Pólya walkers," as he was the first to investigate the properties of walkers on a lattice (Pólya, 1921).

Consider, now, an N-step walk on this lattice. The next-to-last step in this walk must have left the walker at a site adjacent to its final destination. This means that the number of walks that start at site $\vec{x}$ and end at site $\vec{y}$ is equal to the sum of the numbers of $N - 1$-step walks starting at $\vec{x}$ and ending at sites adjacent to $\vec{y}$.[1] This implies the following relationship between $C(N; \vec{x}, \vec{y})$ and the corresponding

[1] Actually, the desired quantity is the number of $N - 1$-step walks that start at $\vec{x}$, land at points adjacent to $\vec{y}$ and then go on to $\vec{y}$ at the Nth step. We assume that any walker that is at an adjacent site can move to y in one step, so the two numbers are the same.

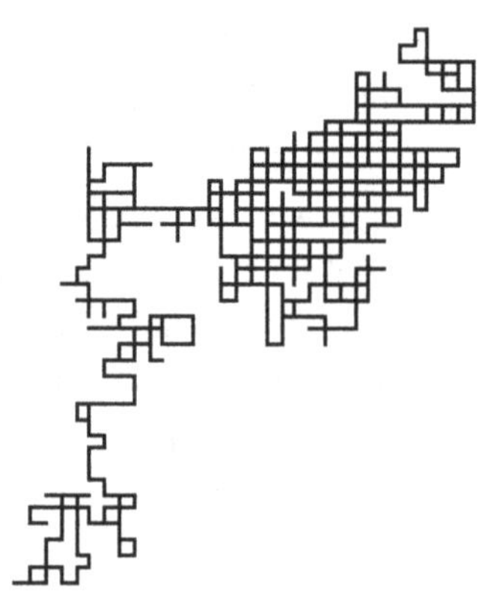

Fig. 2.2. A walker on a two-dimensional square lattice.

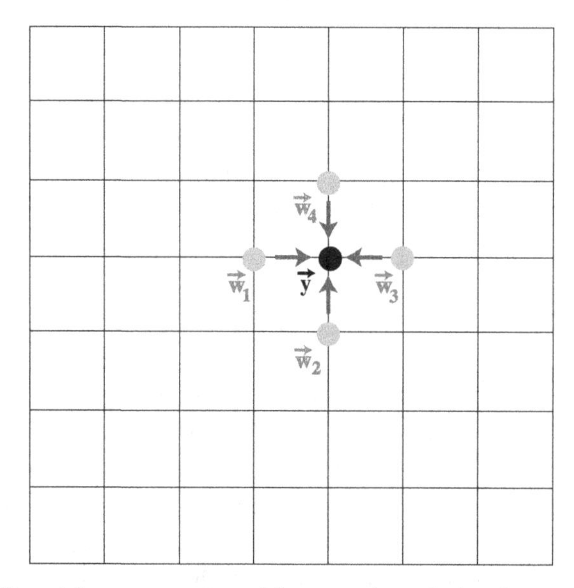

Fig. 2.3. Pictorial representation of the recursion relation for random walks.

quantities for walks starting at $\vec{x}$ and ending at sites $\vec{w}_i$, where the subscript indexes the sites that are nearest neighbors to $\vec{y}$.

$$C(N; \vec{x}, \vec{y}) = \sum_i C(N-1; \vec{x}, \vec{w}_i) \tag{2.7}$$

Figure 2.3 is a pictorial representation of the process quantified in (2.7). Equation (2.7) holds for all positive, non-zero values of N. When $N = 0$, (2.7) is replaced by the identity $C(0, \vec{x}, \vec{y}) = \delta_{\vec{x},\vec{y}}$, where δ is the discrete delta function.

As the next step, we convert (2.7) to a relation between generating functions. We do this by multiplying (2.7) by z^N and summing over N, from $N = 0$ to $N = \infty$.

This yields

$$\sum_{N=1}^{\infty} z^N C(N;\vec{x},\vec{y}) = \sum_{N=1}^{\infty} z^N \left(\sum_i C(N-1;\vec{x},\vec{w}_i)\right)$$
$$= z\sum_i \sum_{N=0}^{\infty} z^N C(N;\vec{x},\vec{w}_i)$$
$$= z\sum_i G(z;\vec{x},\vec{w}_i) \tag{2.8}$$

The left hand side of (2.8) is almost equal to $G(z;\vec{x},\vec{y})$. It is simply missing the term $z^0C(0;\vec{x},\vec{y}) = \delta_{\vec{x},\vec{y}}$. If we add this term to the equation, we end up with

$$G(z;\vec{x},\vec{y}) = z\sum_i G(z;\vec{x},\vec{w}_i) + \delta_{\vec{x},\vec{y}} \tag{2.9}$$

Notice that the new equation involves only the argument z. We are, however, not quite at the desired point. The relation above is non-local in space, in that the left hand side of (2.9) depends on the position vector $\vec{y}$, of the end-point of the walk, while the right hand side of the equation contains the locations, $\vec{w}_i$, of the neighboring sites. This means that it is not yet possible to solve the equation by simple algebraic means. Nevertheless, the recursion relations summed up in (2.7) and (2.9) have a number of interesting and useful features, which we will explore at the end of this section.

Translational symmetry and spatial Fourier transform

The recursion relation for the generating function of random walks yields readily to an analysis based on the translational symmetry of the lattice (Montroll, 1956). To take advantage of this symmetry, we first note that both $C(N;\vec{x},\vec{y})$ and the generating function $G(z;\vec{x},\vec{y})$ depend on the position vectors $\vec{x}$ and $\vec{y}$ *only through their difference*, $\vec{x}-\vec{y}$. This means that we can write the generating function in the form $G(z;\vec{x}-\vec{y})$. Next, we perform a spatial Fourier expansion of (2.9). If

$$g(z;\vec{q}) \equiv \sum_{\vec{x}} G(z;\vec{x}-\vec{y})\mathrm{e}^{\mathrm{i}\vec{q}\cdot(\vec{x}-\vec{y})}, \tag{2.10}$$

then, multiplying (2.9) by $\mathrm{e}^{\mathrm{i}\vec{q}\cdot(\vec{x}-\vec{y})}$ and summing over $\vec{x}$,

$$g(z;\vec{q}) = z\sum_{\vec{x},\vec{w}_i} \mathrm{e}^{\mathrm{i}\vec{q}\cdot(\vec{x}-\vec{y})}G(z;\vec{x}-\vec{w}_i) + 1$$
$$= z\sum_{\vec{x}-\vec{w}_i,\vec{w}_i} \mathrm{e}^{\mathrm{i}\vec{q}\cdot(\vec{x}-\vec{w}_i)}G(z;\vec{x}-\vec{w}_i)\mathrm{e}^{\mathrm{i}\vec{q}\cdot(\vec{w}_i-\vec{y})} + 1$$
$$= z\, g(z;\vec{q})\sum_{\vec{w}_i} \mathrm{e}^{\mathrm{i}\vec{q}\cdot(\vec{w}_i-\vec{y})} + 1$$
$$\equiv z\, g(z;\vec{q})\chi(\vec{q}) + 1. \tag{2.11}$$

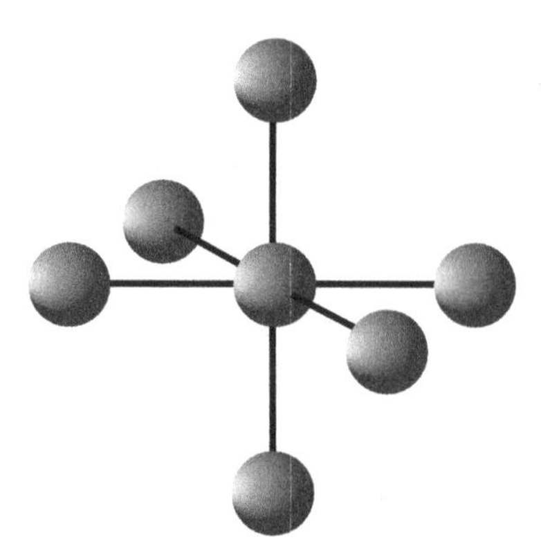

Fig. 2.4. A site and its nearest neighbors in a simple cubic lattice.

The last line introduces the *structure function* of the lattice, $\chi(\vec{q})$. Notice that the extraction of an explicit expression for the generating function $g(z;\vec{q})$ requires nothing more than simple algebra.

Simple cubic lattice

Up to this point, we have not referred to the specific structure of the lattice on which the walker moves. It works for walks in one dimension, where we previously encountered $\chi(q)$, or in d dimensions and a lattice of any symmetry. If the walk takes place on a three-dimensional cubic lattice, then the six nearest neighbors are as shown in Figure 2.4, and the final sum in (2.11) is easy to perform. We find for the structure function

$$\chi(\vec{q}) = 2\left(\cos(q_x a) + \cos(q_y a) + \cos(q_z a)\right) \tag{2.12}$$

The quantity a in the above equation is the distance between adjacent points on the cubic lattice. The sum above is over nearest neighbors to the site $\vec{x} = 0$.

Exercise 2.6

Calculate the structure function, $\chi(\vec{q})$, for body-centered cubic and face-centered cubic lattices. For information that will help you to obtain a solution, see Supplement 2 at the end of this chapter.

If we now perform the algebraic steps necessary to solve (2.11) we obtain

$$g(z;\vec{q}) = \frac{1}{1 - z\chi(\vec{q})} \tag{2.13}$$

And, we're done. The right hand side of (2.13), with $\chi(\vec{q})$ as given by the right hand side of (2.12), contains all the information there is to be had about the number of random walks between any two vertices of a simple cubic lattice – assuming, of course, that all of the walker's steps are along the bonds joining nearest neighbors, and they occur with equal probability. That is, the walk is *unbiased*.

In this way, we have obtained, in a reasonably small number of relatively simple steps, the function that generates an important piece of information regarding the statistics of random walks. The advertised advantages of the generating function are not illusory.

2.1.3 Asymptotic behavior of $C(N;\vec{x},\vec{y})$ for long walks: $N \gg 1$

Now, let's see what we can do with the expression we've obtained. First, we can immediately extract the spatial Fourier transform of $C(N;\vec{x}-\vec{y})$. It is the coefficient of z^N in the power series expansion of the left hand side of (2.13), and expressing $1/(1-z\,\chi(\vec{q}))$ in powers of z is easy:

$$\frac{1}{1-z\,\chi(\vec{q})} = 1 + z\chi(\vec{q}) + \cdots + \left(z\chi(\vec{q})\right)^n + \cdots \tag{2.14}$$

The spatial fourier coefficient

$$c(N;\vec{q}) = \sum_{\vec{x}-\vec{y}} e^{i\vec{q}\cdot(\vec{x}-\vec{y})} C(N;\vec{x}-\vec{y}) \tag{2.15}$$

is given by

$$\begin{aligned} c(N;\vec{q}) &= \left(\chi(\vec{q})\right)^N \\ &= \left(2\left(\cos(q_x a) + \cos(q_y a) + \cos(q_z a)\right)\right)^N \end{aligned} \tag{2.16}$$

We want a result for the number of walks between two given points on the *real* lattice, and the expression on the right hand side of (2.16) only takes us part of the way towards our goal. Note that we take the lattice to possess cubic symmetry and explicitly introduce the nearest neighbor lattice distance, a. The next step is to extract $C(N;\vec{x}-\vec{y})$ from its inverse fourier transform.

$$C(N;\vec{x}-\vec{y}) = \left(\frac{a}{2\pi}\right)^3 \int_{\vec{q}\in \mathrm{BZ}} e^{-i\vec{q}\cdot(\vec{x}-\vec{y})} c(N;\vec{q})\, d\vec{q} \tag{2.17}$$

where the integration over $\vec{q}$ is confined to the first Brillouin zone[2] of the simple cubic lattice, e.g.$-\frac{\pi}{a} < q_i \le \frac{\pi}{a}$, $i = x, y, z$. Now, the integral is not particularly easy to carry out if N is large and one wants an exact result. However, the most important contributions to the integral can be extracted with the use of a few simple

[2] Supplement 2 at the end of this chapter.

tricks. First, we exponentiate a portion of $c(N;\vec{q})$ as follows:

$$\begin{aligned}&\left(\cos(q_x a)+\cos(q_y a)+\cos(q_z a)\right)^N\\&\quad=\exp\left[N\,\ln\left(\cos(q_x a)+\cos(q_y a)+\cos(q_z a)\right)\right]\end{aligned} \tag{2.18}$$

Then, we expand the exponent as a power series in the components of $\vec{q}$:

$$\begin{aligned}&\exp\left[N\,\ln\left(\cos(q_x a)+\cos(q_y a)+\cos(q_z a)\right)\right]\\&\quad=\exp\left[N\,\ln\left(3-\frac{1}{2}\left(a^2\left(q_x^2+q_y^2+q_z^2\right)+O(\vec{q}^{\,4})\right)\right)\right]\\&\quad=3^N\mathrm{e}^{-\frac{N}{6}\left(a^2|\vec{q}|^2\right)+\cdots}\end{aligned} \tag{2.19}$$

We are now ready to proceed.

The Gaussian limit

We will start out by dropping the terms of order $\vec{q}^{\,4}$ and higher in the exponent. The consequences of this are far-reaching. As we will soon see, the fact that the exponent can be truncated at quadratic order in $\vec{q}$ leads to the result that long random walks have *universal* behavior. That is, the statistics of walks with N steps, as exemplified by the quantity $C(N;\vec{x}-\vec{y})$, are, when $N \gg 1$, independent of details, such as on what kind of lattice the walker must take its steps, or even whether or not the walker is confined to a lattice. This is a particular example of the type of universality that has played such an important role in the study of critical-point behavior in equilibrium thermodynamics.

Before discussing in more detail the approximations that have been made in the truncation, we will use the exponential in the last line of (2.19), without the ellipses, to extract the leading order behavior of the quantity $C(N;\vec{x}-\vec{y})$ when $N \gg 1$. The integration we have to perform is over a simple Gaussian. Since the Gaussian decays to zero when $\vec{q}$ has a magnitude significantly greater than $1/\sqrt{N}$, and because we are interested in the limit $N \gg 1$, we can ignore the restriction of $\vec{q}$ to the first Brillouin zone. The Gaussian factor ensures that there is no contribution to the integral of any consequence outside of a small region in the center of the zone. Inserting (2.19) into (2.17) we are left with

$$\begin{aligned}C(N;\vec{x}-\vec{y})&=\left(\frac{a}{2\pi}\right)^3\int 6^N\mathrm{e}^{-\mathrm{i}\vec{q}\cdot(\vec{x}-\vec{y})-\frac{N}{6}\left(a^2|\vec{q}|^2\right)}\,\mathrm{d}\vec{q}\\&=6^N\left(\frac{3}{2\pi N}\right)^{3/2}\mathrm{e}^{-3|\vec{x}-\vec{y}|^2/2Na^2}\end{aligned} \tag{2.20}$$

The expression for the number of random walks between two locations has a Gaussian form. This result is correct as long as the distance from the starting point to the location at which the walker finds itself at last is not too large compared to $\sqrt{N}a$.

The last line in (2.20) is straightforwardly obtained by completing squares in the exponent[3] and utilizing standard results for the Gaussian integral in the integration over $\vec{q}$.

2.1.4 Sources of universality

Let's see just how much of a mistake we are making in discarding the term of order $|\vec{q}|^4$ in (2.19). To keep things simple we will imagine that $\vec{q}$ is a scalar. This does not compromise the validity of our arguments. We now assess the consequences of neglecting the term q^4 in the integral

$$\int_{-\infty}^{\infty} \mathrm{e}^{-\mathrm{i}qr-\frac{a^2N}{6}q^2-bNq^4}\,\mathrm{d}q \tag{2.21}$$

We can do this in a variety of ways. The simplest is to expand the exponent with respect to the quartic term. If we retain the first order term in this expansion, the integral is of the form

$$\begin{aligned}&\int_{-\infty}^{\infty} \mathrm{e}^{-\mathrm{i}qr-\frac{a^2N}{6}q^2}\left(1-bNq^4\right)\mathrm{d}q\\ &=\sqrt{\frac{6\pi}{Na^2}}\mathrm{e}^{\frac{-3r^2}{2aN}}\left[1-\frac{27b}{Na^4}\left(1-6\frac{r^2}{Na^2}+3\left(\frac{r^2}{Na^2}\right)^2\right)\right]\end{aligned} \tag{2.22}$$

The error we have made is, apparently, unimportant as long as the ratio b/a^4 is small compared to N, and we are interested in what happens when N is very large. A more systematic investigation supports the conclusion that we just reached. In the limit $N \gg 1$ the number of walks between two points is given in terms of a Gaussian. Note also that according to (2.22) the Gaussian form for $C(N; \vec{x}, \vec{y})$ holds under the assumption that $r = |\vec{x} - \vec{y}|$ is not large compared to $\sqrt{N}$. This relationship is constistent with $\overline{r} \sim \sqrt{N}$ and $N \gg 1$. As the separation between the end-points approaches the number of steps, a different limiting form applies. The new form reflects the fact that there can be no paths longer than the one left by a walker who has gone in a straight line.

Notice that in the integration we performed in (2.20) we were able to ignore the fact that the lattice has a Brillouin zone. In discarding terms in the exponent beyond those of quadratic order in the wave-vector $\vec{q}$, we also ignored the other consequence of the crystalline structure of the lattice – the detailed dependence on $\vec{q}$ of $\chi(\vec{q})$. In fact, the *only* property of $\chi(\vec{q})$ that we made use of is the fact that it has a power-series expansion in $\vec{q}$, and that the two lowest order terms in

[3] For details, see Supplement 1 at the end of this chapter.

that expansion are the zeroth and second order ones. Any lattice that has this last property will yield a result for $C(N; \vec{x} - \vec{y})$ that, in the limit $N \gg 1$, has the same form as the second part of (2.20).

To recapitulate, the source of the universal Gaussian form for the number of random walks is two-fold. First, there is the fact that the structure function $\chi(\vec{q})$ has the power-series expansion $c_0 + c_2 q^2 + \cdots$. Second, there is the fact that at large number of steps, N, the only terms in that expansion that matter are the two lowest ones. If $\chi(\vec{q})$ expressed as a power series had different terms at lowest order, i.e. $\chi(\vec{q}) = c_0 + c' q^{1.3} + \cdots$, then the number of walks would not be expressible as a Gaussian.

Off-lattice walkers

We can go even further. The walker need not be restricted to a lattice. Imagine that the probability of a step having a length and direction described by the displacement vector $\vec{r}$ is described by a distribution that does not confine the walker to the lattice. If, for example, the probability that $\vec{r}$ lies in the infinitesimal volume $\mathrm{d}^3 r \equiv \mathrm{d}x\, \mathrm{d}y\, \mathrm{d}z$ centered at $\vec{r}_0$ is equal to $p(|\vec{r}_0|)\, \mathrm{d}^3 r \equiv p(r_0)\, \mathrm{d}^3 r$, then we can derive a result for $C(N; \vec{x} - \vec{y})$ in exactly the same way as we did for the case of a walker on a lattice. Following a procedure similar to that taken for walks on a lattice, but allowing the steps to be taken in a continuum, the quantity $\chi(\vec{q})$ will be given by

$$\chi(\vec{q}) = \int \mathrm{e}^{\mathrm{i}\vec{q}\cdot\vec{r}} p(r)\, \mathrm{d}^3 r \tag{2.23}$$

Suppose the probability density $p(r)$ admits of a moment expansion. That is, write

$$\int \mathrm{e}^{\mathrm{i}\vec{q}\cdot\vec{r}} p\left(|\,\vec{r}\,|\right) \mathrm{d}\vec{q} = \int \left(1 + \mathrm{i}\vec{q}\cdot\vec{r} + \frac{1}{2!}\left(\mathrm{i}\vec{q}\cdot\vec{r}\right)^2 + \cdots\right) p(r)\, \mathrm{d}^3 r \tag{2.24}$$

and suppose that the integrals involving the first several terms in the expansion of the exponential converge. Then, because the integral of $\mathrm{i}\vec{q}\cdot\vec{r}\; p(r)$ vanishes by symmetry, we have the same sort of expansion for $\chi(\vec{q})$ as was the case for the walker on a lattice, and the Gaussian form for $C(N; \vec{x} - \vec{y})$ follows.

An important consideration in the specification of $p(r)$ is that this probability is normalized, in that

$$\int p(r)\, \mathrm{d}^3 r = 1 \tag{2.25}$$

This tells us that $\chi(0) = 1$.

Exercise 2.7
Suppose

$$p(r) = \sqrt{\frac{2}{\pi}} r_0^{-3} \mathrm{e}^{-r^2/2r_0^2}$$

Find $\chi(\vec{q})$ and use this result to obtain the probability distribution for the end-to-end distance of an N-step off-lattice random walk.

Worked-out example
Suppose

$$p(r) = \frac{r_0}{\pi^2 (r^2 + r_0^2)^2} \tag{2.26}$$

Find $\chi(\vec{q})$ and use this result to obtain the probability distribution for the end-to-end distance of an N-step off-lattice random walk.

Solution
Here, we have the case of a walker in which the likelihood of very long steps falls off sufficiently slowly that standard results do not follow. In particular $\int r^2 p(r)\, \mathrm{d}^3 r$ is not a finite quantity. As we will see, we do not end up in this case with a Gaussian distribution for walks with N steps in the regime $N \gg 1$.

Our first task is to find the function $\chi(q)$. Making use of (2.23), we have

$$\begin{aligned}\chi(q) &= \int \mathrm{e}^{\mathrm{i}\vec{q}\cdot\vec{r}} \frac{r_0}{\pi^2 (r^2 + r_0^2)^2}\, \mathrm{d}r \\ &= \frac{4}{\pi} \int_0^\infty \frac{\sin qr}{q} \frac{r r_0}{(r^2 + r_0^2)^2}\, \mathrm{d}r \end{aligned} \tag{2.27}$$

The last line in (2.27) follows from the integration over θ and ϕ in

$$\int \mathrm{e}^{\mathrm{i}\vec{q}\cdot\vec{r}} f(r)\, \mathrm{d}^3 r = \int_0^{2\pi} \mathrm{d}\phi \int_0^\pi \sin\theta\, \mathrm{d}\theta \int_0^\infty r^2\, \mathrm{d}r\, \mathrm{e}^{\mathrm{i}qr\cos\theta} f(r) \tag{2.28}$$

The integration over r in (2.27) can be worked out with the use of contour methods. For example, we write

$$\begin{aligned}\frac{4}{\pi} \int_0^\infty \frac{r r_0 \sin qr}{q} \frac{1}{(r^2 + r_0^2)^2}\, \mathrm{d}r &= -\frac{4 r_0}{\pi} \frac{1}{q} \frac{\mathrm{d}}{\mathrm{d}q} \int_0^\infty \frac{\cos qr}{(r^2 + r_0^2)^2}\, \mathrm{d}r \\ &= -\frac{2 r_0}{\pi} \frac{1}{q} \frac{\mathrm{d}}{\mathrm{d}q} \int_{-\infty}^\infty \frac{\mathrm{e}^{\mathrm{i}qr}}{(r^2 + r_0^2)^2}\, \mathrm{d}r \end{aligned} \tag{2.29}$$

To find the result of the integration, one closes the integration contour in the upper half r-plane, around the second order pole at $r = \mathrm{i}r_0$. This yields the result

$$\chi(q) = \mathrm{e}^{-qr_0} \tag{2.30}$$

The next step in the calculation is to reconstruct the probability that an N-step walk takes the walker a distance r from its point of origin. We know that this quantity is given by

$$\begin{aligned} \frac{1}{(2\pi)^3} \int \chi(q)^N \mathrm{e}^{\mathrm{i}\vec{q}\cdot\vec{r}} \mathrm{d}^3 q &= \frac{1}{(2\pi)^3} 4\pi \int_0^\infty \mathrm{e}^{-Nqr_0} \frac{q \sin qr}{r} \, \mathrm{d}q \\ &= \frac{Nr_0}{\pi^2 (r^2 + N^2 r_0^2)^2} \end{aligned} \tag{2.31}$$

The right hand side of the top line in (2.31) follows from (2.28). The integration leading to the last line in (2.31) is easily carried with the use of a table of integrals or by exploiting de Moivre's theorem and replacing the sine function by complex exponentials.

The end result for the probability distribution of the N-step off-lattice walk controlled by the single-step probability distribution (2.26) looks very much like the original probability distribution. It is clear that in this case, the "average" distance of the walker from the point of origin scales like Nr_0. That is, the "average" distance away from its point of origin of the walker in question scales as N, rather than as $N^{1/2}$. This average distance cannot be obtained in terms of a second moment of the distribution $P(N; 0, \vec{r})$ as given by the last line of (2.31) because that distribution function does not have a finite second moment.

2.1.5 Gaussian walks, the continuum limit, and diffusion

There are other ways to derive the Gaussian distribution for long random walks. Those who have some familiarity with the central limit theorem of statistics will recognize the Gaussian form as a special application of that theorem. The form is most revealing in that it also makes clear the connection with diffusion. To see this, recall (2.7), the recursion relation we derived previously for $C(N; \vec{x}, \vec{y})$. This equation can be turned into a recursion relation for the probability that the walker that starts out at $\vec{x}$ and takes N steps will end up at $\vec{y}$. If we call this probability $P(N; \vec{x}, \vec{y})$, then the new recursion relation is

$$P(N; \vec{x}, \vec{y}) = \frac{1}{2d} \sum_{\text{nns}} P(N-1; \vec{x}, \vec{y} - \vec{w}_i) \tag{2.32}$$

where nns means nearest neighbor sites. The above equation holds for unbiased walks on the simple cubic lattice in d dimensions. The unbiased nature of the walker is manifested in the factor $1/2d$ in (2.32); the quantity $2d$ is the coordination number of the lattice, e.g. the number of nearest neighbors. What is implied is that the probability of the walker's landing on any nearest neighbor site in one step is equally likely. It is also clear from (2.32) how to generalize the recursion relationship for *biased* walks, that is walks for which the probability of landing on neighboring sites is p_i instead of $1/2d$, then the correct equation to represent this situation would be

$$P(N;\vec{x},\vec{y}) = \sum_{\text{nns}} P(N-1;\vec{x},\vec{y}-\vec{w}_i)p_i \tag{2.33}$$

Back to the unbiased walkers on a simple cubic lattice: the recursion relation can also be written as a difference equation

$$P(N;\vec{x},\vec{y}) = \frac{1}{6}\sum_{j=1}^{3}\left[P(N-1;\vec{x},\vec{y}-\hat{e}_j a) + P(N-1;\vec{x},\vec{y}+\hat{e}_j a)\right] \tag{2.34}$$

where the $\hat{e}_j$'s are unit vectors along the positive x, y and z axes. Adding and subtracting $P(N-1;\vec{x},\vec{y})$ casts the recursion relation in a very suggestive form:

$$\begin{aligned} P(N;\vec{x},\vec{y}) - P(N-1,\vec{x},\vec{y}) = \frac{1}{6}\sum_{j=1}^{3}\big[&P(N-1;\vec{x},\vec{y}-\hat{e}_j a) \\ &+ P(N-1;\vec{x},\vec{y}+\hat{e}_j a) - 2P(N-1;\vec{x},\vec{y})\big] \end{aligned} \tag{2.35}$$

This final result is exact for the specific random walk we are considering. However, when the number of steps is very large ($N \gg 1$) the above functions are very slowly varying, $P(N-1;\vec{x},\vec{y}) \approx P(N;\vec{x},\vec{y})$, and can therefore be approximated by the leading terms in a Taylor series expansion, i.e.

$$P(N;\vec{x},\vec{y}) - P(N-1;\vec{x},\vec{y}) \approx \frac{\partial}{\partial N}P(N;\vec{x},\vec{y}) \tag{2.36}$$

Carrying this procedure out for both sides of (2.35) transforms the difference equation into a partial differential equation

$$\frac{\partial}{\partial N}P(N;\vec{x},\vec{y}) = \frac{a^2}{6}\nabla^2 P(N;\vec{x},\vec{y}) \tag{2.37}$$

which, as the reader might well realize, is the diffusion equation. This is not an unexpected result. The well-known solution to this equation is given by the left hand side of our (2.20), with the factor 6^N removed. It is now abundantly clear that the approximations leading to the Gaussian form for the number of walks washes out the discrete nature of the lattice and produces the continuum limit. It is worth noting that the generating function itself also satisfies a partial differential equation in the continuum limit. Applying the same arguments as above reduces (2.9) to the following partial differential equation for $G(z;\vec{x},\vec{y})$:

$$za^2\nabla^2 G(z;\vec{x},\vec{y}) - (2dz-1)\,G(z;\vec{x},\vec{y}) = \delta_{\vec{x},\vec{y}} \tag{2.38}$$

which rightfully earns the function G the status of a Green's function. We will have more to say about the Green's function nature of G in a subsequent chapter, where we develop the random walk problem along the lines of a field theory similar to what has been done in statistical mechanics in the study of critical phenomena.

Exercise 2.8
Verify by direct substitution that the function

$$P(N;\vec{x},\vec{y}) = \left(\frac{3}{2}\right)^{3/2} \frac{1}{a^3\pi^{3/2}N^{3/2}} \exp\left(-\frac{3r^2}{2Na^2}\right)$$

where $r = |\vec{x}-\vec{y}|$, is a solution to (2.37). Furthermore, show that this probability distribution is normalized, in that

$$\int P(N;\vec{x},\vec{y})\,\mathrm{d}^3x = 1$$

Exercise 2.9
Derive a diffusion equation for a biased one-dimensional walk. Under what conditions will the probability density be of the Gaussian form?

2.1.6 Summing up

Let's review what we have managed to achieve. Accepting that the number of N-step walks that start and end at specified points is a quantity worth knowing (and it most definitely is), we have shown that the generating function provides a measure of random walk statistics that contains all information with regard to this quantity for a walk with any number of steps. Furthermore, we have shown that this latter mathematical entity admits of a relatively straightforward derivation. Its form has been explicitly derived in the special case of a walker stepping randomly along the edges of a cubic structure. While we have made some restrictive assumptions about the walk – for instance, that it takes place on a certain type of lattice, that it is unbiased – most of the restrictions are easily relaxed within the context of the generating function calculation at no great cost in the form of significant complications to the analysis.

Having obtained an explicit expression for the generating function, we show how to extract the number of N-step walks with specified starting and ending points. Finally, we investigate the asymptotic behavior of this function in the limit of large N, rediscovering features already demonstrated in the previous chapter.

The remainder of this chapter consists of three supplements, in which techniques utilized in the exploitation of the generating function method are reviewed. It is not necessary to read them until those methods are called upon. On the other hand, the reader may wish to be forearmed with some information concerning the mathematics that underlie the discussions to follow.

2.2 Supplement 1: Gaussian integrals

Gaussian functions and Gaussian integrals play an important role in the statistics of random walks. Here, we review some basic results associated with integrals of Gaussians. The first is the well-known formula for the integral of a simple Gaussian:

$$\int_{-\infty}^{\infty} \mathrm{e}^{-ax^2} \mathrm{d}x = \sqrt{\frac{\pi}{a}} \tag{2:S1-1}$$

The next set of integrals are *moments* of the Gaussian distribution. The nth moment of the Gaussian e^{-ax^2} is the integral

$$\int_{-\infty}^{\infty} x^n \mathrm{e}^{-ax^2} \,\mathrm{d}x \tag{2:S1-2}$$

The most efficient way to find the value of the integral above for general values of the power n is, appropriately enough, through a generating function. Consider the integral

$$\int_{-\infty}^{\infty} \mathrm{e}^{-ax^2+bx} \,\mathrm{d}x \tag{2:S1-3}$$

Expanding (2:S1-3) in powers of b, we see that the coefficient of $b^n/n!$ is the nth moment of the Gaussian. This means that (2:S1-3) is a generating function for moments of the Gaussian e^{-ax^2}.

Given this useful fact, let's evaluate the integral in (2:S1-3). We do this by completing squares in the exponent. Writing

$$ax^2 - bx = a\left(x - \frac{b}{2a}\right)^2 - \frac{b^2}{2a} \tag{2:S1-4}$$

we have

$$\int_{-\infty}^{\infty} e^{-ax^2+bx} = \int_{-\infty}^{\infty} e^{-a(x-b/2a)+b^2/4a}\, dx$$
$$= e^{b^2/4a} \int_{-\infty}^{\infty} e^{-ay^2}\, dy$$
$$= e^{b^2/4a} \sqrt{\frac{\pi}{a}} \quad (2:S1\text{-}5)$$

performing the expansion and comparing terms, we see that

$$\int_{-\infty}^{\infty} x^n e^{-ax^2}\, dx = \begin{cases} \sqrt{\frac{\pi}{a}} \frac{n!}{(n/2)!(4a)^{n/2}} & n \text{ is even} \\ 0 & \text{otherwise} \end{cases} \quad (2:S1\text{-}6)$$

2.3 Supplement 2: Fourier expansions on a lattice

2.3.1 On Fourier transforms

The reader is, no doubt, familiar with the expansion of functions in plane waves of the form $e^{i\vec{k}\cdot\vec{r}}$. Here, we recall some of the basic features of such expansions. A function of the variable x defined on the interval $(-\infty, \infty)$, satisfying certain integrability criteria, admits an expansion of the form

$$f(x) = \frac{1}{2\pi} \int_{-\infty}^{\infty} g(k) e^{-ikx}\, dk \quad (2:S2\text{-}1)$$

This transform can be inverted as follows:

$$g(k) = \int_{-\infty}^{\infty} f(x) e^{ikx}\, dx \quad (2:S2\text{-}2)$$

The function $g(k)$ is the fourier transform of the function $f(x)$, and $f(x)$ is the inverse fourier transform of the function $g(k)$. In three dimensions the equations above generalize to

$$f(\vec{r}) = \frac{1}{(2\pi)^3} \int g(\vec{k}) e^{-i\vec{k}\cdot\vec{r}}\, d^3k \quad (2:S2\text{-}3)$$

and

$$g(\vec{k}) = \int f(\vec{r}) e^{i\vec{k}\cdot\vec{r}}\, d^3r \quad (2:S2\text{-}4)$$

respectively.

If the function $f(x)$ is defined on a discrete set of points, (i.e. a lattice), then the expansions above must be modified. Here we present, in an abbreviated and non-rigorous fashion, some results pertaining to the fourier representation of functions defined on a lattice, where the lattice can be either unbounded or of finite extent.

We assume that the reader is acquainted with the standard notation and results of lattice theory. If not, a good introductory text on condensed matter physics ought to provide a suitable background (Ashcroft and Mermin, 1976; Ziman, 1979). For ease of presentation, we restrict our analysis to simple lattices, ones without a basis. These lattices are called Bravais lattices. They can be generated by adding and subtracting integral multiples of three non-coplanar primitive vectors, $\vec{a}_1$, $\vec{a}_2$, and $\vec{a}_3$. These vectors are often referred to as basis vectors, or primitive vectors. All the lattice points are thus connected to a central point, also on the lattice, by position vectors, called *direct lattice vectors*, of the form

$$\vec{R} = n_1\vec{a}_1 + n_2\vec{a}_2 + n_3\vec{a}_3 \tag{2:S2-5}$$

The n_i's take on all positive and negative integral values. The volume in which the lattice sits can be filled by translations through the lattice vectors $\vec{R}$ of small volumes, the smallest of which is called the *primitive unit cell*. The shape of the primitive unit cell is not unique, although its volume is. One of the commonly used unit cells is a parallelepiped whose edges are the basis vectors $\vec{a}_1$, $\vec{a}_2$, and $\vec{a}_3$. The volume, Ω, of the primitive unit cell is then seen to be

$$\Omega = \vec{a}_1 \cdot \vec{a}_2 \times \vec{a}_3 \tag{2:S2-6}$$

For every Bravais lattice, one can construct another lattice, called the *reciprocal lattice*. The basis vectors spanning the reciprocal lattice are denoted $\vec{b}_1$, $\vec{b}_2$, and $\vec{b}_3$. One constructs them from the basis vectors of the direct lattice according to the prescription

$$\vec{b}_i = \frac{2\pi}{\Omega}\vec{a}_j \times \vec{a}_k \tag{2:S2-7}$$

where i, j and k are cyclic permutations of x, y and z. This construction guarantees that $\vec{a}_i \cdot \vec{b}_j = 2\pi\delta_{ij}$. The reciprocal lattice will play a crucial role in the development of plane wave representations of functions defined on a lattice. All the points of the reciprocal lattice can be reached by position vectors of the form

$$\vec{K} = l_1\vec{b}_1 + l_2\vec{b}_2 + l_3\vec{b}_3$$

with l_2, l_2 and l_3 taking on all positive and negative integral values. The vectors $\vec{K}$ are called reciprocal lattice vectors, and by construction, they have the property

$$\vec{K} \cdot \vec{R} = 2\pi \times (\text{an integer}) \tag{2:S2-8}$$

for any $\vec{K}$ and $\vec{R}$ drawn from the reciprocal of the direct lattice, respectively. This follows immediately on recalling that $\vec{a}_i \cdot \vec{b}_j = 2\pi\delta_{ij}$. The primitive unit cell of the reciprocal lattice is called the *first Brillouin zone*. It is straightforward to show that $\Omega_{\mathrm{BZ}} = \vec{b}_1 \cdot \vec{b}_2 \times \vec{b}_3 = (2\pi)^3/\Omega$.

A trivial example, but one pertinent to our previous discussion, is the simple cubic lattice. For a lattice spacing a, we find that the basis vectors

$$\vec{a}_1 = a\hat{\imath} \tag{2:S2-9}$$

$$\vec{a}_2 = a\hat{\jmath} \tag{2:S2-10}$$

$$\vec{a}_3 = a\hat{k} \tag{2:S2-11}$$

generate the cubic lattice. The primitive cell is a cube of side a and volume a^3.

The basis vectors of the associated reciprocal lattice are found to be

$$\vec{b}_1 = \frac{2\pi}{a}\hat{\imath} \tag{2:S2-12}$$

$$\vec{b}_2 = \frac{2\pi}{a}\hat{\jmath} \tag{2:S2-13}$$

$$\vec{b}_3 = \frac{2\pi}{a}\hat{k} \tag{2:S2-14}$$

by straightforward application of (2:S2-7). The Brillouin zone is again a cube of length $2\pi/a$, on each side. The origin of the Brillouin zone is taken to be at the center of the cube, and, of course, its volume is $\Omega = (2\pi)^3/a^3$.

Two other lattices with cubic symmetry are the *body-centered cubic lattice* and the *face-centered cubic lattice*. Three basis vectors for the body-centered cubic (bcc) lattice are

$$\vec{a}_1 = \frac{a}{\sqrt{3}}\left(\hat{\imath} + \hat{\jmath} + \hat{k}\right) \tag{2:S2-15}$$

$$\vec{a}_2 = \frac{a}{\sqrt{3}}\left(\hat{\imath} + \hat{\jmath} - \hat{k}\right) \tag{2:S2-16}$$

$$\vec{a}_3 = \frac{a}{\sqrt{3}}\left(\hat{\imath} - \hat{\jmath} + \hat{k}\right) \tag{2:S2-17}$$

Three basis vectors for the face-centered cubic (fcc) lattice are

$$\vec{a}_1 = \frac{a}{\sqrt{2}}\left(\hat{\imath} + \hat{\jmath}\right) \tag{2:S2-18}$$

$$\vec{a}_2 = \frac{a}{\sqrt{2}}\left(\hat{\imath} + \hat{k}\right) \tag{2:S2-19}$$

$$\vec{a}_3 = \frac{a}{\sqrt{2}}\left(\hat{\jmath} + \hat{k}\right) \tag{2:S2-20}$$

In all cases above, a is the spacing between nearest neighbor vertices on the lattice.

2.3.2 Infinite lattices; Fourier expansions

The discussion above lays the foundation for plane wave expansions of functions defined on any one of the Bravais lattices. Let such a function be denoted $F(\vec{R})$.

Here the function $F(\vec{R})$ is defined for all direct lattice vectors $\vec{R}$ spanning the points of our lattice. A function of this type is $C(N;\vec{x},\vec{y})$, the number of walks beginning at the lattice point $\vec{x}$ and ending at the lattice point $\vec{y}$, with end-to-end displacement vector $\vec{R}=\vec{y}-\vec{x}$, which is also a lattice vector. We desire an appropriate, and, we hope, convergent expansion of such a function in a set of plane waves. As a first step toward achieving this goal, define the function $f(\vec{k})$ as follows:

$$f(\vec{k})=\sum_{\vec{R}} F(\vec{R})\mathrm{e}^{\mathrm{i}\vec{k}\cdot\vec{R}} \tag{2:S2-21}$$

The first thing to note is that the function $f(\vec{k})$ is periodic in reciprocal space e.g.

$$f(\vec{k}+\vec{K})=f(\vec{k}) \tag{2:S2-22}$$

This follows from Equation (2:S2-8). Thus, knowledge of $f(\vec{k})$ in the first Brillouin zone suffices to specify the functions for all values of $\vec{k}$. Another important feature of (2:S2-21) is that the equation can easily be inverted. To accomplish this, multiply both sides of the equation by $\mathrm{e}^{-\mathrm{i}\vec{k}\cdot\vec{R}'}$ and integrate through the volume of the Brillouin zone.

$$\int_{\mathrm{BZ}} \mathrm{e}^{-\mathrm{i}\vec{k}\cdot\vec{R}'} f(\vec{k})\,\mathrm{d}^3k=\sum_{\vec{R}} f(\vec{R})\int_{\mathrm{BZ}} \mathrm{e}^{\mathrm{i}\vec{k}\cdot(\vec{R}-\vec{R}')}\,\mathrm{d}^3k \tag{2:S2-23}$$

Using the following result (see Ashcroft and Mermin, 1976; Ziman, 1979):

$$\begin{aligned}\int_{\mathrm{BZ}} \mathrm{e}^{\mathrm{i}\vec{k}\cdot(\vec{R}-\vec{R}')}\,\mathrm{d}^3k &= \Omega_{\mathrm{BZ}}\delta(\vec{R}-\vec{R}')\\ &= \frac{(2\pi)^3}{\Omega}\delta(\vec{R}-\vec{R}')\end{aligned} \tag{2:S2-24}$$

allows one to insert (2:S2-21) and solve for $F(\vec{R})$. The pair of equations

$$F(\vec{R})=\frac{\Omega}{(2\pi)^3}\int_{\mathrm{BZ}} f(\vec{k})\mathrm{e}^{-\mathrm{i}\vec{k}\cdot\vec{R}}\,\mathrm{d}^3k \tag{2:S2-25}$$

and

$$f(\vec{k})=\sum_{\vec{R}} F(\vec{R})\mathrm{e}^{\mathrm{i}\vec{k}\cdot\vec{R}} \tag{2:S2-26}$$

are the appropriate fourier representation for a function defined on a Bravais lattice specified by lattice vectors $\vec{R}$. These equations are the generalization to the fourier transform equations for functions defined throughout all space.

2.3.3 Plane wave expansions; finite lattices

Our interest in finite size lattices is not purely academic. Walks confined to finite regions of space have wide application to the properties of polymers and will occupy our attention at a later point. The role of boundary conditions is an important feature of constrained walks and could significantly alter their statistical properties, but a discussion of these matters is best deferred until we are confronted with a problem. Here, as a way of establishing an appropriate language useful for analyzing the behavior of walks with bounding surfaces, we introduce lattices of finite extent through the artifact of invoking *periodic* boundary conditions. The finite lattice is imagined to be in the shape of a large parallelepiped of edges N_1a_1, N_2a_2, and N_3a_3, with a total number of lattice points $N_0 = N_1N_2N_3$.

The function $F(\vec{R})$ is now required to be periodic in the displacements $N_1\vec{a}_1$, $N_2\vec{a}_2$, and $N_3\vec{a}_3$. Hence,

$$\begin{aligned} F(\vec{R}) &= F(\vec{R} + N_1\vec{a}_1) \\ &= F(\vec{R} + N_2\vec{a}_2) \\ &= F(\vec{R} + N_3\vec{a}_3) \end{aligned} \tag{2:S2-27}$$

Plane waves of the type $e^{i\vec{k}\cdot\vec{R}}$ are again natural candidates, but, in order to guarantee periodicity, the wave-vector $\vec{k}$ no longer varies continuously within the Brillouin zone. It takes on discrete values. Indeed, the boundary conditions require

$$\vec{k} \cdot N_i\vec{a}_i = 2\pi m_i, i = 1, 2, 3 \tag{2:S2-28}$$

where the m_i takes on integral values.

The vector $\vec{k}$ satisfying the above constraints can be written in terms of the reciprocal basis set as follows

$$\vec{k} = \frac{m_1}{N_1}\vec{b}_1 + \frac{m_2}{N_2}\vec{b}_2 + \frac{m_3}{N_3}\vec{b}_3 \tag{2:S2-29}$$

Vectors of this sort generate a lattice with primitive basis vectors b_1/N_1, b_2/N_2, and b_3/N_3, and primitive cell volume $\Omega_k = \Omega_{\mathrm{BZ}}/N_0$. There are exactly N_0 $\vec{k}$ points within the Brillouin zone. The density of these points is

$$\begin{aligned} \rho_k &= \frac{1}{\Omega_k} \\ &= \frac{N_0}{\Omega_{\mathrm{BZ}}} \\ &= \frac{N_0\Omega}{(2\pi)^3} \\ &= \frac{V}{(2\pi)^3} \end{aligned} \tag{2:S2-30}$$

In the above, V is the volume of the direct lattice. For macroscopically large lattices the number N_0 is of the order of Avogadro's number. The $\vec{k}$ points are infinitesimally close and the wave-vector $\vec{k}$ can therefore be treated as a continuous variable, with $\sum_{\vec{k}} \to \int \rho_k \, \mathrm{d}^3k$. Keeping this in mind, the appropriate plane wave expansion for functions defined on a lattice with periodic boundary conditions are the pair of equations

$$F(\vec{R}) = \frac{1}{N_0} \sum_{\vec{k}} f(\vec{k}) \mathrm{e}^{\mathrm{i}\vec{k}\cdot\vec{R}} \tag{2:S2-31}$$

$$f(\vec{k}) = \sum_{\vec{R}} F(\vec{R}) \mathrm{e}^{-\mathrm{i}\vec{k}\cdot\vec{R}} \tag{2:S2-32}$$

with the discretely valued wave-vector $\vec{k}$ given by (2:S2-29).

Before ending the discussion, one further point needs mentioning, and that is that in order to invert the above equation and solve for the fourier coefficient $f(\vec{k})$ in (2:S2-32) we made use of the property

$$\sum_{\vec{R}} \mathrm{e}^{\mathrm{i}\vec{k}\cdot\vec{R}} = N_0 \delta_{\vec{k},0} \tag{2:S2-33}$$

The reader can most directly see that this relation is true by explicitly doing the sums, which we rewrite as

$$\sum_{n_1=0}^{N_1-1} \sum_{n_2=0}^{N_2-1} \sum_{n_3=0}^{N_3-1} \mathrm{e}^{2\pi\mathrm{i}\left(\frac{m_1}{N_1}n_1+\frac{m_2}{N_2}n_2\frac{m_3}{N_3}n_3\right)}$$
$$= \left(\frac{1-\mathrm{e}^{2\pi m_1 \mathrm{i}}}{1-\mathrm{e}^{2\pi \mathrm{i} m_1/N_1}}\right)\left(\frac{1-\mathrm{e}^{2\pi m_2 \mathrm{i}}}{1-\mathrm{e}^{2\pi \mathrm{i} m_2/N_2}}\right)\left(\frac{1-\mathrm{e}^{2\pi m_3 \mathrm{i}}}{1-\mathrm{e}^{2\pi \mathrm{i} m_3/N_3}}\right) \tag{2:S2-34}$$

Since m_1, m_2, and m_3 are integers not equal to N_1, N_2, and N_3, respectively, the sum is equal to zero unless $m_1 = m_2 = m_3 = 0$, in which case it is just the number of lattice points, N_0. Thus we recover (2:S2-33).

2.4 Supplement 3: asymptotic coefficients of power series

Once we have managed to calculate a generating function, we need to invert the process. That is, we need to determine the coefficients in the power series. Recall the relationship between the generating function $G(z; \vec{x}, \vec{y})$ and $C(N; \vec{x}, \vec{y})$, the latter quantity being the number of N-step walks starting out at point $\vec{x}$ and ending up at point $\vec{y}$:

$$G(z; \vec{x}, \vec{y}) = \sum_{N=0}^{\infty} z^N C(N; \vec{x}, \vec{y}) \tag{2:S3-1}$$

This means that, given $G(z;\vec{x},\vec{y})$, we can find $C(N;\vec{x},\vec{y})$ if we know how to extract the coefficient of z^N in the expansion of $G(z;\vec{x},\vec{y})$ as a power series in z.

In most cases of interest the functional dependence of $G(z;\vec{x},\vec{y})$ on z will not be sufficiently simple that the behavior of coefficients in the power series expansion is immediately evident. In this supplement we will look at a few cases that turn out to be particularly important for applications to random walk problems. We will also introduce the method of steepest descents, a remarkably powerful technique that allows us to extract the leading behavior of high order coefficents in the power series expansion of a large class of functions.

2.4.1 Simple pole

Let's start with one of the simplest examples of a function with an infinite power-series expansion in z: $f(z) = 1/(z_c - z)$. If $|z| < |z_c|$, we have

$$\begin{aligned} \frac{1}{z_c - z} &= \frac{1}{z_c}\left[1 + \frac{z}{z_c} + \left(\frac{z}{z_c}\right)^2 + \cdots\right] \\ &= \frac{1}{z_c}\sum_{n=0}^{\infty}\left(\frac{z}{z_c}\right)^n \end{aligned} \tag{2:S3-2}$$

so the coefficient of z^N is $z_c^{-(N+1)}$.

2.4.2 Two or more simple poles

Now, suppose z_c and z_d are both real, positive numbers and $z_c < z_d$. Furthermore, let

$$f(z) = \frac{a}{z_c - z} + \frac{b}{z_d - z} \tag{2:S3-3}$$

Then, if z is suficiently small($z < z_c$)

$$f(z) = \frac{a}{z_c}\sum_{n=0}^{\infty}\left(\frac{z}{z_c}\right)^n + \frac{b}{z_d}\sum_{n=0}^{\infty}\left(\frac{z}{z_d}\right)^n \tag{2:S3-4}$$

so the coefficient of z^N is $c_N = a z_c^{-(N+1)} + b z_d^{-(N+1)}$.

We can also write

$$c_N = a z_c^{-(N+1)}\left[1 + \frac{b}{a}\left(\frac{z_c}{z_d}\right)^{N+1}\right]$$

Now, let $z_c = z_d(1-\Delta)$, where $\Delta > 0$. Then

$$\begin{aligned} c_N &= a z_c^{-(N+1)}\left(1+\frac{b}{a}(1-\Delta)^{N+1}\right) \\ &= a z_c^{-(N+1)}\left(1+\frac{b}{a}\mathrm{e}^{(N+1)\ln(1-\Delta)}\right) \\ &= a z_c^{-(N+1)}\left(1+\frac{b}{a}\mathrm{e}^{-\delta(N+1)}\right) \end{aligned} \tag{2:S3-5}$$

where $\delta = -\ln(1-\Delta)$ and $\delta > 0$. As N gets larger and larger the second term in brackets in (2:S3-5) vanishes exponentially. Thus, for very large N the coefficient of z^N in $a/(z_c - z) + b/(z_d - z)$ is essentially equal to the coefficient of $a/(z_c - z)$. We will eventually generalize this result as follows:

If the functions $f_a(z)$ and $f_b(z)$ have poles or branch points at z_a and z_b, respectively, and if $z_b > z_a > 0$ (z_a and z_b both real), then, when N is large, the coefficient of z^N in $Af_a(z) + Bf_b(z)$ is, for all practical purposes, equal to the coefficient of z^N in $Af_a(z)$.

2.4.3 Higher order poles and branch points

What about the more general case $f(z) = (z_c - z)^{-\alpha}$, where the exponent α need not be an integer? One way of finding the coefficient of z^N is to use the binomial expansion formula. Another way is the use the following identity:

$$\int_0^\infty t^A \mathrm{e}^{-xt}\,\mathrm{d}t = x^{-A-1}\Gamma(A+1) \tag{2:S3-6}$$

where $\Gamma(A)$ is the gamma function. When A is an integer, n, $\Gamma(n+1) = n!$. With the use of (2:S3-6) we have

$$(z_c - z)^{-\alpha} = \frac{1}{\Gamma(\alpha)}\int_o^\infty t^{\alpha-1}\mathrm{e}^{-t(z_c - z)}\,\mathrm{d}t \tag{2:S3-7}$$

To find the coefficient ofz^N in $(z_c - z)^{-\alpha}$ we expand the right hand side of (2:S3-7) with respect to z. The coefficient of z^N in that expansion is

$$\begin{aligned} \frac{1}{\Gamma(\alpha)}\frac{1}{N!}\int_0^\infty t^{\alpha-1+N}\mathrm{e}^{-tz_c}\,\mathrm{d}t &= \frac{1}{\Gamma(\alpha)}\frac{1}{N!}z_c^{-(N+\alpha)}\Gamma(\alpha+N) \\ &= \frac{\Gamma(\alpha+N)}{\Gamma(\alpha)\Gamma(N+1)}z_c^{-(N+\alpha)} \end{aligned} \tag{2:S3-8}$$

Now, we use Stirling's formula for the gamma function of a large argument

$$\Gamma(N) = \exp[(N-1)\ln(N-1) - (N-1)] \qquad N \gg 1 \tag{2:S3-9}$$

When N is large, the coefficient of interest is

$$\begin{aligned}
&\frac{z_c^{-N-\alpha}}{\Gamma(\alpha)}\exp[(\alpha+N-1)\ln(\alpha+N-1) - (\alpha+N-1) - N\ln N + N] \\
&= \frac{z_c^{-N-\alpha}}{\Gamma(\alpha)}e^{(\alpha-1)\ln(N)} \\
&= \frac{z_c^{-N-\alpha}}{\Gamma(\alpha)}N^{\alpha-1}
\end{aligned} \tag{2:S3-10}$$

2.4.4 Logarithmic singularities

One more complication: suppose the function is of the form $-(z_c - z)^{-\alpha}\ln(z_c - z)$. We obtain the coefficient of z^N by noting that this function is just $d/d\alpha(z_c - z)^{-\alpha}$. Taking the derivative with respect to α of the last term in (2:S3-10), we have for the desired coefficient

$$\begin{aligned}
\frac{d}{d\alpha}\frac{z_c^{-N-\alpha}}{\Gamma(\alpha)}N^{\alpha-1} &= \ln(z_c)\frac{z_c^{-N-\alpha}N^{\alpha-1}}{\Gamma(\alpha)} - \frac{\Gamma'(\alpha)z_c^{-N-\alpha}}{\Gamma(\alpha)^2}N^{\alpha-1} + \ln(N)\frac{z_c^{-N-\alpha}N^{\alpha-1}}{\Gamma(\alpha)} \\
&= \ln(N)\frac{z_c^{-N-\alpha}}{\Gamma(\alpha)}N^{\alpha-1}\left(1 + O\left(\frac{1}{\ln(N)}\right)\right)
\end{aligned} \tag{2:S3-11}$$

By the same token, we can find the coefficient of z^N in $-(z_c - z)^{-\alpha}/\ln(z_c - z)$ by extracting the coefficient of z^N in the integral

$$\int_{-\alpha}^{\infty}(z_c - z)^y\, dy \tag{2:S3-12}$$

Using (2:S3-10) to find the coefficient of z^N in the integrand, one is left with

$$\begin{aligned}
\int_{-\alpha}^{\infty}\frac{y_c^{-N+y}}{\Gamma(-y)}N^{y-1}\, dy &= -\left.\frac{N^{y-1}}{\ln N}\frac{y_c^{-N+y}}{\Gamma(-y)}\right|_{-\alpha}^{\infty} - \frac{1}{\ln N}\int_{-\alpha}^{\infty}N^{y-1}\frac{d}{dy}\left(\frac{y_c^{-N+y}}{\Gamma(-y)}\right)dy \\
&= \frac{N^{\alpha-1}y_c^{-N-\alpha}}{\ln(N)\Gamma(\alpha)}\left(1 + O\left(\frac{1}{\ln N}\right)\right)
\end{aligned} \tag{2:S3-13}$$

The first equality in (2:S3-13) results from an integration by parts. A further integration by parts establishes the second equality.

3

Generating functions II: recurrence, sites visited, and the role of dimensionality

We've now had an introduction to the random walk. There has also been an introduction to the generating function and its utilization in the analysis of the process. Here we investigate some aspects of the random walk, both because they are interesting in their own right and because they further demonstrate the usefulness of the generating function as a theoretical technique in discussing this problem. In particular, we will apply the generating function method to study the problem of "recurrence" in the random walk. That is, we will address the question of whether the walker will ever return to its starting point, and, if so, with what probability. We will find that the spatial dimensions in which the walk takes place play a crucial role in determining this probability. Using an almost identical approach, the number of distinct points visited by the walker will also be determined.

We present the two studies below as a display of the power of the generating function technique in the study of the random walk process. For the skeptic, other examples will be provided in subsequent chapters.

3.1 Recurrence

We begin this chapter with a discussion of the question of recurrence of an unrestricted random walk. We will utilize the generating function, and an important statistical identity, to see how the dimensionality of the random walker's environment controls the probability of its revisiting the site from which it has set out. This study complements our earlier discussion and provides evidence for the power of the generating function approach.

3.2 A new generating function

A useful – but apparently little known[1] – quantity enables one to obtain some key results with remarkably little effort. This quantity is an expanded version of the

[1] Little known among practising solid state physicists, that is.

generating function we've been utilizing, and it refers to a walk that may or may not visit a special site. Suppose the quantity $C(N, M; \vec{x}, \vec{y}, \vec{w})$ is the number of N-step walks that start at the location $\vec{x}$, end at the location $\vec{y}$ and visit the site at location $\vec{w}$ exactly M times in the process of getting from $\vec{x}$ to $\vec{y}$. The quantity of interest is

$$C'(N, t; \vec{x}, \vec{y}, \vec{w}) = \sum_{M=0}^{\infty} C(N, M; \vec{x}, \vec{y}, \vec{w})(1-t)^M \tag{3.1}$$

Clearly, terms in the sum on the right hand side of the above equation for which $M > N + 1$ will not count, as there is no walk that visits a site more times than it leaves footprints. That is $C(N, M; \vec{x}, \vec{y}, \vec{w}) = 0$ if $M > N + 1$. We obtain the coefficients of the power series expansion in $(1-t)^n$ that produces this generating function in the standard way. It is easy to show that

$$C(N, M; \vec{x}, \vec{y}, \vec{w}) = (-1)^M \frac{1}{M!} \frac{\mathrm{d}^M}{\mathrm{d}t^M} C'(N, t; \vec{x}, \vec{y}, \vec{w}) \bigg|_{t=1} \tag{3.2}$$

Note that in this case the value to which the expansion parameter is set is not zero, but one.

3.3 Derivation of the new generating function

We derive the new generating function by introducing a weighting factor. The weighting factor W has the following form for an N-step walk that visits the site s_i at the ith step

$$W = \prod_{i=1}^{N} w(s_i) \tag{3.3}$$

where factor $w(s_i)$ is given by

$$w(s_i) = 1 - t\delta_{s_i, S} \tag{3.4}$$

Here, S is the special site of interest. The overall weighting factor for a given walk is then

$$\prod_{i=1}^{N} \left(1 - t\delta_{s_i, S}\right) \tag{3.5}$$

Suppose $t = 1$. Then, the weighting factor will have the effect of excluding any walk that visits the special site S; all other walks will have a weighting factor of one. This means that if we multiply all walks by the weighting factor above, set $t = 1$ and sum, we will obtain the number of walks that never visit the site S. In the case of N-step walks that start at $\vec{x}$ and end up at $\vec{y}$, this is just $C(N, 0; \vec{x}, \vec{y}, \vec{w})$, where $\vec{w}$ is the position vector of the site S. Suppose we take the derivative of the weighted sum with respect to t, and then set $t = 1$. In that case, we will end up with -1 times the number of walks that visit the site only once. Next, take the nth derivative of the weighted sum over walks with respect to t, multiply by $1/n!$, and then set $t = 1$. This yields $(-1)^n$ times the number of walks that visit the special site exactly n times. This is because each derivative generates a factor equal to $-\delta_{s_j,S}$ and all terms that "escape" the derivative become $(1 - \delta_{s_k,S})$ when $t = 1$. The factor $1/n!$ compensates for the $n!$ ways in which the n derivatives with respect to t operate on the product in (3.5).

Now, we can evaluate the weighted walk by expanding the weighting factor, (3.5), in powers of t.[2] At first order we generate the quantity

$$-t \sum_{i=1}^{N} \delta_{s_i,S} \tag{3.6}$$

When walks are multiplied by this weighting factor and summed, we end up with $-t$ times the sum of all N-step walks that visit the site S at one step, with no restriction on what happens either before or after that step. At second order in the expansion we have

$$t^2 \sum_{i=1}^{N} \sum_{j=1}^{i-1} \delta_{s_i,S}\delta_{s_j,S} \tag{3.7}$$

When walks are multiplied by this second order weighting factor and summed, we end up with t^2 times the sum of all N-step walks that visit the site S twice with no restriction on what happens before, after or between those two visitations. A graphical representation for the expansion in t of the new generating function is shown in Figure 3.1. If the starting and end-point of the walk are fixed, and if the site in question is at the location v, then the sum in Figure 3.1 is given by

$$C(N;\vec{x},\vec{y}) - t \sum_{N_1+N_2=N} C(N_1;\vec{x},\vec{w})C(N_2;\vec{w},\vec{y})$$
$$+t^2 \sum_{\substack{N_1+N_2+N_3=N \\ N_2 \geq 1}} C(N_1;\vec{x},\vec{w})C(N_2;\vec{w},\vec{w})C(N_3;\vec{w},\vec{y}) + \cdots \tag{3.8}$$

The inequality that applies to N_2 in the sum above simply requires the walker to take at least one step before revisiting the site at $\vec{w}$. Otherwise, we would count zero step "walks" in the sum.

[2] We are thus treating t as if it were a small quantity and expanding in it. This technique will be used later for a different expansion parameter in the case of self-avoiding walks, in which case we will generate a virial expansion.

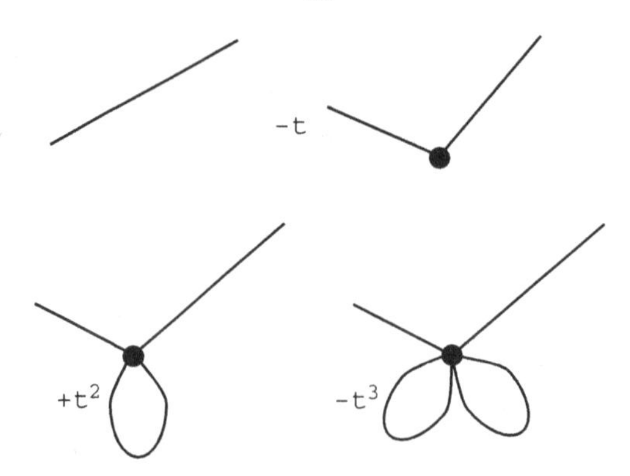

Fig. 3.1. Graphical representation of the virial expansion of the generating function defined in (3.1). The large dots in the diagrams lie at the location $\vec{w}$.

Now, we take the step of multiplying our new function by z^N and summing. This has the effect of removing the convolution over the N_i's, and we obtain

$$\begin{aligned} G(z, t; \vec{x}, \vec{y}, \vec{w}) &= \sum_{N=0}^{\infty} z^N C'(N, t; \vec{x}, \vec{y}, \vec{w}) \\ &= G(z; \vec{x}, \vec{y}) - tG(z; \vec{x}, \vec{w})G(z; \vec{w}, \vec{y}) \\ &\quad + t^2 G(z; \vec{x}, \vec{w})G_1(z; \vec{w}, \vec{w})G(z; \vec{w}, \vec{y}) + \cdots \\ &= G(z; \vec{x}, \vec{y}) - t\frac{G(z; \vec{x}, \vec{w})G(z; \vec{w}, \vec{y})}{1 + tG_1(z; \vec{w}, \vec{w})} \end{aligned} \tag{3.9}$$

The subscript 1 in $G_1(z; \vec{w}, \vec{w})$ indicates that it is a generating function for walks of at least one step.

The quantity

$$\begin{aligned} &(-1)^M \frac{1}{M!} \frac{\mathrm{d}^M}{\mathrm{d}t^M} G(z, t; \vec{x}, \vec{y}, \vec{w})\bigg|_{t=1} \\ &\quad = (-1)^M \frac{1}{M!} \frac{\mathrm{d}^M}{\mathrm{d}t^M} \left[G(z; \vec{x}, \vec{y}) - t\frac{G(z; \vec{x}, \vec{w})G(z; \vec{w}, \vec{y})}{1 + tG_1(z; \vec{w}, \vec{w})} \right]\bigg|_{t=1} \end{aligned} \tag{3.10}$$

is, then, the generating function for all walks that start at $\vec{x}$, end up at $\vec{y}$ and visit the site at $\vec{w}$ exactly M times.

Exercise 3.1

Fill in the steps in the derivation of (3.9).

3.4 Dimensionality and the probability of recurrence

Our result allows us to determine whether or not a random walk is *recurrent*. If it is, then almost all long random walks that start out at a given point will revisit that point. If it is not, then only a finite fraction of those walks do so, and the walk is called *transient*. Recurrent walks are also referred to as "persistent." Pólya (1919) was the first to demonstrate that walks occurring in one and two dimensions return to their starting point with absolute certainty, if they consist of an infinite number of steps and that in higher dimensions the walker has a non-zero probability of never revisiting its starting point, no matter how long its walk. Let's see what (3.9) tells us about the recurrence of random walks on a lattice. We want to know how many of the walks that start out at $\vec{x}$ revisit their point of origin, so we set $\vec{w}$ in (3.9) equal to $\vec{x}$. The generating function takes the form

$$G(z;\vec{x}-\vec{y}) - t\frac{G_1(z;0)G(z;\vec{x}-\vec{y})}{1+tG_1(z;0)} = \frac{G(z;\vec{x}-\vec{y})}{1+tG_1(z;0)} \tag{3.11}$$

The reason for the subscript 1 in the numerator of the second term on the left hand side of (3.11) is that we want to count only those walks that take at least one step from the starting point at x before revisiting that point. Otherwise, we count walks that "revisit" their point of origin after zero steps.

To find out how many N-step walk start out at $\vec{x}$ and end up at $\vec{y}$, never having revisited $\vec{x}$, we need to calculate the coefficient of z^N in the function

$$\frac{G(z;\vec{x}-\vec{y})}{1+G_1(z;0)} = \frac{G(z;\vec{x}-\vec{y})}{G(z;0)} \tag{3.12}$$

The right hand side of the equality in (3.12) follows from the fact that the contribution to the generating function $G(z;0)$ of walks consisting of no steps is exactly one, by convention.

Let's be even less restrictive and ask how many of the walks that start out at $\vec{x}$ and end up *anywhere* never revisit the starting point $\vec{x}$. We simply sum the expression above over all possible end-points $\vec{y}$ – excluding $\vec{x}$ – and see what we have. Using the relation between $G(z;\vec{x}-\vec{y})$ and its spatial Fourier transform, $g(z;\vec{q})$ we have

$$\begin{aligned}\sum_{\vec{y}\neq\vec{x}} \frac{G(z;\vec{x}-\vec{y})}{G(z;0)} &= \frac{g(z;0)}{G(z;0)} - \frac{G(z;0)}{G(z;0)} \\ &= \frac{g(z;0)}{G(z;0)} - 1\end{aligned} \tag{3.13}$$

where we have used the identity

$$g(z,0) = \sum_{\vec{y}} G(z,\vec{x}-\vec{y}) \tag{3.14}$$

for the Fourier coefficient $g(z, \vec{q})$. From Chapter 2, we know that

$$
\begin{aligned}
g(z, \vec{q} = 0) &= \frac{1}{1 - z\chi(\vec{q} = 0)} \\
&\equiv \frac{1}{1 - z/z_c}
\end{aligned}
\tag{3.15}
$$

The last line of (3.15) serves as a definition of the quantity z_c. This tells us that the number of N-step walks that start out at $\vec{x}$ and end up anywhere is the coefficient of z^N in $(1 - z/z_c)^{-1}$, while the number of walks that start at $\vec{x}$ and end up anywhere, not having ever revisited the point of origin, $\vec{x}$, is the coefficient of z^N in

$$
\frac{z_c}{z_c - z} \frac{1}{G(z;0)}
\tag{3.16}
$$

As it turns out, the z-dependence of the generating function $G(z;0)$ in two dimensions or less is different in a very important respect from its behavior in three and higher dimensions. This will lead to fundamentally different results for the "recurrence" of walks in two and one dimensions from what we will find in the case of three-dimensional walks.

Worked-out example

Compute the probability that the walker never revisits its starting point when the walk takes place in one dimension.

Solution

First, note that in one dimension $\chi(q) = \mathrm{e}^{\mathrm{i}q} + \mathrm{e}^{-\mathrm{i}q}$. This means that

$$
\begin{aligned}
z_c &= 1/\chi(q = 0) \\
&= 1/2
\end{aligned}
\tag{3.17}
$$

For one-dimensional walks

$$
\begin{aligned}
G(z, 0) &= \frac{1}{2\pi} \int_{-\pi}^{\pi} \frac{\mathrm{d}q}{1 - z\left(\mathrm{e}^{\mathrm{i}q} + \mathrm{e}^{-\mathrm{i}q}\right)} \\
&= \int_0^{2\pi} \frac{\mathrm{d}q}{1 - z\left(\mathrm{e}^{\mathrm{i}q} + \mathrm{e}^{-\mathrm{i}q}\right)}
\end{aligned}
\tag{3.18}
$$

The integral reduces to an contour integral around the unit circle if one makes the change of variables $\alpha = \mathrm{e}^{\mathrm{i}q}$. The integral in (3.18) is, then

$$
\begin{aligned}
G(z, 0) &= \frac{1}{2\pi} \oint \frac{\mathrm{d}\alpha}{\mathrm{i}\alpha\left(1 - z\left(\alpha + \alpha^{-1}\right)\right)} \\
&= -\frac{1}{2\pi \mathrm{i} z} \oint \frac{\mathrm{d}\alpha}{\alpha^2 - \alpha/2 + 1/z}
\end{aligned}
\tag{3.19}
$$

The path of integration in (3.19) is counter-clockwise around the unit circle.

The poles of the integrand are located at

$$\alpha_{\pm} = \frac{1}{2}\left\{\frac{1}{z} \pm \sqrt{\left(\frac{1}{z}\right)^2 - 4}\right\} \tag{3.20}$$

Since $z < 1, \alpha_-$ is the only pole lying inside the unit circle. Using Cauchy's formula we find

$$\begin{aligned} G(z,0) &= -\frac{2\pi\mathrm{i}}{2\pi\mathrm{i}z}\left(\frac{1}{\alpha_- - \alpha_+}\right) \\ &= \frac{1}{\sqrt{1-4z^2}} \\ &= \frac{z_c}{\sqrt{z_c - z}\sqrt{z_c + z}} \end{aligned} \tag{3.21}$$

The behavior of interest is related to the analytic properties of the function of z that will be constructed in the vicinity of $z = z_c$. If we insert our result for $G(z, 0)$ into the expression (3.13) for the number of walks that do not revisit their point of origin we find that the number of such N-step walks is the coefficient of z^N in

$$\begin{aligned} \frac{g(z;0)}{G(z;0)} &\to \frac{z_c}{z_c - z}\frac{\sqrt{z_c - z}}{z_c} \\ &= \frac{1}{\sqrt{z_c - z}} \end{aligned} \tag{3.22}$$

The next step is to extract the coefficient of z^N in the expression on the last line of (3.22). Making use of the results derived in Supplement 3 of Chapter 2, we find for that coefficient

$$\frac{z_c^{-N}}{\Gamma(1/2)}N^{-1/2} \tag{3.23}$$

The total number of N-step walks goes as z_c^{-N} (the coefficient of z^N in the expansion of $z_c/(z_c - z)$). Dividing by this quantity we find that the probability that a one-dimensional walker "escapes" without revisiting the point from which it started is

$$N^{-1/2}\frac{1}{\Gamma(1/2)} \tag{3.24}$$

which is vanishingly small in the limit $N \to \infty$.

Exercise 3.2

Consider a biased walker, for which the probability p, of a step to the right is not equal to 1/2. Show in this case that the probability of a walker's not revisiting its site of origin is not equal to zero in the limit of an infinite number of steps.

Exercise 3.3

In the case above, show that the probability of escape (that is, the probability of never returning to the point of origin) is equal to one when $p = 1$. There is an obvious intuitive argument for this result, but the solution desired is based on an analysis of the formula for the probability of return utilizing generating functions.

Exercise 3.4

In the case of the biased one-dimensional walk presented in Exercise 3.2, find the general formula for the probability of eventual return to the site of origin when $p \neq 1/2$.

3.5 Recurrence in two dimensions

In the above worked-out example, we found that a one-dimensional walker that has taken enough steps will inevitably revisit its point of origin. In fact, it is possible to show that if the one-dimensional walker takes enough steps, it will eventually visit any given site. This will also turn out to be the case in two dimenensions. And it is to two dimensions that we now turn.

As in the case of the one-dimensional walk, the key to the calculation of the probability of eventual return to the point of origin for walks in two dimensions is the extraction of the coefficient of z^N in the power-series expansion of the right hand side of (3.13). As already noted, when N is large, the dominant contribution comes from the first term on the right hand side of (3.13), because of the singular behavior of $g(z;0)$. First, recall that $g(z;0) = z_c/(z_c - z)$ (see (3.15)). In the case of a walk on the d-dimensional version of a simple cubic lattice with nearest neighbor distance equal to one,[3] $\chi(\vec{q}) \to 2d$ as $\vec{q} \to 0$. Thus, the above relationship holds with $z_c = 1/2d$.

The quantity of interest is, then, the coefficient of z^N in

$$\frac{z_c}{(z_c - z)} \frac{1}{G(z;0)} \tag{3.25}$$

In general

$$G(z;0) = \frac{1}{(2\pi)^d} \int \frac{d^d q}{1 - z\chi(\vec{q})} \tag{3.26}$$

where

$$\chi(\vec{q}) = c_1 - c_2|\vec{q}|^2 + O(|\vec{q}|^4) \tag{3.27}$$

[3] In two dimensions, this is a square lattice.

where $c_1 = 2d$ and $c_2 = 1/2$ for a simple cubic lattice. Truncating the structure function $\chi(\vec{q})$ to lowest order in $|\vec{q}|$ allows us to integrate over the angular coordinates in q-space, yielding the following simplified form for $G(z;0)$

$$
\begin{aligned}
G(z;0) &= \frac{1}{(2\pi)^d} \int \frac{1}{c_1} \frac{\mathrm{d}^d q}{z_\mathrm{c} - z + \left(c_2/c_1^2\right)|\vec{q}|^2} \\
&= \frac{K_d}{(2\pi)^d} \int \frac{1}{c_1} \frac{q^{d-1}\,\mathrm{d}q}{(z_\mathrm{c} - z) + c_2 q^2/c_1^2}
\end{aligned}
\tag{3.28}
$$

with the quantity K_d being the result of the angular integration, given by

$$
K_d = \frac{2\pi^{d/2}}{\Gamma(d/2)} \tag{3.29}
$$

Exercise 3.5

Equation (3.29) follows from the equation for the area, A, of a d-dimensional sphere of radius r:

$$
A = K_d r^{d-1}
$$

Derive (3.29) by evaluating the integral for the surface area of such a sphere:

$$
\begin{aligned}
A &= \int \cdots \int \delta\left(\sqrt{x_1^2 + \cdots + x_d^2} - r\right) \mathrm{d}x_1\,\mathrm{d}x_2 \ldots \mathrm{d}x_d \\
&= \int \cdots \int 2r\,\delta\left(x_1^2 + \cdots + x_d^2 - r^2\right) \mathrm{d}x_1\,\mathrm{d}x_2 \ldots \mathrm{d}x_d
\end{aligned}
$$

The result of the integral over $q = |\vec{q}|$ clearly depends on the dimensionality of q-space. When $d = 2$, we have

$$
\begin{aligned}
G(z;0) &\propto \int_0^{q_0} \frac{q\,\mathrm{d}q}{z_\mathrm{c} - z + c_2 q^2/c_1^2} \\
&= \frac{c_1^2}{2c_2} \ln\left(\frac{z_\mathrm{c} - z + c_2 q_0^2/c_1^2}{z_\mathrm{c} - z}\right).
\end{aligned}
\tag{3.30}
$$

Equation (3.30) tells us that, as $z \to z_\mathrm{c}$, $G(z;0) \propto -\ln(z_\mathrm{c} - z)$, so the number of two-dimensional non-recurring random walks is proportional to the coefficient of z^N in $1/(z_\mathrm{c} - z)\ln(z_\mathrm{c} - z)$. The asymptotic form of this coefficient is worked out

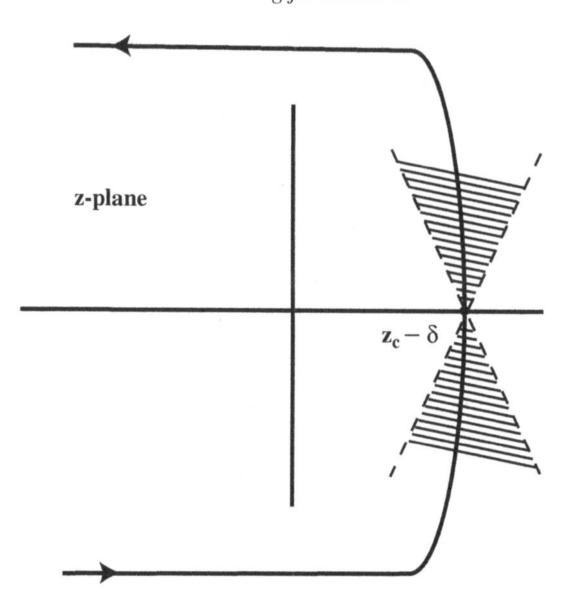

Fig. 3.2. The path of steepest descents.

in Supplement 3 of Chapter 2. Here, we will show how it can be obtained with the use of steepest descents.[4]

The coefficient of interest is given by Cauchy's formula as

$$\frac{1}{2\pi i}\oint \frac{1}{(z_c - z)\ln(z_c - z)}\frac{dz}{z^{N+1}} \equiv \oint h(z)e^{f(z)}\,dz$$

where $h(z) = 1/\ln(z_c - z)$, and $f(z) = -\ln(z_c - z) - (N+1)\ln(z)$. The function $f(z)$ has an extremum on the positive real axis at $z = z_c(1 - 1/N) = z_c - \delta$, with $f''(z_c) = 1/\delta^2 > 0$. Thus, the contour of integration can be nicely distorted to pass through the steepest descents path which, for the case at hand, is parallel to the imaginary axis passing through the extremum point $z_c - \delta$ (see Figure 3.2), yielding a value of a constant times $z_c^{-(N+1)}/\ln N$ for the integral.

The total number of walks leaving $\vec{x}$ is equal to z_c^{-N}. The number of non-recurring walks is smaller than the total number of walks by the factor $1/\ln N$. As $N \to \infty$, non-recurring walks become an infinitesimally small subset of all walks.

3.6 Recurrence when the dimensionality, d, lies between 2 and 4

When the dimensionality falls between 2 and 4 we can reduce the integral for $G(z; 0)$ as follows:

[4] See the supplement at the end of Chapter 1.

$$G(z;0) \sim \int_0^{q_0} \frac{q^{d-1}\,\mathrm{d}q}{z_c - z + c_2 q^2/c_1^2} \tag{3.31a}$$

$$= \frac{c_1^2}{c_2}\int_0^{q_0} \frac{q^{d-3}\left(z_c - z + c_2 q^2/c_1^2\right)\mathrm{d}q}{z_c - z + c_2 q^2/c_1^2} - \frac{c_1^2}{c_2}\int_0^{q_0} \frac{(z_c - z)q^{d-3}\,\mathrm{d}q}{z_c - z + c_2 q^2/c_1^2} \tag{3.31b}$$

$$= \frac{c_1^2}{c_2}\frac{q_0^{d-2}}{d-2} - \frac{c_1^2}{c_2}\int_0^{\infty} \frac{(z_c - z)q^{d-3}\,\mathrm{d}q}{z_c - z + c_2 q^2/c_1^2} + \frac{c_1^2}{c_2}\int_{q_0}^{\infty} \frac{(z_c - z)q^{d-3}\,\mathrm{d}q}{z_c - z + c_2 q^2/c_1^2} \tag{3.31c}$$

The reason that (3.31a) does not contain an equality is that we have approximated the denominator in the integral by the two lowest powers of q that appear in it. This tells us in general how the integral behaves, and we are able to properly reproduce the leading order singularity structure of $G(z;0)$ at $z = z_c$. However, the function itself does not emerge from the integration with complete accuracy.

In the second integral in (3.31c) we can rescale the integration variable q by defining $q = Qa(z_c - z)^{1/2}/c_2^{1/2}$. Furthermore, let the final contribution to (3.31c) be equal to $R(z_c - z)$. Then,

$$G(z;0) \propto \frac{c_1^2}{c_2}\frac{q_0^{d-2}}{d-2} - \left(\frac{a^2}{c_2}\right)^{d/2}(z_c - z)^{(d-2)/2}\int_0^{\infty}\frac{Q^{d-3}\,\mathrm{d}Q}{1+Q^2} + R(z_c - z) \tag{3.32}$$

The result on the right hand side of (3.32) tells us that the generating function $G(z;0)$ behaves non-analytically as a function of the fugacity z. There is, to be precise, a mathematical singularity in $G(z;0)$ at $z = z_c$. However, when $2 < d < 4$, the singularity appears as a positive, but non-integral, power of the quantity $z_c - z$. The first derivative with respect to z of the generating function is infinite at $z = z_c$. On the other hand, when $2 < d < 4$, the quantity $G(z_c;0)$ is finite. The number of walks that start at $\vec{x}$ and end up anywhere, never having revisited the point of origin is then, in the limit of very large N, equal to the coefficient of z^N in the power-series expansion of

$$\frac{z_c}{z_c - z}\frac{1}{G(z_c;0)} \tag{3.33}$$

That coefficient is $z_c^{-N}/G(z_c;0)$. Dividing by the total number of N-step walks, we find that the probability of a walker's starting out at $\vec{x}$ and ending up anywhere – never having revisited the point of origin – is equal to $G(z_c;0)^{-1}$.

In order to obtain precise results for this probability, we must attempt an accurate evaluation of the quantity $G(z_c;0)$.

Exercise 3.6
In the vicinity of $z = z_c$, we now know from (3.32) that a walk in three dimensions will have a generating function $G(z; 0)$ of the form

$$G(z; 0) = B_0 - B_1(z_c - z)^{1/2}$$

This means that to leading order in N, the number of walks that never revisit their point of origin is a finite fraction of all walks that start at that same point. Use the above expression to find the most significant correction to this leading order result when N is large but finite.

3.7 The probability of non-recurrence in walks on different cubic lattices in three dimensions

As noted above, a walker on a three-dimensional lattice will, with a probability that is *not* equal to zero, "escape" to infinity without ever having revisited the lattice point from which it started. We can make use of the formulas that we now have at our disposal to find out what this probability is in the case of walkers on various lattices. We will concentrate here on the three types of cubic lattices that were introduced earlier in this book: the simple cubic lattice, the body-centered cubic lattice and the face-centered cubic lattice.

Recall what we just found for the probability that an N-step walker, having left its point of origin, will never revisit that point. This probability is equal to

$$G(z_c; 0)^{-1} \tag{3.34}$$

Now,

$$\begin{aligned} G(z_c; 0) &= \frac{1}{(2\pi)^d} \int \frac{d^d q}{1 - z_c \chi(\vec{q})} \\ &= \frac{1}{(2\pi)^d} \int \frac{d^d q}{1 - \chi(\vec{q})/\chi(0)} \end{aligned} \tag{3.35}$$

The structure factor $\chi(\vec{q})$ is, in the three lattice cases we will talk about here, as given below.

- Simple cubic (sc)

$$\chi(\vec{q}) = 2\left(\cos q_x + \cos q_y + \cos q_z\right) \tag{3.36}$$

- Body-centered cubic (bcc)

$$\chi(\vec{q}) = 8 \cos q_x \cos q_y \cos q_z \tag{3.37}$$

- Face-centered cubic lattice (fcc)

$$\chi(\vec{q}) = 4\left(\cos q_x \cos q_y + \cos q_y \cos q_z + \cos q_x \cos q_z\right) \tag{3.38}$$

The three-fold integrations that lead to the probabilities of escape without return to the starting point are now known as Watson integrals (Watson, 1939). Their evaluation is detailed in the literature (Hughes, 1995). The results of the integrations can be expressed in terms of gamma functions and elliptic integrals. After some highly non-trivial analysis, one obtains the numerical results

$$G_{\rm sc}(z_{\rm c}; 0) \approx 1.516\,386\ldots \tag{3.39}$$

$$G_{\rm bcc}(z_{\rm c}; 0) \approx 1.393\,203\ldots \tag{3.40}$$

$$G_{\rm fcc}(z_{\rm c}; 0) \approx 1.344\,661\ldots \tag{3.41}$$

This means that the probability that a walker on a simple cubic lattice will never revisit its point of origin is approximately 0.659 463. The corresponding probabilities for the body-centered and face-centered cubic lattices are ≈0.71 777 and ≈0.743 682, respectively.

Exercise 3.7
In the case of a three-dimensional simple cubic lattice, construct an expression for the fraction of walks that escape without revisiting their point of origin. If the expression is in terms of an integral over q-space, it is not necessary to perform the integration.

Exercise 3.8
In the case of a three-dimensional simple cubic lattice, construct an expression for the fraction of walks that escape after visiting the point of origin exactly m times. Call that fraction f_m. Show that your expression leads to the equality

$$\sum_{m=0}^{\infty} f_m = 1$$

3.8 The number of sites visited by a random walk

Here we will apply some of the results just derived to see how many distinct sites are visited by a random walker, on average. To do this, we will start by finding out what fraction of all the walks of n steps end up on a site that was never visited before. We will then use the answer to this question to find out how many new sites are visited by a walker of arbitrary length. The connection between the two

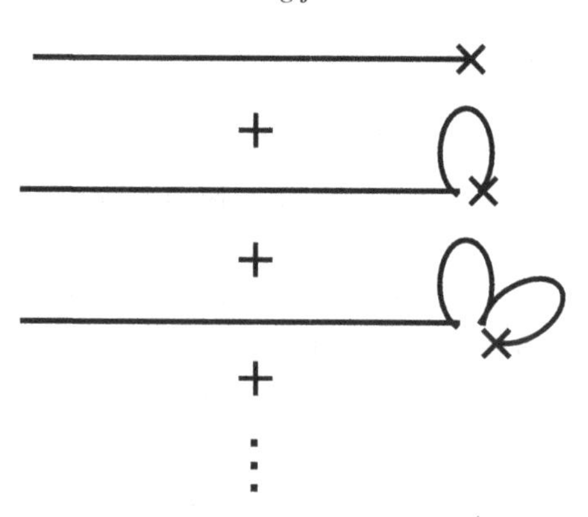

Fig. 3.3. Graphical representation of the set of terms that tell us how many walks that end up at a given site have never visited that site before.

quantities is fairly straightforward. Suppose a walker that has taken n steps ends up, on average, on Δ_n sites that have never been visited before. This means that at the nth step, Δ_n sites are, on the average, added to the total number visited by the walker. To find the total number of sites, we simply add up the new sites visited at each step leading up to the last one. Thus, if δ_n is the total number of distinct sites visited by the walker in n steps, then

$$\delta_n = \sum_{n'=0}^{n} \Delta_{n'} \tag{3.42}$$

Clearly, $\Delta_0 = 1$ and $\Delta_n = \delta_n - \delta_{n-1}$. Now, the average number of new sites visited at the nth step is just the probability that the walker has landed on a site not previously visited. This probability is related to what is known as the probability of *first passage* (Barber and Ninham, 1970). To see how this goes, define $F_n(\vec{x} - \vec{y})$ as the probability that the walker, starting at $\vec{x}$, reaches $\vec{y}$ for the first time at its nth step. Then, the sum $\sum_{\vec{y} \neq \vec{x}} F_n(\vec{x} - \vec{y})$ is the probability that the walker reaches *any* site for the first time at the nth step. This is another way of saying that the walker has reached a "new" site at the nth step. Thus,

$$\Delta_n = \sum_{\vec{y} \neq \vec{x}} F_n(\vec{x} - \vec{y}) \tag{3.43}$$

The connection is thus established. The next stage of our calculation is the determination of the probability, Δ_n. Here is how we do it. We make use of the method that was utilized to calculate recurrence in a random walk to find out how many walks that end at a given site have never visited that site before. This quantity is depicted in Figure 3.3. What we do is eliminate from the walks that end at the site of interest all walks that have visited that site previously, in the same way as we accomplished that elimination when we investigated the question of recurrence.

Calling the site of origin A and the site at which the walk ends up B, the sum represented by the diagrams in Figure 3.3 is given by

$$G(z;\vec{r}_{\mathrm{B}}-\vec{r}_{\mathrm{A}})-G(z;\vec{r}_{\mathrm{B}}-\vec{r}_{\mathrm{A}})G_1(z;0)+\cdots=G(z;\vec{r}_{\mathrm{B}}-\vec{r}_{\mathrm{A}})\frac{1}{1+G_1(z;0)} \quad (3.44)$$

As previously noted, the quantity $G_1(z;0)$ is related to the generating function for walks that start and end at the same step. The difference between G_1 and the standard generating function is that there must be at least one step in G_1. That is

$$G_1(z;0)=\sum_{N=1}^{\infty}z^N C(N;0) \quad (3.45)$$

where $C(N;0)$ is the sum of all N-step walks that start and end at the same point. This tells us that the spatial Fourier transform of G_1 is given by[5]

$$g_1(z;\vec{q})=\frac{z\chi(\vec{q})}{1-z\chi(\vec{q})} \quad (3.46)$$

It also tells us that

$$1+G_1(z;0)=G(z;0) \quad (3.47)$$

If we now sum over all possible end-points, we obtain the generating function for all walks that end up at a point that has not been previously visited. Summing over all end-points picks out the $\vec{q}=0$ Fourier transform of the generating function G, which means that the generating function that tells us how many walks end at a newly visited site is

$$\begin{aligned} g(z;0)\frac{1}{1+G_1(z;0)} &= \frac{1}{1-z\chi(0)}\frac{1}{1+G_1(z;0)} \\ &= \frac{z_{\mathrm{c}}}{z_{\mathrm{c}}-z}\frac{1}{1+G_1(z;0)} \end{aligned} \quad (3.48)$$

Now, let's evaluate the function $G_1(z;0)$. We have

$$\begin{aligned} G_1(z;0) &= \frac{1}{(2\pi)^d}\int\frac{z\chi(\vec{q})}{1-z\chi(\vec{q})}\,\mathrm{d}^d q \\ &= \frac{1}{(2\pi)^d}\int\frac{1}{1-z\chi(\vec{q})}\,\mathrm{d}^d q-\frac{1}{(2\pi)^d}\int\mathrm{d}^d q \\ &= \frac{1}{(2\pi)^d}\int\frac{1}{1-z\chi(\vec{q})}\,\mathrm{d}^d q-1 \end{aligned} \quad (3.49)$$

We've done this integral before. Using results that we have already, we find the following.

[5] For convenience, the Ω_{BZ} is set equal to $(2\pi)^d$.

(1) In one dimension

$$G_1(z;0) \propto (z_c - z)^{-1/2} \tag{3.50}$$

(2) In two dimensions

$$G_1(z;0) \propto \ln(z_c - z) \tag{3.51}$$

(3) In three dimensions

$$G_1(z;0) \propto \int d^d q \left(\frac{1}{1 - \chi(\vec{q})/\chi(0)} - 1 \right) - \mathcal{K}(z_c - z)^{1/2} + \cdots \tag{3.52}$$

The integral in (3.52), which represents the $z = z_c$ limit of the quantity G_1, is finite. In one and two dimensions, that limit yields an infinite result for G_1.

Let's see what we get for the asymptotic behavior of the generating function for the number of walks that end at a never-before-visited site. In one dimension, as $z \to z_c$, the generating function of interest goes as

$$\frac{1}{z_c - z} \frac{1}{(z_c - z)^{-1/2}} \to (z_c - z)^{-1/2} \tag{3.53}$$

We find (see Supplement 3 at the end of Chapter 2) that the coefficient of z^n in $(z_c - z)^{1/2}$, when n is large, goes as a constant time $\sqrt{2/\pi} n^{-1/2} z_c^{-n}$. If we divide by the number of all n-step walks, which goes strictly as z_c^{-n}, then, we have a probability of landing on a new site that decays away as $n^{-1/2}$ for large n. To find the average number of sites visited in n steps, we sum over this quantity for all n' up to n. The quantity of interest is, then,

$$\sqrt{\frac{2}{\pi}} \sum_{n'=1}^{n} \left(n'\right)^{-1/2} = 2\sqrt{\frac{2}{\pi}} n^{1/2} \tag{3.54}$$

We obtain the right hand side of (3.54) by approximating the sum on the left hand side by an integral, which is reasonable as long as n is large.

In two dimensions, we find the limiting behavior of the generating function for walks that end at a newly visited site is

$$\frac{1}{(z_c - z)\ln(z_c - z)} \tag{3.55}$$

Again, from Supplement 3 of Chapter 2, we find the coefficient of z^n in this sum, and we find that when n is large, this coefficient goes as $\pi z_c^{-n}/\ln n$. The sum that tells us how many distinct sites are visited, on the average, by the two-dimensional walker after n steps is, then, of the form

$$\pi \sum_{n'=2}^{\infty} \frac{1}{\ln n'} = \pi \frac{n}{\ln n} \tag{3.56}$$

The lower limit of the sum has been set equal to two so as to avoid the singularity at $n' = 1$. This singularity is an artifact of the use of an expression that is asymptotically correct for large argument in a regime in which it does not apply. The right hand side of (3.56) follows from the replacment of the sum on the right hand side by an integral, and then by an integration by parts. To be specific, we use

$$\int_1^n \frac{dn'}{\ln n'} = \left. \frac{n'}{\ln n'} \right|_1^n + \int_1^n \frac{dn'}{(\ln n')^2}$$
$$= \frac{n}{\ln n} + \cdots \tag{3.57}$$

where the ellipsis in (3.57) represents terms that are asymptotically smaller than the leading order one displayed there.

All this means that the number of distinct sites visited by a two-dimensional walker grows as the number of steps taken, divided by the log of the number of steps. When n is extremely large, this ratio is a negligible fraction of n. The results derived here are in agreement with the asymptotic formulas for δ_n derived originally with other methods (Montroll and Weiss, 1965).

Finally, in the case of the three-dimensional walker, the generating function of interest has the form, in the vicinity of $z = z_c$

$$\frac{z_c}{z_c - z} \frac{1}{G(z_c; 0)} \tag{3.58}$$

The coefficient of z^n for this walk goes as $z_c^{-n}/(G(z_c; 0))$. If we divide by the number of all n-step walks, we end up with the n-independent ratio $1/(G(z_c; 0))$. This tells us that the number of sites visited by an n-step walk goes as $n/(G(z_c; 0))$. In one and two dimensions, the number of distinct sites visited by a walker who goes on a long walk is an infinitesimal fraction of the total number of steps taken. In three dimensions, the number of distinct sites visited is, asymptotically, a determinable and finite fraction of the total number of steps taken in the course of the walk.

Though the results just obtained for recurrent walks and the number of distinct sites visited can also be arrived at by other means (Barber and Ninham, 1970), our goal was to demonstrate that these results follow relatively quickly and easily with the use of generating functions.

Exercise 3.9

Find the mean number of distinct sites visited by an N-step walk on a body-centered cubic lattice when N is very large. If your answer is in the form of an integral over q-space, it is not necessary to evaluate that integral.

Exercise 3.10

For the case of a biased one-dimensional walk, in which the probability of a step to the right, p, is not equal to 1/2, find the mean number of new sites visited by an N-step walk when N is very large. The leading order term in your result will be proportional to N. Can you justify the constant of proportionality qualitatively?

Exercise 3.11

In the case of an N-step walk on a three-dimensional, simple cubic lattice, the next-to-leading order term in the result for the mean number of distinct sites visited when N is very large goes as a power of N. What is that power?

Exercise 3.12

In the case of a simple cubic lattice, find out how many sites are visited, approximately and on the average, in an N-step walk, where N is large.

4

Boundary conditions, steady state, and the electrostatic analogy

The modification of random walk statistics resulting from the imposition of constraints on the walkers will be repeatedly visited in this book. In fact, several chapters will be dedicated to the treatment of walkers that are forbidden to step on points they have previously visited. This kind of constraint represents the influence of a walk on itself. As we will see, calculations of the properties of the self-avoiding random walk require the use of fairly sophisticated techniques. The payoff is the ability to extract the fundamental properties of a model that accurately represents the asymptotic statistics of important physical systems, particularly long linear polymers.

The discussion referred to above will take place in later chapters. Here, we focus on other constraints, which embody the influence of a static environment on a walker. In the first part of this chapter, we will address the effect on a random walker of the requirement that it remain in a specified region of space. Specifically, we will look at walkers confined to a half-space, a linear segment in one dimension and a spherical volume in three dimensions. Then we will look at the case of walkers that are confined to the region outside a spherical volume. This case turns out to be relevant to issues relating to the intake of nutrients and the excretion of wastes by a simple, stationary organism, such as a cell. In fact, the final section of this chapter utilizes random walk statistics to investigate the limits on the size of a cell that survives by ingesting nutrients that diffuse to it from its surroundings.

In some important cases the study of the effects of environmental influences on random walk statistics is facilitated if the walkers in question are members of an ensemble that is continuously replenished at various points in the system. Because of this replenishment, the number of such walkers at a given point does not change with time. In other words, the system has achieved *steady state*. This stratagem allows for the calculation of time-averaged quantities in many random walk problems. We will derive the equation that is satisfied by the function that describes the number of steady state walkers in a couple of ways. The equation is, in the approximations that are applicable here, equivalent to the equation for the electrostatic potential due to a collection of charges. That is, the calculation of the number of walkers in steady state is equivalent to the solution of the equation for the electrostatic potential due to a set of charges. In addition, we will see how boundaries that surround the region in which the walkers propagate can mimic the effects of the boundaries enclosing the region containing static charges and the electrostatic potential that they generate. This allows us to make use of the powerful techniques that have been developed to calculate electrostatic potentials (Jackson, 1999) to answer interesting questions about steady state random walks and diffusion. Finally, we will make use of the mathematical connection between steady state walks and electrostatics to look at diffusion in a region that surrounds a compact volume, such as a sphere. We will obtain results that are relevant to the process by which simple organisms absorb nutrients and eliminate waste.

4.1 The effects of spatial constraints on random walk statistics

The first question we ask is: what is the change in the total number of walks when a particular kind of barrier is present? Such a barrier will eliminate all walkers that pass through it. It represents a particularly effective (if ruthless) means of keeping walkers out of a "forbidden" region. This kind of partition has direct relevance to the actual, physical, barrier that surrounds the region to which a polymer may be confined. As we will see, the real confining surface that keeps a polymer in a particular region has the same mathematical properties as the absorbing boundary that does away with walkers who step across it.

4.1.1 The image walker

To be specific, imagine a set: of walkers who are free to wander in the half-space $x > 0$. However, any walker that steps on the boundary at $x = 0$ finds its path eliminated. One way of accomplishing this elimination is with the device of the "image walker" (see Figure 4.1). This walker starts out at a point that is the mirror image with respect to the $x = 0$ plane of the point of origin of the original walker. The way in which one determines the number of allowed walks between the confined walker's starting point and some ultimate destination is by counting all paths that the walker might have taken between those two points – *including* paths that enter the forbidden region $x \leq 0$ – and then subtracting the number of walks that image walkers can take from their starting point to the same final point. This subtraction has the effect of eliminating all of the original walker's paths that have passed over the boundary. The way in which the elimination is achieved is illustrated in Figure 4.1. The number of allowed walks, which we denote by $C'(N; \vec{r}, \vec{r}_0)$, is, then,

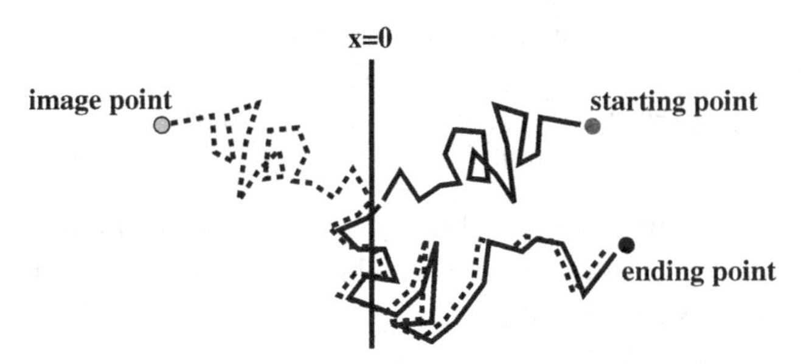

Fig. 4.1. Illustrating the way in which an image walker eliminates a path by the original walker that passes over the boundary at $x = 0$. For every path which crosses the boundary and returns to the ending point, there is a path starting at the image point that can reach the same ending point in the same number of steps, N. This is not the case for walks that do not cross the boundary. Note that the portion of the image walker's path that precedes its first contact with the boundary is the mirror image of the corresponding portion of the real walker's path and that after that first contact the image walker shadows the real one.

given by:

$$C'(N; \vec{r}, \vec{r_0}) = C(N; \vec{r} - \vec{r_0}) - C(N; \vec{r} - \vec{r_I}) \tag{4.1}$$

On the right hand side of (4.1), the function $C(N; \vec{r} - \vec{r}_0)$ is the number of unrestricted N-step walks that start at the point of origination, $\vec{r}_0$, and that end at the point $\vec{r}$, while $C(N; \vec{r} - \vec{r}_{\rm I})$ is the number of N-step walks starting at the image point $\vec{r}_{\rm I}$ that end up at $\vec{r}$. If $\vec{r}_0 = (x_0, y, z)$, then $\vec{r}_{\rm I} = (-x_0, y, z)$. In terms of the explicit expression that has been developed for the number of unrestricted walks in three dimensions, we have for the quantity C'

$$C'(N; \vec{r}, \vec{r}_0) = \mu^N \left(\frac{2\pi N}{3}\right)^{-3/2} \left[\mathrm{e}^{-3|\vec{r}-\vec{r}_0|^2/2N} - \mathrm{e}^{-3|\vec{r}-\vec{r}_{\rm I}|^2/2N}\right] \tag{4.2}$$

The quantity μ is the connectivity constant. For an unconstrained walk on a simple cubic (sc) lattice, $\mu = 6$. It is relatively straightforward to verify that the effect of the constraint that the walk remain in the region $x > 0$, as enforced by the image walker, has little effect on the total number of walks unless the number of steps in the walk, N, is sufficiently great that $N \gtrsim x_0^2$. We can make this a little more explicit by summing over all possible y and z coordinates of the final points. Using the fact that the Gaussian function $\mathrm{e}^{-\alpha(x^2+y^2+z^2)}$ can be written as a product of functions depending on only one of the three coordinates, we find that the integrals over y and z of both contributions to the right hand side of (4.2) are readily evaluated. The end result of this procedure is the following expression for the number of allowed walks starting at $\vec{r}_0$ and ending up anywhere in the plane $x = x_F$

$$\mu^N \left(\frac{2\pi N}{3}\right)^{-1/2} \left[e^{-3(x_F - x_0)^2/2N} - e^{-3(x_F + x_0)^2/2N}\right] \tag{4.3}$$

Suppose we would like to assess the *net* effect of the absorbing barrier on the number of walks. We can sum over all possible locations of the walker after it has taken N steps. This is an integral over x_F, from 0 to ∞. The result is

$$\Gamma(N; x_0) = \mu^N \left(\frac{2\pi N}{3}\right)^{-1/2} \int_0^\infty \mathrm{d}x_F \left[e^{-3(x_F - x_0)^2/2N} - e^{-3(x_F + x_0)^2/2N}\right] \tag{4.4a}$$

$$= \mu^N \left(\frac{2\pi N}{3}\right)^{-1/2} \left[\int_0^{2x_0} e^{-3(x_F - x_0)^2/2N}\, \mathrm{d}x_F + \int_{2x_0}^\infty e^{-3(x_F - x_0)^2/2N}\, \mathrm{d}x_F - \int_0^\infty e^{-3(x_F + x_0)^2/2N}\, \mathrm{d}x_F\right] \tag{4.4b}$$

$$= \mu^N \left(\frac{2\pi N}{3}\right)^{-1/2} \int_{-x_0}^{x_0} \mathrm{d}x\, e^{-3x^2/2N} \tag{4.4c}$$

(4.4c) follows from changes of integration variable in the terms in brackets in (4.4b).

Let's look at the final expression on the right hand side of (4.4c). The integral is an error function (Abramowitz and Stegun, 1970). If the limits of integration are sufficiently large that the absolute value of the exponent in the integrand is sizable, then the integral can be replaced by the complete Gaussian integral, and the end result for the total number of walks is μ^N, just as if there were no barrier in place. This limit is achieved under the condition $x_0^2 \gg 2N$. In other words, if the absorbing wall is far enough away from the walker's starting point that the walker is unlikely to encounter the wall in the course of its wanderings, then the wall will exert a relatively mild effect on the number of walks.

On the other hand, suppose that the opposite limit holds, $x_0^2 \ll 2N$. Then, the right hand side of (4.4c) is well-approximated by the expression

$$\mu^N \left(\frac{2\pi N}{3}\right)^{-1/2} \times 2x_0 \tag{4.5}$$

In the limit that the number of steps, N, is large, the number of allowed walks becomes a small fraction of the number of unrestricted walks. Dividing (4.5) by the number of unrestricted walks, we find that the reduction in the total number is $\sqrt{6x_0^2/\pi N}$.

Finally, we can sum over all initial points, to see what effect the absorbing wall has on the total number of walks that start anywhere. We start by recasting the right hand side of (4.4c) as follows:

$$\begin{aligned}
&\mu^N \left(\frac{2\pi N}{3}\right)^{-1/2} \int_{-x_0}^{x_0} \mathrm{d}x \mathrm{e}^{-3x^2/2N} \\
&= \mu^N \left(\frac{2\pi N}{3}\right)^{-1/2} \left[\int_{-\infty}^{\infty} \mathrm{d}x \mathrm{e}^{-3x^2/2N} - 2\int_{x_0}^{\infty} \mathrm{d}x \mathrm{e}^{-3x^2/2N}\right] \\
&= \mu^N - 2\mu^N \left(\frac{2\pi N}{3}\right)^{-1/2} \int_{x_0}^{\infty} \mathrm{e}^{-3x^2/2N}\, \mathrm{d}x \qquad (4.6)
\end{aligned}$$

The first term on the right hand side of (4.6) is just the total number of unrestricted walks that start out at x_0. The second term represents the reduction in the total number of walks that results from the presence of the absorbing barrier. We now sum over all possible starting points, by integrating over x_0, to determine the net effect of the boundary on walks starting anywhere in the half-space $x > 0$. This integration is readily performed

$$\begin{aligned}
\int_0^{\infty} \mathrm{d}x_0 \int_{x_0}^{\infty} \mathrm{e}^{-3x^2/2N}\, \mathrm{d}x &= x_0 \int_{x_0}^{\infty} \mathrm{e}^{-3x^2/2N}\, \mathrm{d}x \Bigg|_{x_0=0}^{x_0=\infty} + \int_0^{\infty} x_0 \mathrm{e}^{-3x_0^2/2N}\, \mathrm{d}x_0 \qquad (4.7a) \\
&= \frac{1}{2}\int_0^{\infty} \mathrm{e}^{-3y/2N}\, \mathrm{d}y \qquad (4.7b) \\
&= \frac{N}{3} \qquad (4.7c)
\end{aligned}$$

(4.7a) represents the results of an integration by parts. To obtain (4.7b), we implemented a change of integration variables.

Combining (4.7) with (4.6) we find that the net reduction in the number of walks is given by

$$\mu^N \left(\frac{2N}{3\pi}\right)^{1/2} \qquad (4.8)$$

The absorbing wall has the effect of substantially eliminating all walks that start within a distance $\sim\sqrt{N}$ of it.

Integration of the first term on the right hand of (4.6) yields the total number of walks that there would be if no absorbing wall were present. We will see that the reduction in the total number of walkers because of an absorbing wall can have dramatic effects when the volume to which the walker is confined is finite.

4.1.2 Confinement to a one-dimensional region

The random walker may face further restrictions. There may be walls on more than one side, forcing the walk to take place in a smaller portion of space. For example, a one-dimensional walker may find itself confined by absorbing walls to the region $0 \leq x \leq L$. Two questions naturally arise in this case:

(1) What are the consequences of this confinement?
(2) How does one perform calculations leading to the answer to the question above?

Clearly, a successful response to the first query depends crucially on how well one is able to address the second. It is tempting to consider the use of image walkers to effect a cancellation of unwanted walks. In fact, this stratagem can be implemented. Unfortunately, it is necessary to introduce an infinite number of image walkers in order to carry it out. As a result, one is faced with awkward summations. A much more efficient approach involves the use of the generating function, and the introduction of an appropriate basis set of functions.

Recall that the generating function obeys the equation (2.9). This equation was simplified by the introduction of the spatial Fourier transform of the generating function. The transformed generating function was defined in terms of plane waves, $e^{i\vec{k}\cdot\vec{r}}$. These functions express in the simplest form the translational symmetry of the configuration. However, plane waves are not the most suitable basis set when the environment is inconsistent with periodic boundary conditions. In the case at hand, because of the presence of absorbing walls at $x = 0$ and $x = L$, the generating function must equal zero at those two boundaries.[1] A set of functions that incorporate the translational invariance of the recursion relation in the interior of the allowed volume of space, and that also respect the action of the absorbing walls, is the following

$$\psi(x, y, z) = e^{ik_y y + ik_z z} \sin qx \tag{4.9}$$

The wave-vector component q is quantized according to the prescription

$$q = \frac{n\pi}{L}; \quad n = 1, 2, 3 \ldots \tag{4.10}$$

If (4.10) is satisfied, then it is easy to demonstrate that the function $\psi(x, y, z)$ is zero at both $x = 0$ and $x = L$.

Now, let's simplify the situation, and focus on a one-dimensional walker confined as above. The recursion relation is as presented in Chapter 1

$$G(z; x_1, x_2) = z\left(G(z; x_2 + \Delta, x_1) + G(z; x_2 - \Delta, x_1)\right) + \delta_{x_1, x_2} \tag{4.11}$$

Here, Δ is the length of the step taken. For a walker on a lattice, Δ is the lattice spacing. Expanding the generating function in terms of the sine waves introduced

[1] It is superficially reasonable that the generating function is equal to zero at an absorbing barrier. The detailed argument leading to this condition is simple but not entirely trivial. See the supplement at the end of this chapter; for a more detailed discussion, see (Weiss, 1994).

above, we find that (4.11) is replaced by the following equation for the Fourier amplitude $g(z; q)$:

$$g(z;q) = 2z\cos q\Delta g(z;q) + \sqrt{\frac{2}{L}}\sin qx_1 \tag{4.12}$$

The L-dependent factor on the right hand side of (4.12) results from the normalization of the sinusoidal mode. Solving this equation for $g(z; q)$, we obtain

$$g(z;q) = \sqrt{\frac{2}{L}}\frac{\sin qx_1}{1 - 2z\cos q\Delta} \tag{4.13}$$

To make life even simpler, we will expand the denominator in (4.13) in powers of q and retain only those terms in the power-series expansion that are quadratic or lower. This approximation is tantamount to replacing the difference equation (4.11) by a partial differential equation; see Chapter 2, Equation (2.37). We then find

$$g(z;q) = \sqrt{\frac{2}{L}}\frac{\sin qx_2}{(1 - 2z) + z\Delta^2 q^2} \tag{4.14}$$

Reconstituting the generating function in real space, we find

$$G(z;x_1,x_2) = \frac{2}{L}\sum_{n=0}^{\infty}\left[\sin\left(\frac{n\pi x_1}{L}\right)\sin\left(\frac{n\pi x_2}{L}\right)\right]\Big/\left[1 - 2z + z\Delta^2\left(\frac{n\pi}{L}\right)^2\right] \tag{4.15}$$

If this expression for $G(z; x_1, x_2)$ looks familiar to you – it should. We already demonstrated in Chapter 2 that the generating function satisfies, in the continuum limit, a Green's function-type equation, see (2.38). And, (4.15) is a solution to that equation using eigenfunctions appropriate to absorbing-wall boundary conditions.

Exercise 4.1
Derive an expression for the generating function, $G(z; x_1, x_2)$, of walkers confined to the half-space $x > 0$.

Exercise 4.2
Using the generating function obtained above show that (4.3) results.

Let's use the result (4.15) to calculate the total number of one-dimensional walks in the region $0 \le z \le L$, by integrating over both the final position of the walk, x_1

and the walk's initial location x_2. Making use of the result

$$\int_0^L \sin\left(\frac{n\pi}{L}\right) \mathrm{d}x = \frac{L}{n\pi}(1 - \cos n\pi)$$
$$= \begin{cases} 0 & n = 2m \\ 2L/n\pi & n = 2m + 1 \end{cases} \tag{4.16}$$

we find the total number of walks in the confined region is

$$\frac{8L}{\pi^2} \sum_{m=0}^{\infty} \frac{1}{(2m+1)^2} \frac{1}{1 - 2z + z\Delta^2((2m+1)\pi/L)^2} \tag{4.17}$$

As we are interested in the total number of N-step walks, we extract the coefficient of z^N in (4.17). This procedure generates the factor

$$\left(2 - \Delta^2\left(\frac{(2m+1)\pi}{L}\right)^2\right)^N = \exp\left[N \ln\left(2 - \Delta^2\left(\frac{(2m+1)\pi}{L}\right)^2\right)\right]$$
$$\to 2^N \exp\left[-N\Delta^2\left(\frac{(2m+1)\pi}{L}\right)^2 \Big/ 2\right] \tag{4.18}$$

We then obtain the following result for the total number of one-dimensional random walks that start anywhere – and that end anywhere – in the region $0 \leq x \leq L$

$$2^N \frac{8L}{\pi^2} \sum_{m=0}^{\infty} \frac{1}{(2m+1)^2} \exp\left[-N\Delta^2\left(\frac{(2m+1)\pi}{L}\right)^2 \Big/ 2\right] \tag{4.19}$$

Exercise 4.3

Show where image walkers are placed in order to reproduce the effect of absorbing boundaries on walkers confined to the region $0 < x < L$.

Exercise 4.4

Verify by explicit calculation that (4.12) follows from the substitution of a sine-wave expansion of the generating function in (4.11).

4.1.3 Limiting cases

The expression (4.19) does not reduce to a closed form in terms of elementary functions. However, it is possible to see what happens in interesting limits. For instance, if the number of steps is sufficiently small that $\Delta^2 N/L^2 \ll 1$, we can replace the exponential in (4.19) by one, and we obtain for the total number of walks confined to the region $0 \leq x \leq L$

$$2^N \frac{8L}{\pi^2} \sum_{m=0}^{\infty} \frac{1}{(2m+1)^2} \tag{4.20}$$

The sum in (4.20) is readily performed if we take advantage of the result (Gradshteyn *et al.*, 2000)

$$\sum_{n=1}^{\infty} \frac{1}{n^2} = \frac{\pi^2}{6} \tag{4.21}$$

Then, we write the sum over odd integers as the difference between the sum over all integers and the sum over even integers

$$\begin{aligned}
\sum_{m=0}^{\infty} \frac{1}{(2m+1)^2} &= \sum_{n=1}^{\infty} \frac{1}{n^2} - \sum_{l=1}^{\infty} \frac{1}{(2l)^2} \\
&= \sum_{n=1}^{\infty} \frac{1}{n^2} - \frac{1}{4} \sum_{k=0}^{\infty} \frac{1}{k^2} \\
&= \frac{3}{4} \sum_{n=0}^{\infty} \frac{1}{n^2} \\
&= \frac{3}{4} \frac{\pi^2}{6} \\
&= \frac{\pi^2}{8}
\end{aligned} \tag{4.22}$$

Inserting this result into (4.20), we obtain for the total number of N-step one-dimensional walks in a sufficiently large domain

$$2^N L \tag{4.23}$$

The total number of N-step walks in the domain is just equal to the total number of N-step unrestricted walks multiplied by the volume of the domain. This latter factor simply takes into account the fact that we count walks that start anywhere in the domain.

On the other hand, in the opposite limit, $N\Delta^2/L^2 \gg 1$, the exponentials play a decisive role in the evaluation of the sum over m in (4.19). In fact, the exponential gives rise to a rapid damping out of the terms in the sum. The first term dominates, and we are left with

$$2^N \frac{8L}{\pi^2} \mathrm{e}^{-\pi^2 \Delta^2 N/L^2} \tag{4.24}$$

The effect of the boundary is to introduce an exponential decay into the statistics of the confined walk. There is an explanation for this factor based on entropic considerations, due to de Gennes (1979). One argues that the reduction in the number of allowed walks ought to be expressible in the form e^S, where S is an entropy-like function. It is known that the entropy of a large system scales linearly with the size of the system. In the case of the random walk, the "size" is the number of steps in the walk. Thus, $S \propto N$. On the other hand, the entropy ought to depend, in a dimensionless way, on the size of the region to which the walker is confined. That is, S must be a function of L. There is another length in the system: the natural extension of the one-dimensional walk. The square of this length goes as $\Delta^2 N$. It is, thus, reasonable to assume that there is an entropy reduction factor going like $e^{-\alpha N \Delta^2/L^2}$, where α is a numerical constant that cannot be inferred from general arguments such as this one.

Another way to understand the factor is in terms of "leakage" of random walkers out of the confined region. Imagine that a single walker starts out at the point of origin of the walk. Instead of deciding where to step next, the walker splits up into a set of descendants, one for each possible step. For instance, the one-dimensional walker splits into two at each step. One of the descendants moves to the right, and the other descendant moves to the left. To determine the number of walks after N steps, one counts the number of descendants. The leakage mentioned above is a process that competes with the proliferation of walkers. When descendants of the original walker impinge on the absorbing boundary, they are eliminated from the population. To calculate how many N-step walks take place in the region, it suffices to determine the outcome of the combined processes of proliferation and depletion.

The fission of walkers at each step multiplies the population by two. The net effect of depletion at the boundary can also be qualitatively assessed. The loss of walkers at the boundary leads to a reduction of the population in the immediate vicinity of the edge of the region. The loss is proportional to the "flux" of walkers across the boundary, where the flux is proportional to the slope of the distribution. That this is so follows from the fact that the number of walkers arriving from a neighboring point scales with the number of walkers that were at that point at the moment immediately preceding the last step. If there are more walkers to the right of a given location than to the left, then more walkers will be passing by that location moving right to left than left to right. The excess in left-moving walkers will increase proportionately to the difference between the population of the site to the right and the site to the left. Quantitatively, we write

$$J(x) = -\alpha \frac{dn(x)}{dx} \tag{4.25}$$

where $J(x)$ is the flux of walker and α is a constant.[2] The minus sign indicates that

[2] A more detailed discussion of this relation will be given later in the chapter. See Section 4.2.1 onwards.

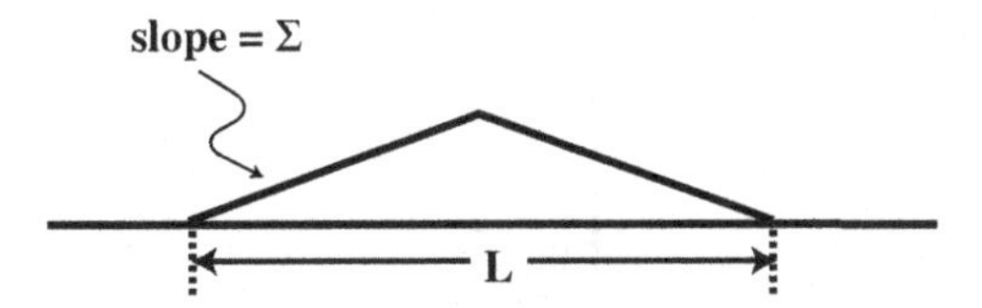

Fig. 4.2. Density of walkers in a finite region. Note that the density falls off at the edge of the allowed region, where the absorbing boundary depletes the population of walkers. This figure is an accurate representation of the distribution of a set of walkers that is continually replenished through the introduction of new walkers at the center, while depletion takes place as walkers are absorbed at the edges.

the random walkers tend to move from more heavily populated regions to regions in which there is a relative deficit of walkers. The next step is to relate the slope of the distribution at the boundary to the total number of walkers contained in the finite region. Suppose that the slope of the distribution of walkers is Σ. Then, the total number of walkers in a region of width L is approximately L times the magnitude of the distribution. This magnitude goes as the slope times the width of the region (see Figure 4.2). This means that the total number of walkers, n_{w}, is equal, to within a constant of proportionality, to ΣL^2, or that $\Sigma = \kappa n_{\mathrm{w}}/L^2$, where κ is a proportionality constant.

In light of the above discussion, if $n_{\mathrm{w}}(N)$ is the number of walkers after N steps, then the number of walks after $N + 1$ steps is given by

$$n_{\mathrm{w}}(N) = n_{\mathrm{w}}(N-1) \times 2 \times \left(1 - \frac{\alpha\kappa}{L^2}\right) \tag{4.26}$$

Iterating this relationship we find

$$\begin{aligned} n_{\mathrm{w}}(N) &= 2^N \left(1 - \frac{\alpha\kappa}{L^2}\right)^N n_{\mathrm{w}}(0) \\ &= 2^N \exp\left[N \ln\left(1 - \frac{\alpha\kappa}{L^2}\right)\right] \\ &\approx 2^N \mathrm{e}^{-\kappa\alpha N/L^2} \end{aligned} \tag{4.27}$$

The constants κ and α follow from detailed calculations, which in the case at hand lead to the more precise formula (4.24).

4.1.4 Confinement in a sphere

As a final investigation of the effects of spatial constraints on random walk statistics, we will consider a three-dimensional walk confined to a sphere of radius R. Here, the appropriate basis set is the combination, in spherical coordinates (Morse and

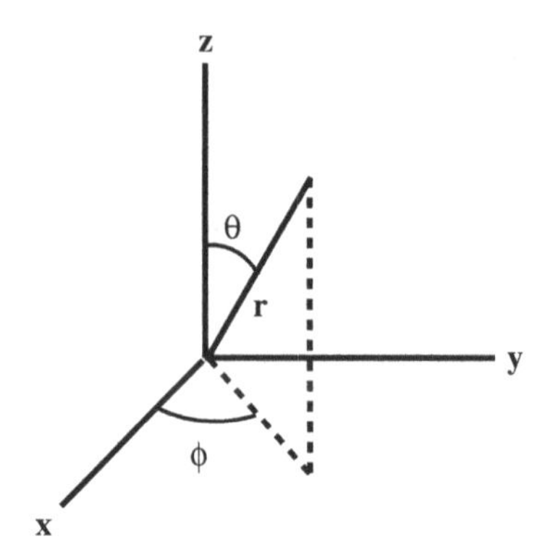

Fig. 4.3. The relationship between Cartesian and spherical coordinates.

Feshbach, 1953)

$$j_l(r)Y_l^m(\theta, \phi) \tag{4.28}$$

where $j_l(r)$ is a spherical Bessel function and $Y_l^m(\theta, \phi)$ is a spherical harmonic (Abramowitz and Stegun, 1970). The relationship between the Cartesian coordinate system (x, y, z) and the spherical coordinates (r, θ, ϕ), is illustrated in Figure 4.3. The solution to the three-dimensional version of (2.38) is given in terms of the functions in (4.28) by

$$\sum_{l=0}^{\infty} \sum_{|m|\leq l} \sum_{n_l=1}^{\infty} C_{n_l} \frac{j_l(k_{n_l} r_1) j_l(k_{n_l} r_2) Y_l^m(\theta_1, \phi_1) Y_l^m(\theta_2, \phi_2)}{1 - 6z + z\Delta^2 k_{n_l}^2} \tag{4.29}$$

The quantity C_{n_l} represents the normalization of the basis set $j_l Y_l^m$. The allowed values of k_{n_l} are determined by the properties of the spherical Bessel functions. The quantity k_{n_l} is adjusted to that $j_l(k_{n_l} R) = 0$. In this way, the boundary conditions appropriate to an absorbing wall at the outer bounding surface of the sphere are reflected in the basis set used.

We will focus on the total number of walks inside the sphere, which means that we will sum (actually integrate) formula (4.29) over all spatial coordinates. The spherical harmonics have the property that

$$\int_0^{2\pi} \mathrm{d}\phi \int_0^{\pi} \sin\theta \, \mathrm{d}\theta \; Y_0^0(\theta, \phi) Y_l^m(\theta, \phi) = \delta_{l,0}\delta_{m,0} \tag{4.30}$$

The integral over the angular coordinates thus leave us with the following expression to evaluate

$$\frac{8\pi}{R} \sum_{n_0} \int_0^R r_1^2 \, \mathrm{d}r_1 \int_0^R r_2^2 \, \mathrm{d}r_2 \frac{2}{R} \frac{j_0(k_0 r_1) j_0(k_0 r_2)}{1 - 2z + z\Delta^2 k_{n_0}^2} \tag{4.31}$$

The spherical Bessel function $j_0(x)$ is given by

$$j_0(x) = \frac{\sin x}{x} \tag{4.32}$$

This means that the k_{n_0}'s are given by

$$k_{n_0} = \frac{n_0 \pi}{R} \tag{4.33}$$

where the integers n_0 range from 1 to ∞. Integrating over the two radial coordinates, we are left with the following sum

$$\frac{8R^3}{\pi} \sum_n \frac{1}{n^2} \frac{1}{1 - 6z + z\Delta^2 \pi^2 n^2 / R^2} \tag{4.34}$$

We are interested in the coefficient of z^N in the above expression. Once again, we expand with respect to z. We find for the number of N-step walks originating and terminating anywhere in the spherical region for which the walker never leaves that region

$$6^N \frac{8R^3}{\pi} \sum_{n=1}^{\infty} \frac{1}{n^2} \mathrm{e}^{-\Delta^2 \pi^2 N / R^2} \tag{4.35}$$

As in the case of the one-dimensional confined walk, we do not have a formula that reduces to a simple, closed form expression. Again, as in the one-dimensional case, we can obtain limiting behavior. When $N\Delta^2/R^2 \ll 1$, we can ignore the exponential in (4.35). The number of walks in that region is, then, equal to

$$6^N \frac{8R^3}{\pi} \sum_{n=1}^{\infty} \frac{1}{n^2} = 6^N \left(\frac{4}{3} \pi R^3 \right) \tag{4.36}$$

The right hand side of (4.36) follows from (4.21). Note that we recover in this limit the result that the total number of random walks in the sphere is equal to the total number of completely unrestricted random walks multiplied by the volume of the sphere. This is entirely consistent with the result that was obtained in the equivalent limit for one-dimensional random walks confined to a finite domain.

In the limit that $N\Delta^2/R^2 \gg 1$, the sum is once again over a set of rapidly decreasing terms. The leading contribution is the first one:

$$\frac{8R^3}{\pi} \left(6\mathrm{e}^{-\Delta^2 \pi^2 / R^2} \right)^N \tag{4.37}$$

Once again, there is the N-independent prefactor and the exponential decay term.

This term derives from the mechanisms mentioned in the discussion of the one-dimensional case.[3]

Exercise 4.5

Derive an expression for the total number of N-step walks available to a walker confined to a cube of side length L.

4.2 Random walk in the steady state

Our focus in this section will be somewhat different. We will be interested in time averages, rather than what takes place at some particular point in the history of the random walk. We can address this problem by considering an ensemble of walkers, rather than a single one. We start by reformulating the problem in terms of what happens to a group of walkers. As discussed previously, we measure time in discrete intervals. At the Nth interval, the number of walkers at location $\vec{r}$ will be depicted by the function $W(N;\vec{r})$. Let's assume that the walkers move around on a cubic lattice, and we'll also assume that the walk is unbiased. There is a recursion relation between $W(N;\vec{r})$ and the quantity at the next time step. Here is the recursion relation:

$$W(N+1;\vec{r}) = \frac{1}{6}\sum_{\vec{\rho}_i} W(N;\vec{\rho}_i) + S(\vec{r}) \tag{4.38}$$

The $\vec{\rho}_i$'s in (4.38) are the locations of the sites nearest to the site at $\vec{r}$. We assume a simple cubic lattice, and the factor $1/6$ in (4.38) reflects the fact that each walker has six choices for the direction in which it is to step. The final term on the right hand side of (4.38) is the source term, and it represents the constant supply of walkers that is provided at the various points at which the walkers originate.

It is in this latter point that our reformulation represents the greatest change from previous discussions. We will allow walkers to appear at different instants in time. In fact, we will now imagine a steady supply of walkers, materializing at a continuous rate at one or more points in space. There may be some difficulty in imagining this scenario, unless the walkers appear at the boundary of the region under consideration, in which case they will have entered from that region's surroundings. However, if the walkers are molecules, this kind of production is possible if chemical reactions are taking place, of which the walkers in question are the end-products. From a mathematical point of view, we are creating a steady state that reproduces the results of averaging over a long time interval.

[3] For an additional discussion of the above situation, see (Barber and Ninham, 1970).

Now, in the steady state, the number of walkers at a particular point in space is independent of time, which means that we can replace $W(N;\vec{r})$ by the N-independent quantity $w(\vec{r})$. The recursion relation is, then, a requirement on this expression:

$$w(\vec{r}) = \frac{1}{6}\sum_{\vec{\rho}_i} w(\vec{\rho}_i) + S(\vec{r}) \tag{4.39}$$

As we've done before, we'll perform a Taylor-series expansion of the terms in the sum on the right hand side of (4.39). Retaining terms up to second order in the derivative of $w(\vec{r})$, we end up with

$$w(\vec{r}) = w(\vec{r}) + \frac{a^2}{3}\nabla^2 w(\vec{r}) + S(\vec{r}) \tag{4.40}$$

The quantity a is the distance between the lattice point at $\vec{r}$ and any one of its six nearest neighbors. That there are no linear terms follows from the reflection invariance of the square lattice to which the walkers are restricted. We've already seen that a lattice is not a necessity for the derivation of equations like (4.40). Cancelling the $w(\vec{r})$'s on the two sides of (4.40), we are left with the following equation for the steady state distribution of random walkers on the lattice:

$$-\nabla^2 w(\vec{r}) = \frac{3}{a^2} S(\vec{r}) \tag{4.41}$$

Except for the constant of proportionality, this is identical to the equation coupling the electrostatic potential, $\phi(\vec{r})$, and the charge distribution, $\rho(\vec{r})$, that gives rise to it.

Thus, *steady state diffusion, or the steady state random walk with sources, maps onto Poisson's equation for the electrostatic potential.*

In order to fully exploit the relationship between the two problems, we must also review the effects of boundaries. We've already looked at the influence of absorbing boundaries. At such bounding surfaces, the distribution of walkers vanishes. This is equivalent, in the electrostatic analogy, to having the potential $\phi(\vec{r})$ constant and equal to zero. In other words, absorbing boundaries play the same mathematical role in steady state diffusion as conducting boundaries do in electrostatics. There is another important type of boundary, and that is one that reflects the walkers that impinge on it. The appropriate boundary condition will be taken up after we rederive (4.41) with the use of another approach that is more intuitive, but identical in effect, to the recursion-relation-based derivation of that Poisson-like equation for the distribution of walkers in the steady state.

4.2.1 Fick's law and the random walk

The second approach to the question of the distribution of random walkers in the steady state utilizes the notion of a current density of walkers and builds on the argument in Section 4.1.3. Let's imagine that we have a collection of walkers meandering through a continuum. We'll assume that there is an overall organized movement of the walkers in any given point in space, superimposed on the random motion. As it turns out, this organized motion is driven by spatial imbalances in the concentration of the walkers, and it is an inevitable consequence of the random nature of the movement of individual walkers. If we describe this overall motion in terms of a current density, $\vec{j}(\vec{r}, t)$, then, the density of walkers, which we'll denote by $c(\vec{r}, t)$, satisfies the continuity equation

$$\frac{\partial c(\vec{r}, t)}{\partial t} = -\vec{\nabla} \cdot \vec{j}(\vec{r}, t) + S(\vec{r}, t) \tag{4.42}$$

This equation expresses the fact that a change in the number of walkers in a given region will either result from an excess of current into the region or from any external source that might be "materializing" walkers into that region.

The current density is the result of walkers that wander in from neighboring regions. Because the walkers step in random directions, we expect that the current density that occurs will be the result of walkers "spilling" into regions with a lower concentration of walkers from regions of higher concentration. We can quantify this by writing

$$\vec{j}(\vec{r}, t) = -\beta \vec{\nabla} c(\vec{r}, t) \tag{4.43}$$

This relationship, which actually follows from an analysis that takes a form similar to our recursion relations, is known as *Fick's first law*.[4] The parameter β is empirical. It can be obtained if we know more about the details of the random walk process, or it can be determined from measurements on the system in which the actual walk is taking place. Combining (4.42) and (4.43), we obtain

$$\frac{\partial c(\vec{r}, t)}{\partial t} = \beta \nabla^2 c(\vec{r}, t) + S(\vec{r}, t) \tag{4.44}$$

This is a continuum version of the recursion relation that we've used to obtain the generating function. For the time being, we'll focus on the steady state, but it is possible to construct variants of the equations that we've used to obtain the generating functions for random walks from (4.44). Making the assumption that the source term is time-independent, and removing the time dependence from the function $c(\vec{r}, t)$, we are left with the steady state equation

$$-\beta \nabla^2 c(\vec{r}) = S(\vec{r}) \tag{4.45}$$

[4] It is apparent from (4.43) that the appropriate boundary condition for a wall that perfectly reflects walkers instead of absorbing them is that the net flux of walkers into the wall is zero, or, equivalently, $\hat{n} \cdot \vec{\nabla} c = 0$, where $\hat{n}$ is the unit vector perpendicular to the wall.

Again, we have derived an equation that is mathematically identical to Poisson's equation for the electrostatic potential.

Worked-out example

For a steady stream of one-dimensional walkers emanating from the point x_0, which lies between two absorbing boundaries located at $x = 0$ and at $x = L$, find the concentration of walkers at any point $0 < x < L$.

Solution

The one-dimensional version of the equations satistifed by the distribution of walkers is

$$\frac{\partial c(x,t)}{\partial t} = -\frac{\partial j(x,t)}{\partial x} + s_0\delta(x - x_0) \tag{4.46a}$$

$$j(x,t) = -\beta\frac{\partial c(x,t)}{\partial x} \tag{4.46b}$$

where (4.46a) is the one-dimensional version of the continuity equation and (4.46b) is the one-dimensional version of Fick's first law. In the steady state, the equation satisfied by the concentration, $c(x)$ is

$$\beta\frac{d^2c(x)}{dx^2} = -s_0\delta(x - x_0) \tag{4.47}$$

This equation tells us that the first derivative of the steady state concentration is constant throughout the region $0 < x < L$, except for the point $x = x_0$, at which the slope of $c(x)$ suffers a discontinuity, the concentration itself remaining continuous. Furthermore, we know that the concentration obeys the boundary conditions $c(0) = c(L) = 0$. The discontinuity in the slope is given by

$$\left.\frac{dc(x)}{dx}\right|_{x_0^+} - \left.\frac{dc(x)}{dx}\right|_{x_0^-} = -\frac{s_0}{\beta} \tag{4.48}$$

The solution with all these characteristics is

$$c(x, x_0) = \begin{cases} \dfrac{s_0}{\beta} x \dfrac{L - x_0}{L} & x < x_0 \\ \dfrac{s_0}{\beta} x_0 \dfrac{L - x}{L} & x > x_0 \end{cases} \tag{4.49}$$

Exercise 4.6

Show that the probability of a walker being absorbed at $x = 0$ in the example above is given by

$$P(x_0) = \frac{L - x_0}{L}$$

Note that as $L \to \infty$, $P(x_0) \to 1$. In other words, the walker never escapes to infinity.

4.2.2 Solution for $c(\vec{r})$ in the presence of a localized source

We'll start with the simplest configuration of sources: one that is localized at a single point in space, which we'll place at the origin of our coordinate system. Then,

$$S(\vec{r}) = s_0 \delta(\vec{r}) \tag{4.50}$$

We'll take as our steady state equation, the one we've derived from Fick's first law:

$$\begin{aligned} -\beta \nabla^2 c(\vec{r}) &= S(\vec{r}) \\ &= s_0 \delta(\vec{r}) \end{aligned} \tag{4.51}$$

By analogy to electrostatics, we know, more or less immediately, that the solution to this equation is

$$c(\vec{r}) = \frac{s_0}{4\pi\beta} \frac{1}{|\vec{r}|} \tag{4.52}$$

The distribution falls off as a power law in the distance from the source. Note that this is very different from the Gaussian distribution that we derived, repeatedly, earlier on. That distribution had an essential time-dependence, expressed in terms of N, the number of steps that the walker had taken. The situation to which the former distribution applies is one in which there is a source that supplies a walker, or a set of walkers, at an instant in time and then "turns off." By contrast, for (4.52) walkers materialize at the origin at a constant rate.

Connection between the steady state and Gaussian Distributions

It is instructive to see how the result for the steady state distribution of walkers emanating from a point source can be derived from the familiar Gaussian distribution. In the latter case, the walkers are placed at a particular point in space at an instant of time. This corresponds to a source of the form

$$S(\vec{r}, t) = S_0\delta(\vec{r})\delta(t) \tag{4.53}$$

The distribution of walkers will have the form

$$c_G(\vec{r}, t) \propto t^{-3/2}e^{-r^2/at} \tag{4.54}$$

for times t greater than zero. In the steady state situation, the source term is constantly "on," which means that the distribution is obtained by integrating the expression in (4.54) for over all positive values of t. Thus, we have for the steady state distribution

$$c_S(\vec{r}) \propto \int_0^\infty t^{-3/2}e^{-r^2/at}\,dt \tag{4.55}$$

As long as $r \neq 0$, this integral converges. We can extract the r-dependence of $c_S(\vec{r})$ by scaling it out of the integration. Let $t = Tr^2$. Then we have

$$\begin{aligned} c_S(\vec{r}) &\propto \frac{1}{r}\int_0^\infty T^{-3/2}e^{-1/aT}\,dT \\ &= \frac{K}{r} \end{aligned} \tag{4.56}$$

with K an r-independent quantity. The Coulomb's-law-like form of the steady state distribution has been recovered.

4.2.3 Steady state distribution of walkers near an absorbing wall

Let's consider a few interesting cases. Suppose we have a point source of walkers in the vicinity of an absorbing wall. We know that the concentration of walkers falls to zero at the wall, and we know that the point source has the same mathematical effect on the distribution of walkers as a point charge has on an electrostatic potential. Translating this into a problem in electrostatics, we are looking at the problem of a point charge in the vicinity of a plane conducting surface. This is solved with the use of an image charge. Figure 4.4 shows what the distribution of charges looks like. The image source lies on the other side of the absorbing wall. This tells us that the steady state distribution of walkers will be given by a concentration, $c(\vec{r})$, going as

$$c(\vec{r}) \propto \frac{1}{|\vec{r} - \vec{r}_s|} - \frac{1}{|\vec{r} - \vec{r}_i|} \tag{4.57}$$

where $\vec{r}_s$ is the position vector of the source and $\vec{r}_i$ is the position vector of the image. Suppose that the absorbing wall is in the x, z plane, and the point source

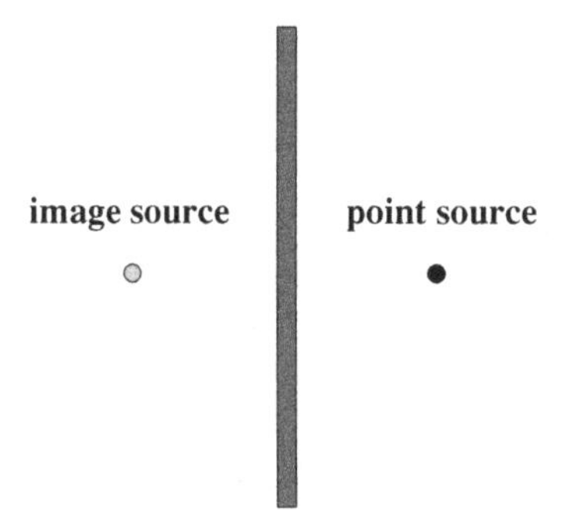

Fig. 4.4. The image charge configuration.

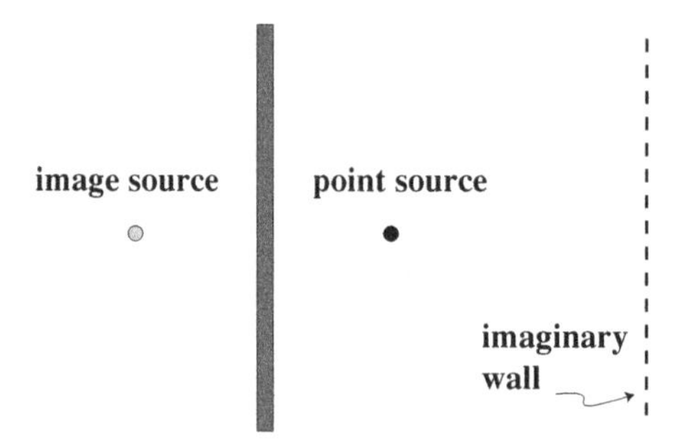

Fig. 4.5. The image charge distribution, with the imaginary wall through which the current is calculated.

is along the x axis a distance l from the wall. Then, $\vec{r}_s = \hat{x}l$, while $\vec{r}_i = -\hat{x}l$. The distribution of walkers has the form

$$\begin{aligned} c(x, y, z) &\propto \frac{1}{\sqrt{(x-l)^2 + y^2 + z^2}} - \frac{1}{\sqrt{(x+l)^2 + y^2 + z^2}} \\ &\rightarrow \frac{2lx}{\left(x^2 + y^2 + z^2\right)^{3/2}} \end{aligned} \tag{4.58}$$

where the last line holds when $r \gg l$. This "dipole field" dependence of the number of walkers at the location $\vec{r}$ represents a serious depletion with respect to the steady state distribution, which goes as $1/r$, when there is no absorbing wall. In fact, it can be demonstrated that the wall has the effect of eliminating almost all the walkers that emanate from the source. One way to see this is to calculate the total current of walkers through an imaginary wall to the right of the source, shown as a vertical dashed line in Figure 4.5. Recall that the current density is proportional to the gradient of the concentration of walkers. To calculate the total flux of walkers through the imaginary wall, we take the partial derivative of $c(\vec{r})$ with respect to x.

We find for the current denstiy

$$
\begin{aligned}
j_x(x, y, z) &\propto -\frac{\partial}{\partial x}\frac{2lx}{\left(x^2+y^2+z^2\right)^{3/2}} \\
&= 2l\left(\frac{3x^2}{\left(x^2+y^2+z^2\right)^{5/2}} - \frac{1}{\left(x^2+y^2+z^2\right)^{3/2}}\right) \qquad (4.59)
\end{aligned}
$$

We've used the form that works when the imaginary wall is far away from the real one. Now, if we integrate this component of the current density over the imaginary wall, we find ourselves taking an integral that is proportional to

$$
\begin{aligned}
&\int_0^\infty \left[\frac{3x^2}{\left(x^2+R^2\right)^{5/2}} - \frac{1}{\left(x^2+R^2\right)^{3/2}}\right] R\,\mathrm{d}R \\
&= \frac{1}{2}\int_0^\infty \left[\frac{3x^2}{\left(x^2+w\right)^{5/2}} - \frac{1}{\left(x^2+w\right)^{3/2}}\right] \mathrm{d}w \\
&= \left\{-\frac{x^2}{\left(x^2+w\right)^{3/2}} + \frac{1}{\left(x^2+w\right)^{1/2}}\right\}\Bigg|_0^\infty \\
&= 0 \qquad (4.60)
\end{aligned}
$$

In the end, no walkers escape from the absorbing wall, no matter how far out they manage to walk.

4.2.4 Walkers near an absorbing sphere

We can also ask what happens to walkers in the vicinity of an absorbing sphere? First, we'll look at the case of a collection of walkers that are supplied by a set of sources that are infinitely far away from the sphere. The sources are such that the distribution of walkers is uniform at an infinite distance from the sphere. To find what the concentration of walkers is at an arbitrary distance from the sphere, which has a radius equal to r_0, we search for a solution of Laplace's equation (because there are no sources in the region of interest) that is equal to zero at the surface of the sphere and that is a constant infinitely far away from the sphere. Placing the origin at the center of the sphere, we can also demand spherical symmetry about that origin. The solution of interest is then readily intuited. It is

$$
c(\vec{r}) = c_0\left(1 - \frac{r_0}{r}\right) \qquad (4.61)
$$

where r is the distance from the center of the sphere, and the expression on the right hand side of (4.61) holds when $r > r_0$.

We can use this formula to calculate the rate at which the walkers are absorbed by the sphere. To do this, we calculate the current density of walkers by taking the

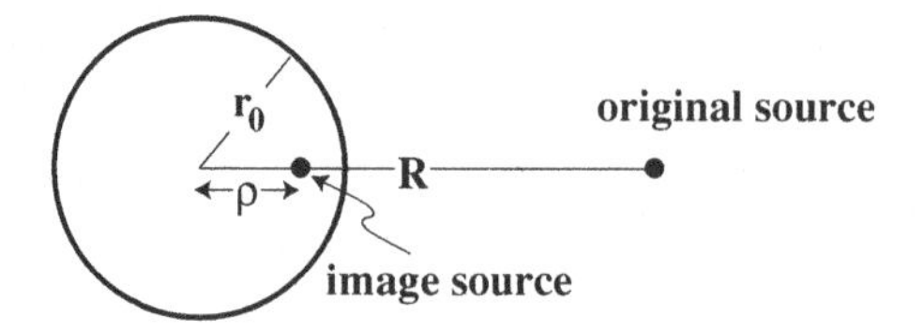

Fig. 4.6. The source and the image source in the case of an absorbing sphere.

gradient of the right hand side of (4.61), and then by calculating the current flux into the sphere. From (4.43), we have

$$\vec{j}(\vec{r}) = -\beta\hat{r}\frac{\partial}{\partial r}c_0\left(1 - \frac{r_0}{r}\right)$$
$$= -\hat{r}\frac{\beta c_0 r_0}{r^2} \tag{4.62}$$

This tells us that the flux of walkers into the sphere is equal to

$$\beta c_0 \frac{r_0}{r_0^2} \times 4\pi r_0^2 = 4\pi\beta c_0 r_0 \tag{4.63}$$

This means that the number of walkers that are absorbed by the sphere scales linearly with the radius of the sphere. If the sphere represents a cell, and the walkers are nutrients in the broth in which the cell sits, the rate at which the cell takes in those nutrients is proportional to its linear size. On the other hand, if the sphere really is a cell, it has metabolic requirements that scale as its volume – in other words, as r_0^3. Ultimately, those requirements will overwhelm the cell's ability to absorb nutrients through diffusion, as the size of the cell increases. In the case of an immobile cell, these considerations place a limit on the maximum size that it can be. In general, the fact that metabolic needs will exceed the rate at which nutrients can be gathered as they diffuse inwards will mandate a different strategy for the acquisition of biological fuel for any organism that is larger than a certain size (Berg, 1993).

It is also possible to calculate the distribution of walkers when there is a point source outside of an absorbing sphere (Figure 4.6). In this case, we make use of a modification of the image charge. If the source is a distance R away from the center of a sphere of radius r_0, then the image source is a distance $\rho = r_0^2/R$ from the center of the sphere, and the "strength" of the source is equal to $-r_0/R$ times the strength of the original, external source. It is possible to calculate how many walkers escape from the sphere by taking the integral of the current flux through a surface that surrounds both the source and the absorbing sphere. We can short-circuit this calculation by noting that a version of Gauss's law holds here, which tells us that the net flux through a surface surrounding a set of sources is proportional to the total strength of those sources. In this case, the total strength is equal to the strength of the original source plus that of the image source, which is the strength of the original source multiplied by $1 - r_0/R$. The fraction of the total number of walkers that emanate from the source that also escape from the sphere is $(R - r_0)/R$.

What we have also achieved here is a derivation of the probability of escape by a three-dimensional walker from a spherical region in the vicinity of its point of origin.

4.2.5 *The analogy with capacitance*

There is a mathematical connection between an absorbing surface into which matter diffuses from a distance and the element of a capacitor. Because of the relationship between the concentration and the electrostatic potential, we can imagine an analogy in which the surface is that of a conductor, surrounded by a spherical shell some distance away. If the potential difference between the two is $\Delta\phi$, and the charge on the inner surface is Q, then the capacitance of the system is $c = Q/\Delta\phi$. The connection between charge and the electrostatic potential is $\rho(\vec{r}) = -4\pi\nabla^2\phi(\vec{r})$. The total charge on the inner surface is then given by Gauss's law:

$$Q = -\frac{1}{4\pi}\int \vec{\nabla}\phi(\vec{r})\cdot \mathrm{d}\vec{S} \tag{4.64}$$

Making use of the relationships we've already established between electrostatic quantities and those in steady state diffusion, we can state that the following holds

$$\frac{-\int \vec{j}(\vec{r})\cdot \mathrm{d}\vec{S}}{c_\infty} = 4\pi\beta C \tag{4.65}$$

where c_∞ is the concentration of walkers far away from the absorbing surface and C is the capacitance of a capacitor consisting of the absorbing surface surrounded, at a great distance, by a spherical shell (see Figure 4.7). For reference, recall that the capacitance of a sphere is given by

$$C_{\text{sphere}} = r_0 \tag{4.66}$$

where r_0 is the sphere's radius. This and (4.65) yields (4.63) for the total current absorbed by the sphere, where we have replaced c_0 by c_∞.

4.2.6 *A sphere covered with receptors*

This scenario allows us to consider what happens if the nutrients are not absorbed uniformly throughout the sphere. Suppose, instead, that the sphere has a number of absorbing sites, or receptors, distributed across its surface. Imagine that there are n of those receptors, and that the radius of a given site is a. We'll also assume

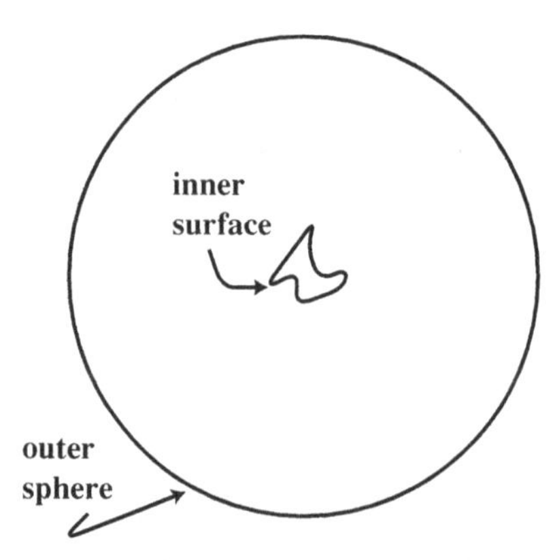

Fig. 4.7. A capacitor consisting of an inner surface surrounded by a spherical shell. The radius of the surrounding shell is much greater than the size of the inner surface.

that $na \ll r_0$, where r_0 is the sphere's radius. This means that the total surface area accounted for by the receptors is a small fraction of the surface area of the sphere. Let's represent this collection of receptors as a network of small conducting surfaces, arranged in the shape of a spherical shell. This network is utilized as one of the elements in a capacitor. What is the capacitance of the resulting capacitor?

First, note that the distance between the receptors is the order of $r_0/\sqrt{n}$. Then, notice that the potential a distance r away from the sphere, where $r \gg r_0/\sqrt{n}$, is going to be the same as the electrostatic potential generated by a uniform distribution of the receptors, smeared out over the sphere. This is because at such a distance, the difference between a set of discrete charges and a uniformly distributed charge is negligible, as far as the electrostatic potential goes. If a charge of Q/n is placed on each receptor, then the electrostatic potential of this array of charges goes as $Q/(r + r_0)$. The electrostatic potential near the sphere is, then, well-approximated by Q/r_0. This means that the capacitance of the spherical arrangement is substantially equal to the capacitance of a uniform sphere. Making use of the electrostatic analogy, we find that the collection of receptors will absorb nutrients at the same rate as if the entire surface of the sphere were a receptor.

We can be a little more explicit about the potential immediately above the surface of the network. The electrostatic potential right next to one of the small components of the network due to the charge carried by that component will go as Q/na, because each of the components carries a charge equal to Q/n, and each has an effective size equal to a. On the other hand, the potential due to all the other components will be essentially the same as if the charges on them were uniformly distributed over the surface of the sphere. This potential is equal to $Q(n-1)/nr_0$. If n is large enough that $na \gg r_0$, which is possible for sufficiently large n, since all we require is that $a \ll r/\sqrt{n}$, the potential is dominated by the second contribution, which, in the limit of large n, goes to Q/r_0. The potential at a point near the surface of the sphere that is far away from one of the small components compared to its size will be absolutely dominated by the second term. Thus, to a very good approximation, the electrostatic potential in the immediate vicinity of the network is the same as the electrostatic potential right next to a sphere carrying a uniform charge equal to Q.

For a more extended discussion of the issues addressed in the last two sections see (Berg, 1993) and (Berg and Purcell, 1977).

4.3 Supplement: boundary conditions at an absorbing boundary

As noted in Section 4.1.2, the distribution of walkers falls to zero at a boundary that eliminates, or absorbs, walkers that cross over it. This boundary condition may seem eminently reasonable.[5] However, a rigorous derivation is not entirely trivial. In this supplement, we show how the condition comes about. We focus our attention on the one-dimensional situation; the argument readily generalizes to walks in a higher dimensional region.

We start with the equation obeyed by the generating function for the one-dimensional walk on a lattice:

$$G(z; x, y) = z\,(G(z; x, y - \Delta) + G(z; x, y + \Delta)) + \delta_{x,y} \tag{4:S-1}$$

The quantity $G(z; x, y)$ is the generating function for one-dimensional walks starting at y and ending at y, where y and x are restricted to lie on a one-dimensional array of points, which means that both x and y take on the values $n\Delta$, where $n \geq 0$. There is an absorbing boundary immediately to the left of the point $n = 0$, which means that all walkers that attempt the step to $y = -\Delta$ are eliminated. The restriction on n means that the equation (4:S-1) holds only when $y \geq 1$. When $y = 0$, the equation satisfied by the generating function is

$$G(z; 0, y) = zG(z; x, \Delta) + \delta_{x,\Delta} \tag{4:S-2}$$

This equation holds because there are no walkers at $y = -n\Delta$.

We can restore the applicability of (4:S-1) to all points in the region occupied by surviving walkers by *requiring* that $G(z; x, -\Delta)$ is equal to zero. In that case, the equation satisfied by the generating function $G(z; x, y)$ is (4:S-1) at *all* points $n\Delta$ with $n \geq 0$, supplemented by the boundary condition

$$G(z; x, -\Delta) = 0 \tag{4:S-3}$$

[5] See (Weiss, 1994) and references therein for a fuller discussion of boundary conditions.

This means that the boundary condition appropriate to an absorbing barrier at the perimeter of a region occupied by random walkers is that the generating function (or the density of walkers) is equal to zero *immediately on the other side of the boundary.* In the continuum limit, in which we ignore details such as the actual distance covered by a walker in a single step, this boundary condition is approximated by the requirement that the walker density – or generating function – vanishes at the exact location of the barrier.

5

Variations on the random walk

We now discuss the ways in which the statistics of the random walk may be affected by variations in the rules governing the walk. As we will see, the methods that have been developed in the previous chapters are still useful. However, the variations brought about by the nature of the environment and on the rules governing the walk give rise to interesting new behavior, and to the possibility of analyzing important physical and probabilistic phenomena.

5.1 The biased random walk

We'll start by considering a walk in which the walker prefers to take a step in a particular direction. In other words, the walk is no longer isotropic, but is rather *biased*. For simplicity, we'll look at the case of a one-dimensional walker that prefers to go either to the right or to the left. In this case, the probability of taking a step to the right will be p and the probability of taking a step to the left is $1 - p$, where $p \neq 1/2$. The probability that the walker has taken n steps to the right and $N - n$ steps to the left after an N-step walk, is, as before, given in terms of the binomial distribution. Recall Exercise 1.1, in Chapter 1. We find

$$P(N; n, N - n) = p^n (1 - p)^{N-n} \frac{N!}{n!(N - n)!} \tag{5.1}$$

We can also obtain this probability with the use of a recursion relation. We start by relating $P(N + 1; x, y)$, the probability that the walker that has started out at x ends up at y after having taken $N + 1$ steps, to the probabilities that in N steps the walker ended up immediately to the left or immediately to the right of its final destination. This recursion relation is

$$P(N + 1; x, y) = pP(N; x, y - l) + (1 - p)P(N; x, y + l) \tag{5.2}$$

We now play the same game as in Chapter 1. We Taylor expand the probabilities on the right hand side of (5.2):

$$P(N+1;x,y) = P(N;x,y) + l(2p-1)\frac{\partial}{\partial y}P(N;x,y) + \frac{l^2}{2}\frac{\partial^2}{\partial y^2}P(N;x,y) + \cdots \tag{5.3}$$

We'll solve this equation by evaluating the Fourier transform of both sides, taking advantage of the fact that this probability depends only on the distance from x to y. Multiplying both sides by $e^{iq(y-x)}$ and integrating, we find for the equation satisfied by the Fourier-transformed quantity $P(N;q)$

$$P(N+1;q) = P(N;q)\left(1 + il(2p-1)q - \frac{l^2q^2}{2} + O(q^3)\right) \tag{5.4}$$

The solution to this equation is

$$P(N;q) = \left(1 + il(2p-1)q - \frac{l^2q^2}{2} + \cdots\right)^N P(0;q) \tag{5.5a}$$

$$= \exp\left[N\ln\left(1 + il(2p-1)q - \frac{l^2q^2}{2} + \cdots\right)\right]P(0;q) \tag{5.5b}$$

$$= \exp\left[Nil(2p-1)q - 2Nq^2l^2p(1-p) + O(q^3)\right]P(0;q) \tag{5.5c}$$

We now neglect terms in the exponent in (5.5c) that are beyond quadratic in q. This presents us with a Fourier-transformed probability that is a Gaussian in q. The inverse Fourier transform will also be a probability, this time in the distance between the initial and end-points. The initial distribution of walkers, $P(N; y-x)$, is given by

$$P(0; y-x) = \delta(y-x) \tag{5.6}$$

a Dirac delta function. The Fourier transform of this function is independent of q. This means that the inverse Fourier transform of (5.5c), with the terms beyond $O(q^2)$ in the exponent neglected, is given by

$$\sqrt{\frac{1}{8\pi Np(1-p)l^2}}\exp\left[-(x-(2p-1)Nl)^2/8Nl^2p(1-p)\right] \tag{5.7}$$

This is a Gaussian whose peak moves to the right a distance $(2p-1)l$ at each step and whose width increases as the square root of N (see Figure 5.1 for an explanation of this behavior). The statistics of the biased walk are those of a Gaussian with a moving peak. Note that the width of the peak also goes as the combination $\sqrt{p(1-p)}$. The width vanishes as the probability of a step to the right or to the left

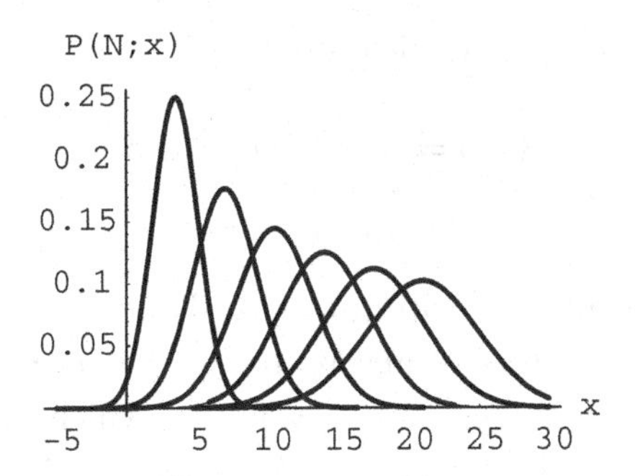

Fig. 5.1. The evolution of the probability distribution, $P(N;x)$ of a one-dimensional walk that is biased to the right. The plots are for various values of N. The original distribution is a delta function centered at the origin. As N increases the distributions move to the right, and they also spread, as shown in this figure.

goes to 1. Of course, this makes perfect sense. In either of those limits there is no randomness in the walk. The walker proceeds to the left or the right with perfect certainty. All the members of a collection of such walkers will end up at exactly the same point after taking N steps from a common origin.

It should be noted that this feature would be lost if we were to go directly from (5.3) to the diffusion equation by writing $P(N+1;x,y) = P(N;x,y) + \partial P(N;x,y)/\partial N$. In that case, the probability would still be centered about $(2p-1)l$, but the width of the peak would no longer go as $\sqrt{p(1-p)}$, but, instead, this factor would be replaced by 1/2. The reason for this becomes clear in a more careful derivation of the diffusion equation in the continuum limit, when N is large. When this is done, one finds that the diffusion equation is only valid in the limit $p \to 1/2$, or for walks that are weakly biased.

Exercise 5.1

Demonstrate the correctness of the statement above, that the diffusion equation is valid only if p is close to 1/2.

5.1.1 A biased walk in three dimensions

In looking at a biased walk in three dimensions, we will make use of another way to approach the possibility of biasing in a random walk. The key here lies in the modification of the structure function $\chi(\vec{q})$. In the case of the unbiased walk, the structure function is the Fourier transform of a function that is centered about $\vec{r} = 0$, and that has a simple peak there. We can put a bias in the walk by translating this function. That is, suppose that $\chi(\vec{q})$ is given by[1]

$$\chi(\vec{q}) = \int e^{-i\vec{q}\cdot\vec{r}} P(\vec{r})\, d^3r \equiv \chi_0(\vec{q}) \tag{5.8}$$

Here, $P(\vec{r})$ is the probability density for the walker to take a step resulting in a displacement equal to $\vec{r}$. A way of biasing the walk is to replace $P(\vec{r})$ by $P(\vec{r} - \vec{r}_B)$. This probability density describes a walk for which the mean value of the displacement after each step is $\vec{r}_B$, instead of zero. Then

$$\begin{aligned}\chi(\vec{q}) &= \int e^{-i\vec{q}\cdot\vec{r}} P(\vec{r} - \vec{r}_B)\, d^3r \\ &= e^{-i\vec{q}\cdot\vec{r}_B} \int e^{-i\vec{q}\cdot(\vec{r}-\vec{r}_B)} P(\vec{r} - \vec{r}_B)\, d^3r \\ &= e^{-i\vec{q}\cdot\vec{r}_B} \chi_0(\vec{q}) \end{aligned} \tag{5.9}$$

The probability that the walker will end up at the location $\vec{r}$ after taking N steps from its point of origin at $\vec{r} = 0$ is

$$\begin{aligned}\frac{1}{(2\pi)^3} \int d^3q\, \chi(\vec{q})^N e^{i\vec{q}\cdot\vec{r}} &= \frac{1}{(2\pi)^3} \int d^3q\, \chi_0(\vec{q})^N e^{i\vec{q}\cdot(\vec{r}-N\vec{r}_B)} \\ &= P(N; \vec{r} - N\vec{r}_B) \end{aligned} \tag{5.10}$$

where $P(N;\vec{r})$ is the probability that the unbiased walk will end up at the location $\vec{r}$ after N steps. Notice that in this case, the width of the distribution is not affected by the fact that the walker moves preferentially in a given direction. This is because one way of thinking about the new rule governing the walker is to say that it takes a purposeful step in a given direction followed by a purely random step. The new walk is a superposition of a purposeful walk and a random one, in which the random component obeys the same rules as in the case of the unbiased walk. Such a walk describes the motion of small particles diffusing in a moving medium that carries them along at a fixed velocity. For instance, one could imagine a cloud of diffusing dye molecules in water that flows steadily in one direction.

5.2 The persistent random walk

The "pure" random walker has no recollection of the previous steps that it took as it decides where to go next. In an important variation of the random walk, the walker's choice of a direction at a given step is influenced by how it moved in the time leading up to that choice. This kind of walker has a memory. Focusing on a

[1] The structure function, $\chi(\vec{q})$, is defined for lattice walks in (2.11) and for off-lattice walks in (2.23).

walker in one dimension, let's assume that the walker has a tendency to move in the same direction as it has already gone. That is, we'll assume that, if the walker took a step to the right previous to the one it is about to take, then the subsequent step is more likely to be to the right than to the left, and conversely for a walker that has arrived at its present position by having taken a step to the left. We'll quantify this tendency by assigning a probability Q for taking a step in the same direction as the immediately preceding one and a probability $1 - Q$ for the walker's reversing its direction at the step it is about to take. Here we encounter a random process which is still Markovian, but is of second order (Feller, 1968). That is, the process depends not only on the previous event, but also on the one that precedes it. In order to analyze this variant of the random walker we'll define a new set of probabilities. Let's denote by $L(N; x, y)$ the probability that a walker that starts out at x and takes N steps ends up at y, the last step having been to the left. By the same token, $R(N; x, y)$ is the same probability, the walker now having taken its last step to the right. The recursion relation is now two-fold. The recursion relation for $L(N; x, y)$ is

$$L(N+1; x, y) = QL(N; x, y+l) + (1-Q)R(N; x, y+l) \tag{5.11}$$

The recursion relation for $R(N; x, y)$ is

$$R(N+1; x, y) = (1-Q)L(N; x, y-l) + QR(N; x, y-l) \tag{5.12}$$

As the first stage in the solution of these two recursion relations, we Fourier transform both sides of (5.11) and (5.12). The equations for the transformed functions are then

$$L(N+1; q) = \mathrm{e}^{-\mathrm{i}ql}\left(QL(N; q) + (1-Q)R(N; q)\right) \tag{5.13}$$

$$R(N+1; q) = \mathrm{e}^{\mathrm{i}ql}\left((1-Q)L(N; q) + QR(N; q)\right) \tag{5.14}$$

We can summarize the two equations above in matrix form. Writing the two probabilities as the entries in a column vector, we end up with the following single equation

$$\begin{aligned}\begin{pmatrix} L(N+1; q) \\ R(N+1; q) \end{pmatrix} &= \begin{pmatrix} Q\mathrm{e}^{-\mathrm{i}ql} & (1-Q)\mathrm{e}^{-\mathrm{i}ql} \\ (1-Q)\mathrm{e}^{\mathrm{i}ql} & Q\mathrm{e}^{\mathrm{i}ql} \end{pmatrix} \begin{pmatrix} L(N; q) \\ R(N; q) \end{pmatrix} \\ &\equiv \mathbf{T} \begin{pmatrix} L(N; q) \\ R(N; q) \end{pmatrix}\end{aligned} \tag{5.15}$$

Here, $\mathbf{T}$ is the two-by-two matrix on the right hand side of the first line of (5.15). The solution for the quantities L and R is

$$\begin{pmatrix} L(N; q) \\ R(N; q) \end{pmatrix} = \mathbf{T}^N \begin{pmatrix} L(0; q) \\ R(0; q) \end{pmatrix} \tag{5.16}$$

A bit of algebra suffices to determine the eigenvalues of the matrix $\mathbf{T}$. We find

$$\lambda_1 = Q\cos ql + \sqrt{(1-Q)^2 - Q^2\sin^2 ql} \tag{5.17}$$

$$\lambda_2 = Q\cos ql - \sqrt{(1-Q)^2 - Q^2\sin^2 ql} \tag{5.18}$$

There are various ways of working out the properties of the persistent walk. One of the easiest utilizes the generating function of the walk. Given the relationship in (5.15), we are able to evaluate the sum leading to the generating function for $L(N;q)$ and $R(N;q)$. If we define

$$l(z;q) = \sum_{N=0}^{\infty} L(N;q)z^N \tag{5.19}$$

and similarly for $r(z;q)$, then

$$\begin{aligned}\begin{pmatrix} l(z;q) \\ r(z;q) \end{pmatrix} &= \sum_{N=0}^{\infty} \mathbf{T}^N z^N \begin{pmatrix} L(0;q) \\ R(0;q) \end{pmatrix} \\ &= (1-\mathbf{T}z)^{-1}\begin{pmatrix} L(0;q) \\ R(0;q) \end{pmatrix}\end{aligned} \tag{5.20}$$

Here

$$(1-\mathbf{T}z)^{-1} = \begin{pmatrix} 1 - zQe^{iql} & z(1-Q)e^{-iql} \\ z(1-Q)e^{iql} & 1 - zQe^{-iql} \end{pmatrix} \Big/ D \tag{5.21}$$

where

$$\begin{aligned} D &= 1 - 2zQ\cos ql + z^2(2Q-1) \\ &= (1 - z\lambda_1)(1 - z\lambda_2) \end{aligned} \tag{5.22}$$

The problem of determining the probability of a certain type of walk is reduced to finding the coefficient of z^N in the expansion of the generating function.

Give what we now know about the asymptotic behavior of power series, we are in a position to extract the long-time statistics of the one-dimensional persistent random walk. The main thing that helps us is the fact that those statistics are dominated by the singularity in the generating function that is closest to the origin. This singularity is at $z = 1/\lambda_1$, λ_1 being the larger of the two eigenvalues of the matrix $\mathbf{T}$. For small values of the wave-vector q,

$$\begin{aligned} \lambda_1 &= Q\left(1 - \frac{q^2l^2}{2}\right) + (1-Q) - \frac{Q^2}{1-Q}\frac{q^2l^2}{2} + O\left(q^4\right) \\ &= 1 - \frac{Q}{1-Q}\frac{q^2l^2}{2} + O\left(q^4\right) \end{aligned} \tag{5.23}$$

One of the key quantities of interest is the total number of walks. The Fourier transform of the total number of N-step walks is equal to the sum $L(N;q) + R(N;q)$. Given (5.20), we have for the generating function of the total number of walks $l(z;q) + r(z;q)$. If we are interested in the asymptotic statistics of the persistent walk, we only need look for the leading order coefficient of z^N in the power-series expansion of $(1 - z\mathbf{T})^{-1}$. This is controlled by the expansion of the denominator, D, in (5.21), where D is given by (5.22). Now, we can expand the product $1/(1 - z\lambda_1)(1 - z\lambda_2)$ in partial fractions as follows:

$$\frac{1}{(1 - z\lambda_1)(1 - z\lambda_2)} = \frac{1}{\lambda_1 - \lambda_2}\left(\frac{\lambda_1}{1 - z\lambda_1} - \frac{\lambda_2}{1 - z\lambda_2}\right) \tag{5.24}$$

The coefficient of z^N in the expansion of this quantity is equal to

$$\frac{1}{\lambda_1 - \lambda_2}\left(\lambda_1^{N+1} - \lambda_2^{N+1}\right) \tag{5.25}$$

This is dominated by the Nth power of the larger of the two eigenvalues, λ_1. Given (5.23), we see that the dependence on N and q of the spatial Fourier transform of the total number of N-step walks is dominated by the quantity

$$\begin{aligned}\left(1 - \frac{Q}{1 - Q}\frac{q^2l^2}{2}\right)^N &= \exp\left[N \ln\left(1 - \frac{Q}{1 - Q}\frac{q^2l^2}{2}\right)\right] \\ &\to \exp\left[-N\frac{Q}{1 - Q}\frac{q^2l^2}{2}\right]\end{aligned} \tag{5.26}$$

which tells us that the number of N-step random walks that end up a distance x from their point of origin goes as

$$\begin{aligned}&\frac{1}{2\pi}\int_{-\infty}^{\infty} \exp(-iqx)\exp[-NQq^2l^2/(2(1 - Q))]\, dq \\ &= \sqrt{\frac{1 - Q}{2\pi NQl^2}}\exp\left[-x^2\frac{1 - Q}{2NQl^2}\right]\end{aligned} \tag{5.27}$$

Note that in the limit $Q = 1$ the width of the distribution on the right hand side of (5.27) goes to infinity, in that the expression loses its dependence on the end-point x. What this tells us is that the Gaussian limit is never achieved in this instance. The case $Q = 1$ corresponds to a walk that is not really random, as it will continue indefinitely in the direction in which it has set out. What happens when $Q = 1$ is that the actual distribution of walkers takes the form of a delta function, centered on the location of the steadily advancing walker.

On the other hand, when $Q = 0$, in which limit the walker inevitably changes its direction, the width of the distribution of walkers, as predicted by the right hand side of (5.27), goes to zero. Of course, this limit is also pathological, in that all randomness has vanished from the process. The walker, fated to reverse its direction at every step, never gets anywhere.

In between those two limits, when the number of steps, N, is large enough, the distribution of walkers takes on an asymptotically Gaussian distribution, with a width that depends non-trivially on the likelihood that the walk changes its direction.

It is possible to extend the analysis here to a walker in more than one dimension. The explicit form that the equations take in dimensions greater than or equal to two is considerably more complicated, but yields to analysis in the continuum limit. We will take up this case later in the chapter.

It is possible to collapse the dependence of the Gaussian in (5.27) on the probability Q into a redefinition of the step length that the walker takes. If we make the following replacements

$$l \to nl \tag{5.28}$$

$$N \to N/n \tag{5.29}$$

where

$$n = \frac{Q}{1-Q} \tag{5.30}$$

Then

$$\exp\left[-x^2 \frac{1-Q}{2NQl^2}\right] \to \exp[-x^2/2Nl^2] \tag{5.31}$$

and the Gaussian distribution for the persistent walk looks just like the Gaussian distribution for an ordinary random walk. What we imagine is a walker that takes n steps before deciding to take a step that is, with perfect unbiased randomness, to the left or the right. We call the distance covered in these n steps, $l_{\mathrm{p}} = nl$, the *persistence length*. The effect of persistence in this case is to alter the effective length of each step. The notion of persistence length carries over to other random walk processes and to many systems that are modeled by random walks.

5.2.1 Another approach to the persistent random walk

While the above development is complete, in principle, it is possible to take another tack in the calculation of the statistics of the persistent random walk (Weiss, 1994). We start by recalling the recursion relations for the quantities $L(N;q)$ and $R(N;q)$

$$L(N+1;q) = \mathrm{e}^{-\mathrm{i}ql}\,(QL(N;q) + (1-Q)R(N;q)) \tag{5.32}$$

$$R(N+1;q) = \mathrm{e}^{\mathrm{i}ql}\,((1-Q)L(N;q) + QR(N;q)) \tag{5.33}$$

Eliminating $L(N;q)$ in (5.32) and (5.33), and then solving for $L(N+1;q)$ we find

$$L(N+1;q) = \frac{1-2Q}{1-Q} e^{-iql} R(N;q) + \frac{Q}{1-Q} e^{-2iql} R(N+1;q) \tag{5.34}$$

Now if we replace N by $N+1$ in (5.33) and use (5.34) to eliminate $L(N+1;q)$ from the resulting equation, we end up with the following recursion relation

$$R(N+2;q) = 2Q\cos ql R(N+1;q) + (1-2Q)R(N;q) \tag{5.35}$$

The eigenvalue equation that this recursion relation leads to can be derived from the matrix form of the recursion relation (5.35):

$$\begin{pmatrix} R(N+2;q) \\ R(N+1;q) \end{pmatrix} = \begin{pmatrix} 2Q\cos ql & 1-2Q \\ 1 & 0 \end{pmatrix} \begin{pmatrix} R(N+1;q) \\ R(N;q) \end{pmatrix} \tag{5.36}$$

The eigenvalue of the matrix in (5.36) is the solution of the equation

$$\lambda^2 - 2Q\cos ql\lambda + (1-2Q) = 0 \tag{5.37}$$

This equation is identical to the equation satisfied by the eigenvalues of the matrix **T**.

Another way to obtain the eigenvalue equation is to posit for $R(N;q)$ the form

$$R(N;q) = A\lambda^N \tag{5.38}$$

Substituting this presumed form into the recursion relation (5.35), we end up with (5.37) for the appropriate λ.

Exercise 5.2

Using the above ansatz for $R(N;q)$ derive the results found previously for $R(N,q) + L(N,q)$.

5.2.2 Persistent walks in the diffusion limit: the one-dimensional persistent walk and the Telegrapher's equation

Up to now, the discussion of the properties of the persistent walk has assumed that the walkers in question take discrete steps. The behavior of an individual walker was utilized to produce results for the distribution of a collection of walkers with the use of a recursion relation. Here, we will start with the notion of a distribution walkers. The walkers under consideration will not take steps from one location to another but will rather travel with a constant speed, changing the direction of their motion at random points of time as they move along. Our calculation of the statistical properties of a collection of this kind of walker will result in explicit predictions for their distribution, as a function of space and time.

We start with walkers in one dimension. A walker moves either to the left or the right, in both cases with a speed equal to v. This means that a full description of the statistics of a set of such walkers consists of two distributions, one referring to "left-moving" walkers and the other to "right-moving" walkers. Let's call the first distribution $L(x, t)$ and the second $R(x, t)$. The recursion relation relevant to a collection of this type of walker relates the distributions at a given location, x, and a given time, t, to the distributions at a slightly earlier time. Let the difference between the two times in question be δt. Then, if there were no interconversion between left- and right-moving walkers, the recursion relation for the distributions would be

$$R(x, t + \delta t) = R(x - v\delta t, t) \tag{5.39}$$

$$L(x, t + \delta t) = L(x + v\delta t, t) \tag{5.40}$$

The first equation above tells us that the right-moving walkers that are at the location x at time $t + \delta t$ were, at time t at the location $x - v\delta t$. The second tells us that the left-moving walkers that are at x at time $t + \delta t$ were at $x + v\delta t$ at time t. Expanding (5.39) and (5.40) to first order in δt, we end up with the following differential equations for the two distributions:

$$\frac{\partial R(x, t)}{\partial t} + v\frac{\partial R(x, t)}{\partial x} = 0 \tag{5.41}$$

$$\frac{\partial L(x, t)}{\partial t} - v\frac{\partial L(x, t)}{\partial x} = 0 \tag{5.42}$$

The most general solution to (5.41) is $R(x, t) = f(x - vt)$, with f any function, and the most general solution to (5.42) is $L(x, t) = f(x + vt)$. There is no randomness; the distributions simply follow the walkers to which they refer, moving to the right or left with a speed equal to v.

The walks become random when left- and right-movers interconvert. This process is encoded in the equations for the two distributions as follows. We rewrite the two equations above as

$$\frac{\partial R(x, t)}{\partial t} + v\frac{\partial R(x, t)}{\partial x} = \gamma\left(L(x, t) - R(x, t)\right) \tag{5.43}$$

$$\frac{\partial L(x, t)}{\partial t} - v\frac{\partial L(x, t)}{\partial x} = \gamma\left(R(x, t) - L(x, t)\right) \tag{5.44}$$

This system of equations comprises the Telegrapher's equation.

Exercise 5.3

Show that (5.43) and (5.44) follow by taking an appropriate limit of the discrete version of these equations.

The right hand sides of (5.43) and (5.44) represent the process by which a right-mover can, with probability γ per unit time, convert into a left-mover and vice versa. This will eventually cause the distributions to look like those for a random walk. However, there will be some memory of the ballistic propagation that underlies the random motion. This memory is eventually expressed in terms of a persistence length, directly related to the mean distance a walker propagates before suffering a reversal in its motion.

To process the equations of motion, we define sum and difference distributions:

$$\Sigma(x,t) = R(x,t) + L(x,t) \tag{5.45}$$

$$\Delta(x,t) = R(x,t) - L(x,t) \tag{5.46}$$

Then, the equations become

$$\frac{\partial \Sigma}{\partial t} + v\frac{\partial \Delta}{\partial x} = 0 \tag{5.47}$$

$$\frac{\partial \Delta}{\partial t} + v\frac{\partial \Sigma}{\partial x} = -2\gamma\Delta \tag{5.48}$$

The function $\Sigma(x,t)$ is the quantity of interest, as it is equal to the total number of walkers reaching the location x at time t.

Exercise 5.4

Find the equation satisfied by the function $\Sigma(x,t)$. This equation is the one-dimensional version of the Telegrapher's equation.

Calculation of the distribution $\Sigma(x,t)$

As noted above, our ultimate goal is the distribution $\Sigma(x,t)$, the distribution of all walkers, both right- and left-movers. The first step in the calculation of this distribution is to take the spatial Fourier and temporal Laplace transforms of the above equations. Let $\sigma(q,\lambda)$ be the transformed function for Σ and δ be the transformed function of Δ. The transformed equations are

$$\lambda\sigma(q,\lambda) + iqv\delta(q,\lambda) = \int_{-\infty}^{\infty} \Sigma(x,0)e^{iqx}\,dx \tag{5.49}$$

$$(\lambda + 2\gamma)\delta(q,\lambda) + iqv\sigma(q,\lambda) = \int_{-\infty}^{\infty} \Delta(x,0)e^{iqx}\,dx \tag{5.50}$$

Assuming that at $t = 0$, the distribution contains equal amounts of left- and right-movers, and that all are concentrated at $x = 0$, the right hand side of (5.49) is equal to one, and the right hand side of (5.50) is equal to zero. Solving (5.50) for $\delta(q,\lambda)$:

$$\delta(q,\lambda) = -iqv\frac{\sigma(q,\lambda)}{\lambda + 2\gamma} \tag{5.51}$$

Substituting into (5.50), we end up with

$$\left[\left(\lambda^2 + 2\gamma\lambda\right) + (qv)^2\right]\sigma(q,\lambda) = (\lambda + 2\gamma) \tag{5.52}$$

The reconstitution of the distribution is carried out with two integrations. We start with the integration over q, which is relatively straightforward. Assuming, for the time being that λ is real and positive, which it is in the standard definition of the Laplace transform:

$$f(\lambda) = \int_0^{\infty} F(t)e^{-\lambda t}\,dt \tag{5.53}$$

we have for the Laplace transform of the spatial distribution $\Sigma(x,t)$

$$\frac{1}{2\pi}\int_{-\infty}^{\infty} \frac{(\lambda + 2\gamma)e^{-iqx}}{(qv)^2 + \lambda^2 + 2\gamma\lambda}\,dq = \frac{1}{2}\frac{\lambda + 2\gamma}{v}\frac{e^{-\sqrt{\lambda^2+2\lambda\gamma}|x|/v}}{\sqrt{\lambda^2 + 2\gamma\lambda}} \tag{5.54}$$

The final stage in the derivation of the form of the distribution is the inverse Laplace transform. This is obtained in the standard way (Jeffreys, 1972). The inverse of (5.53) is

$$F(t) = \frac{1}{2\pi i}\int_{\epsilon - i\infty}^{\epsilon + i\infty} f(\lambda)e^{\lambda t}\,d\lambda \tag{5.55}$$

Here, ϵ is a positive, real infinitesimal. That is, one integrates with respect to λ along, and just to the right of, the imaginary axis. The integrand has a branch cut on the negative λ axis, from $\lambda = -2\gamma$ to $\lambda = 0$ (See Figure 5.2). The quantity $\Sigma(x,t)$ is then given by

$$\Sigma(x,t) = \frac{1}{4\pi i}\int_{-i\infty}^{i\infty} \frac{\lambda + 2\gamma}{v}\frac{e^{-\sqrt{\lambda^2+2\lambda\gamma}|x|/v}}{\sqrt{\lambda^2 + 2\gamma\lambda}}e^{\lambda t}\,d\lambda \tag{5.56}$$

If $x > vt$, the contour in (5.56) can be closed in the right half-plane, in which

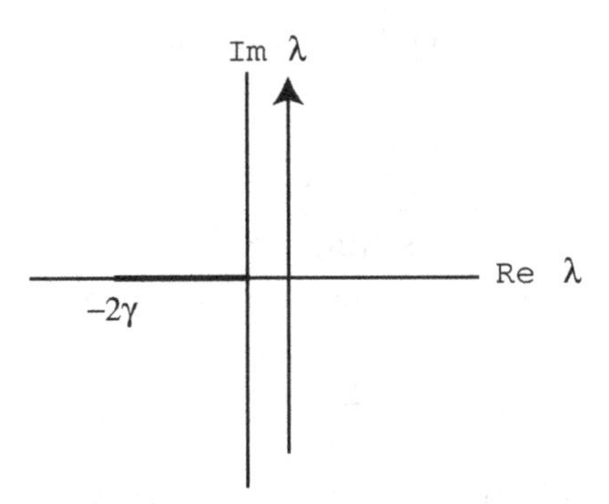

Fig. 5.2. The cut in the integrand, and the contour along the imaginary λ axis.

there are no non-analyticities, and the integration yields the answer zero. On the other hand, when $x < t$, the contour is closed in the left half-plane, and it can be contracted until it wraps around the branch cut. The integral is then reduced with the use of a change of variables. If one writes

$$\lambda = -\gamma(1 + \sin\phi) \tag{5.57}$$

The integral becomes

$$\frac{1}{4\pi}\frac{2\gamma + \partial/\partial t}{v}\left[e^{-\gamma t}\int_0^{2\pi} e^{-\gamma t \sin\phi}\cos\left(\gamma x\cos\left(\phi\right)/v\right) d\phi\, \Theta(vt - x)\right] \tag{5.58}$$

where the Heaviside step function $\Theta(vt - x)$ tells us that the result of the integration is equal to zero when $x > vt$, as noted above. According to a standard formula (Gradshteyn *et al.*, 2000):

$$\int_0^{2\pi} e^{a\sin\phi}\cos(b\cos\phi)\, d\phi = 2\pi I_0\left(\sqrt{a^2 - b^2}\right) \tag{5.59}$$

where I_0 is a modified Bessel function. This tells us that the "sum" distribution, $\Sigma(x, t)$ is given by

$$\frac{1}{2}\left(2\gamma + \frac{\partial}{\partial t}\right)\left[e^{-\gamma t} I_0\left(\gamma\sqrt{t^2 - x^2/v^2}\right)\Theta(vt - x)\right] \tag{5.60}$$

The final result for the number of walkers in the "sum" distribution is

$$\Sigma(x, t) = \frac{1}{2}\left[\left(2\gamma + \frac{\partial}{\partial t}\right) e^{-\gamma t} I_0\left(\gamma\sqrt{t^2 - x^2/v^2}\right)\Theta(vt - x) \right.$$
$$\left. + e^{-\gamma t}\left(\delta(x - vt) + \delta(x + vt)\right)\right] \tag{5.61}$$

To clarify the meaning of the delta functions in (5.61), we go back to the case in which there is no interconversion. The equation that is satisfied by the

Laplace- and Fourier-transformed sum distribution, $\sigma(q, \lambda)$, is

$$\sigma(q, \lambda) = \frac{\lambda}{\lambda^2 + q^2 v^2} \tag{5.62}$$

which is obtained from (5.52) by setting $\gamma = 0$ in that equation and solving for $\sigma(q, \lambda)$. Taking the inverse Fourier transform with respect to q, we are left with

$$\frac{\mathrm{e}^{-\lambda|x|/v}}{2v} \tag{5.63}$$

The inverse Laplace transform with respect to λ is, then, the integral along the imaginary λ axis

$$\begin{aligned} \frac{1}{2\pi \mathrm{i}} \int_{-\mathrm{i}\infty}^{\mathrm{i}\infty} \mathrm{e}^{-\lambda|x|/v} \mathrm{e}^{\lambda t} \frac{\mathrm{d}\lambda}{2v} &= \frac{1}{4\pi v} \int_{-\infty}^{\infty} \mathrm{e}^{\mathrm{i}\omega(t-|x|/v)} \, \mathrm{d}\omega \\ &= \frac{1}{2} \delta(|x| - vt) \\ &= \frac{1}{2} (\delta(x - vt) + \delta(x + vt)) \end{aligned} \tag{5.64}$$

The solution is a pair of delta functions, moving out to the right and left with a speed equal to v. The delta functions in the distribution in (5.61) contain walkers that have propagated ballistically without having undergone any interconversion. Once a walker interconverts, it joins the "trailing" distribution described by the first contribution to the right hand side of that equation. As time goes on, this trailing distribution settles into a Gaussian form, the width of which increases at a rate that can be associated with the propagation of a random walker with a persistence length calculable from the quantities v and γ.

Figure 5.3 displays a set of distributions of ensembles of persistent walkers at various times. The walkers in the ensembles are assumed to have all started at $x = 0$ at time $t = 0$. Furthermore, each ensemble consists initially of equal numbers of left- and right-moving walkers. The distributions are shown in the half-space $x > 0$. The portions of the distributions lying in the half-space $x < 0$ are the mirror image of those portions shown in Figure 5.3. We have set $v = 1$ and $\gamma = 1$ in calculating the graphs.

Exercise 5.5

Show that the distribution of one-dimensional walkers tends to the standard Gaussian form as time, t, becomes sufficiently large. You will need the asymptotic formula

$$I_0(z) \to \frac{\mathrm{e}^z}{\sqrt{2\pi z}} \qquad z \text{ is large}$$

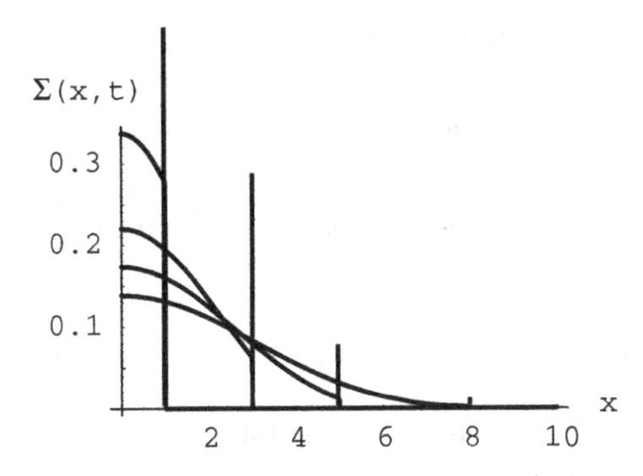

Fig. 5.3. Sum distribution functions, as given by (5.61), for various times, in the half-space $x > 0$. The distributions are symmetric with respect to reflection about the origin. The delta functions at the leading edges of the distribution are indicated in the figure. The speed, v, with which the walkers move is set equal to one, and the times for which the distributions are depicted are $t = 1, 3, 5$ and 8. The rate of interconversion, γ, is equal to one.

Exercise 5.6

How does the width of the distribution, as a function of x, depend on time, t and γ? Use your result to intuit a persistence length.

5.2.3 Persistent walks in two dimensions

We are now going to look at what happens in two dimensions. Here, things are a bit more complicated, because, in part, there are more than two distributions. In fact, there are an infinite number of them, corresponding to the infinite number of directions in which a walker can be going in two dimensions. We'll characterize the direction of the walk by the angle θ. The equation for the distribution function $f(\vec{r}, t, \theta)$ is

$$\frac{\partial f(\vec{r}, t, \theta)}{\partial t} + v \cos\theta \frac{\partial f(\vec{r}, t, \theta)}{\partial x} + v \sin\theta \frac{\partial f(\vec{r}, t, \theta)}{\partial y} = -\Gamma f(\vec{r}, t, \theta) \qquad (5.65)$$

The terms on the left hand side of (5.65) describe the evolution of a distribution in which the particles move ballistically. If the right hand side were equal to zero, then the solution to this equation would have the general form

$$f(\vec{r}, t, \theta) = f_0(\vec{r} - \hat{x}vt\cos\theta - \hat{y}vt\sin\theta, \theta) \qquad (5.66)$$

That the right hand side of (5.66) is a solution to (5.65) with $\Gamma = 0$ is readily verified by substitution.

The quantity Γ on the right hand side of (5.65) is an operator, not a number, and it represents the "mixing" effect of randomizing events. The right hand side of

(5.65) is, more explicitly, given by

$$\int \mathrm{d}\theta' \, \tau(\theta, \theta') f(\vec{r}, t, \theta') \tag{5.67}$$

Let's call the Fourier- and Laplace-transformed distribution $F(\vec{q}, \lambda, \theta)$. Then, (5.65) becomes

$$\left(\lambda + \mathrm{i}v\vec{q} \cdot \hat{n} + \Gamma\right) F(\vec{q}, \lambda, \theta) = f(\vec{q}, t = 0, \theta) \tag{5.68}$$

The right hand side of (5.68) is the Fourier-transformed distribution at $t = 0$, and the quantity $\hat{n}$ is the unit vector with x component equal to $\cos\theta$ and y component equal to $\sin\theta$.

5.2.4 The explicit form of the operator Γ

It is assumed that, at a rate γ, the distribution relaxes to a uniform one. That is, at each point at which the random walker decides to change its direction, it does so completely randomly. The way we write the operator in this case is

$$\Gamma = \gamma - \gamma |0\rangle\langle 0| \tag{5.69}$$

Here, $|0\rangle$ represents the normalized uniform distribution, which is independent of θ. As is clear, we are making use of the Dirac bra and ket notation, in order to take advantage of its compactness and ease of manipulation. The normalization is $\langle 0|0\rangle = 1$. The form (5.69) ensures that $\Gamma|0\rangle = 0$. In other words, the direction-independent portion of the distribution of walkers remains unchanged. All dependence on the direction in which the walker moves washes out as scattering takes place.

Note that if it were not for the second term on the right hand side of (5.69), then what operates on $F(\vec{q}, \lambda, \theta)$ would be simply multiplicative. The solution of that equation would then be

$$F(\vec{q}, \lambda, \theta) = \frac{f(\vec{q}, t = 0, \theta)}{\lambda + \mathrm{i}v\vec{q} \cdot \hat{n} + \gamma} \tag{5.70}$$

and the solution for the distribution $f(\vec{r}, t, \theta)$ would then be the inverse Fourier and Laplace transform of the right hand side of (5.70). Actually, the general solution of the equation for the distribution in this case is

$$f(\vec{r}, t, \theta) = f_0(\vec{r} - v\hat{n}t, \theta)\mathrm{e}^{-\gamma t} \tag{5.71}$$

Again, the unit vector $\hat{n}$ is given by

$$\hat{n} = \hat{x}\cos\theta + \hat{y}\sin\theta \tag{5.72}$$

Our general task is to find the inverse of the operator in parentheses on the left hand side of (5.68). This operator can be written in the form

$$A - \gamma|0\rangle\langle 0| \tag{5.73}$$

with

$$A = \lambda + \mathrm{i}v\vec{q}\cdot\hat{n} + \gamma \tag{5.74}$$

The inverse of this operator is

$$A^{-1} + \frac{\gamma A^{-1}|0\rangle\langle 0|A^{-1}}{1 - \gamma\langle 0|A^{-1}|0\rangle} \tag{5.75}$$

Exercise 5.7

Show that the operator in (5.75) is the inverse of the operator in (5.73).

Suppose, now that the initial distribution is uniform and is concentrated at the location $\vec{r} = 0$: $f(\vec{q}, t = 0, \theta) = |0\rangle$. Suppose also that we are interested in the uniform portion of the distribution, which is directly relevant to the total number of persistent walkers at a given point in space at a given time. We are then interested in the amplitude

$$\langle 0|\left(A^{-1} + \frac{\gamma A^{-1}|0\rangle\langle 0|A^{-1}}{1 - \gamma\langle 0|A^{-1}|0\rangle}\right)|0\rangle = \frac{\langle 0|A^{-1}0|\rangle}{1 - \gamma\langle 0|A^{-1}|0\rangle} \tag{5.76}$$

Equation (5.76) is the symbolic version of the solution in which we are interested. To construct it explicitly, we need to evaluate the "expectation value" $\langle 0|A^{-1}|0\rangle$. This is straightforward to set up. In two dimensions, we have

$$\begin{aligned}\langle 0|A^{-1}|0\rangle &= \frac{1}{2\pi}\int_0^{2\pi} \frac{\mathrm{d}\theta}{\lambda + \mathrm{i}v(q_x\cos\theta + q_y\sin\theta) + \gamma} \\ &= \frac{1}{\sqrt{(\lambda+\gamma)^2 + q^2v^2}}\end{aligned} \tag{5.77}$$

Substituting this result into (5.76), we obtain for the Fourier- and Laplace-transformed expression for the uniform distribution of persistent walkers

$$\frac{1}{\sqrt{(\lambda+\gamma)^2 + q^2v^2} - \gamma} \tag{5.78}$$

The full expression for the quantity of interest is the double integral

$$\frac{1}{2\pi\mathrm{i}}\int_0^\infty \mathrm{d}q \int_{-\mathrm{i}\infty}^{\mathrm{i}\infty} \mathrm{d}\lambda \; qJ_0(qr)\mathrm{e}^{\lambda t}\frac{1}{\sqrt{(\lambda+\gamma)^2 + q^2v^2} - \gamma} \tag{5.79}$$

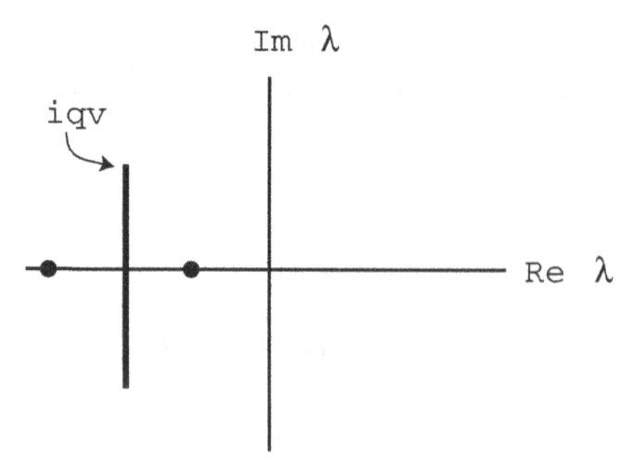

Fig. 5.4. The analytic structure in the complex λ-plane of the integrand in (5.79). Note the poles and the branch cut.

To evaluate the integral over λ in (5.79), we need to investigate the analytical structure of the function in (5.78). As a function of λ, it has two poles and a branch cut (Figure 5.4). To find the poles, we solve the equation

$$\sqrt{(\lambda+\gamma)^2+q^2v^2}-\gamma=0 \tag{5.80}$$

The solution to this equation is

$$\lambda=-\gamma\pm\sqrt{\gamma^2-q^2v^2} \tag{5.81}$$

If we take the contributions from those two poles into account, we are left with the integration

$$\int_0^\infty 2q\gamma J_0(qr)\mathrm{e}^{-\gamma t}\frac{\sinh\left(\sqrt{\gamma^2-q^2v^2}\,t\right)}{\sqrt{\gamma^2-q^2v^2}}\,\mathrm{d}q \tag{5.82}$$

Again, consulting a table of integrals (Gradshteyn *et al.*, 2000), we find

$$\int_0^\infty xJ_0(cx)\frac{\sin\left(a\sqrt{b^2+x^2}\right)}{\sqrt{b^2+x^2}}\,\mathrm{d}x = \begin{cases}\sqrt{\frac{\pi}{2}}b^{1/2}\left(a^2-c^2\right)^{-1/4}J_{-1/2}\left(b\sqrt{a^2-c^2}\right) & c<a\\ 0 & c>a\end{cases} \tag{5.83}$$

To match up with the integral that we want to do, we make the correspondences $c\to r, a\to vt, b\to \mathrm{i}\gamma$. This leaves us with the result

$$2\gamma\sqrt{\frac{\pi}{2}}\mathrm{e}^{-\gamma t}\frac{\cosh\left(\gamma\sqrt{v^2t^2-r^2}\right)}{v\sqrt{v^2t^2-r^2}}\Theta\left(v^2t^2-r^2\right) \tag{5.84}$$

Again, the function Θ is the Heaviside step function. Note the "causality" requirement it enforces. No walker progresses further than if it were propagating ballistically.

We now have to look at the branch cut integration. This branch cut runs from $\lambda = -\gamma - iqv$ to $\lambda = -\gamma + iqv$. If we make the change of variables $\lambda = -\gamma + iqv\sin\theta$, the integration over λ becomes

$$\frac{1}{2\pi}\int_0^\infty dq \int_0^{2\pi} qJ_0(qr)e^{-\gamma t}\frac{e^{iqvt\sin\theta}qv\cos\theta}{qv\cos\theta - \gamma}\, d\theta \tag{5.85}$$

The integral simplifies if we make use of the identity

$$\frac{1}{qv\cos\theta - \gamma} = -\int_0^\infty e^{wq\cos\theta - w\gamma}\, dw \tag{5.86}$$

Inserting this into (5.85), we end up with the double integral

$$\begin{aligned}
&\frac{1}{2\pi}\int_0^{2\pi} d\theta \int_0^\infty dw\, qJ_0(qr)e^{-\gamma t}e^{-w\gamma}\left(-\frac{d}{dw}\right)\exp\left(iqvt\sin\theta + wqv\cos\theta\right)\\
&= \int_0^\infty dw\, qJ_0(qr)e^{-\gamma t}e^{-w\gamma}\left(-\frac{d}{dw}\right)J_0\left(qv\sqrt{t^2-w^2}\right)\\
&= qJ_0(qr)e^{-\gamma t}J_0(qvt) - \gamma e^{-\gamma t}\int_0^\infty qJ_0(qr)e^{-w\gamma}J_0\left(qv\sqrt{t^2-w^2}\right)dw
\end{aligned} \tag{5.87}$$

Using

$$\int_0^\infty qJ_0(qx)J_0(qy)\, dq = \frac{2\pi}{x}\delta(x-y) \tag{5.88}$$

we find for the number of persistent walkers in two dimensions that are a point a distance r from their common point of origin at a time t:

$$\begin{aligned}
&2\pi\frac{e^{-\gamma t}}{r}\delta(r-vt) - \frac{2\pi\gamma}{r}e^{-\gamma t}\int_0^\infty \delta\left(r - v\sqrt{t^2-w^2}\right)e^{-\gamma w}\, dw\\
&= 2\pi\frac{e^{-\gamma t}}{r}\delta(r-vt) - 2\pi\frac{\gamma}{v\sqrt{v^2t^2-r^2}}e^{-\gamma t}e^{-\gamma\sqrt{v^2t^2-r^2}/v}\Theta(vt-r)
\end{aligned} \tag{5.89}$$

The structure of this distribution is similar to what we found in one dimension. The analytic procedure utilized above carries over to three dimensions, as we will see in the next section.

5.2.5 The three-dimensional persistent walk

The procedure here is the same as in the case of the two-dimensional persistent walk. The Laplace- and Fourier-transformed distribution is of the general form

$$A^{-1} + \frac{\gamma A^{-1}|0\rangle\langle 0|A^{-1}}{1 - \gamma\langle 0|A^{-1}|0\rangle} \tag{5.90}$$

and if we start with a distribution that is uniform and we are interested in the contribution to the total distribution of its uniform component, the quantity of interest is

$$\frac{\langle 0|A^{-1}|0\rangle}{1 - \gamma\langle 0|A^{-1}|0\rangle} \tag{5.91}$$

In the case at hand, the expectation value in the above equations is given by

$$\langle 0|A^{-1}|0\rangle = \frac{1}{4\pi}\int \frac{d\Omega}{\lambda + \gamma + iv\vec{q}\cdot\vec{n}} \tag{5.92}$$

The integral in (5.92) is actually over the surface of a three-dimensional sphere. We can perform the integration by going to spherical coordinates, $\hat{n} = \hat{x}\sin\theta\cos\phi + \hat{y}\sin\theta\sin\phi + \hat{z}\cos\theta$. We know that the integral will depend only on the magnitude of $\vec{q}$, so we can have that vector point entirely in the z direction, without loss of generality. The integral in (5.92) then becomes

$$\begin{aligned} \frac{1}{2}\int_0^{\pi} \frac{\sin\theta}{\lambda + \gamma + ivq\cos\theta}\, d\theta &= \frac{1}{2vq}\ln\left[\frac{\lambda + \gamma + ivq}{\lambda + \gamma - ivq}\right] \\ &= \frac{1}{vq}\arctan\frac{vq}{\lambda + \gamma} \end{aligned} \tag{5.93}$$

The spatial distribution of persistent walkers, integrated over direction, as a function of time is the inverse Fourier and Laplace transform of the function

$$\begin{aligned} &\left(\frac{1}{vq}\arctan\frac{vq}{\lambda+\gamma}\right)\Big/\left(1 - \frac{\gamma}{vq}\arctan\frac{vq}{\lambda+\gamma}\right) \\ &= \frac{1}{vq}\arctan\frac{vq}{\lambda+\gamma} + \left[\frac{\gamma}{(vq)^2}\left(\arctan\frac{vq}{\lambda+\gamma}\right)^2\right]\Big/\left[1 - \frac{\gamma}{vq}\arctan\frac{vq}{\lambda+\gamma}\right] \end{aligned} \tag{5.94}$$

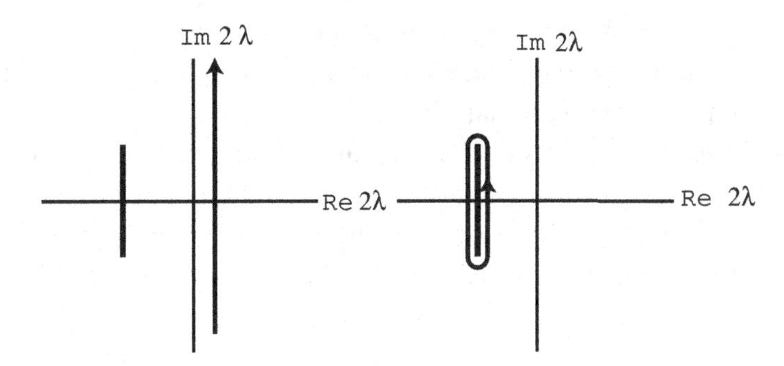

Fig. 5.5. The contour of integration given the singularity structure of the function in (5.93).

After performing the angular integration over the wavenumber q, we are left with the following double integral for the distribution

$$\int_0^\infty \frac{q \sin qr}{q} \, dq \frac{1}{2\pi \mathrm{i}} \int_{-\mathrm{i}\infty}^{\mathrm{i}\infty} d\lambda \; \mathrm{e}^{\lambda t} \left\{ \frac{1}{vq} \arctan \frac{vq}{\lambda + \gamma} \right.$$
$$\left. + \left[\frac{\gamma}{(vq)^2} \left(\arctan \frac{vq}{\lambda + \gamma} \right)^2 \right] \Big/ \left[1 - \frac{\gamma}{vq} \arctan \frac{vq}{\lambda + \gamma} \right] \right\} \tag{5.95}$$

Let us look at the leftmost term in curly brackets in (5.95). The integral over λ is controlled by the singularity structure in the integrand in the complex λ-plane. This consists of a branch cut running from $\lambda = -\gamma - \mathrm{i}vq$ to $\lambda = -\gamma + \mathrm{i}vq$. Consider the function on the last line of (5.93). When $\lambda = -\gamma + \epsilon + \mathrm{i}xvq$, the function is equal to $(1/vq)(\pi/2 - \mathrm{i}\ln((1+x)(1-x))$, this if $-1 < x < 1$. On the other hand, if λ is immediately to the left of the branch cut, $\lambda = -\gamma - \epsilon + \mathrm{i}xvq$, then the function is equal to $(1/vq)(-\pi/2 - \mathrm{i}\ln((1+x)(1-x))$. We integrate up the right hand sides of the branch cut and down the left hand side (see Figure 5.5). Given that the real part of the function is the same on both sides of the branch cut, and that the integrations on either side are in the opposite sense, we end up with the following result for the branch cut integration

$$\frac{1}{2\pi} \frac{1}{vq} \int_{-vq}^{vq} \pi \mathrm{e}^{\mathrm{i}xt} \mathrm{e}^{-\gamma t} \, dx \frac{\sin qr}{r} \tag{5.96}$$

Integration over the variable x yields

$$\frac{1}{2\pi} \frac{2\pi}{vq} \mathrm{e}^{-\gamma t} \frac{\sin vqt}{t} \frac{\sin qr}{r} q \tag{5.97}$$

and, finally, integration over q yields

$$\mathrm{e}^{-\gamma t} \int_0^\infty \frac{1}{vtr} \sin qr \sin qvt \, dq = \frac{\mathrm{e}^{-\gamma t}}{rvt} \frac{\pi}{2} \delta(r - vt) \tag{5.98}$$

This is the exponentially decaying contribution to the distribution consisting of walkers that have not made the plunge and taken a step in a direction different from the one in which they originally started out.

The terms in square brackets in (5.95) require a bit more care. First off, there is a slightly more complicated singularity structure. The function has the branch cut described above. Also, as a function of λ, there is a pole on the negative λ axis. This pole is at the location at which $1 = (\gamma/vq)\arctan(vq/\lambda + \gamma)$. Solving this equation for λ:

$$\lambda = qv \cot \frac{qv}{\gamma} - \gamma \tag{5.99}$$

This pole exists as long as $0 < q < \gamma\pi/2v$. Closing around this pole, and correctly evaluating the residue there entails taking the derivative

$$\frac{\mathrm{d}}{\mathrm{d}\lambda}\left(1 - \frac{\gamma}{vq}\arctan\frac{vq}{\lambda+\gamma}\right)\bigg|_{\lambda = qv\cot\frac{qv}{\gamma} - \gamma} = \frac{\gamma \sin^2\left(\frac{qv}{\gamma}\right)}{(qv)^2} \tag{5.100}$$

Making use of this result, we are left with the following integration over the variable $x = vq/\gamma$

$$\left(\frac{\gamma}{v}\right)^2 \int_0^{\pi/2} \frac{x^3}{r}\frac{1}{\sin^2 x} \mathrm{e}^{-\gamma t}\mathrm{e}^{\gamma x t \cot x} \sin(\gamma x r/v)\, \mathrm{d}x \tag{5.101}$$

The next contribution comes from the branch cut illustrated in Figure 5.5. A careful consideration of the structure of the integral leads to the following expression for the integration around this contour

$$\int_{-\infty}^{\infty} \mathrm{d}q \int_{-1}^{1} \mathrm{d}x \left\{ \frac{1}{2\pi}\left[\frac{\gamma}{(qv)^2}\left(\frac{\pi}{2} - \frac{\mathrm{i}}{2}\ln\left(\frac{1+x}{1-x}\right)\right)^2\right] \Big/ \left[1 - \frac{\gamma}{qv}\left(\frac{\pi}{2} - \frac{\mathrm{i}}{2}\ln\left(\frac{1+x}{1-x}\right)\right)\right] \times q^2 v \frac{\sin qr}{r} \mathrm{e}^{\mathrm{i}qvxt}\mathrm{e}^{-\gamma t} \right\} \tag{5.102}$$

Note the change in the range of the integration over q. Now, we perform the integration over q. The analytic structure of the integrand as a function of q is very straightforward. There is a simple pole at the zero of the denominator in the integrand in (5.102). However, there is a convergence issue, having to do with the behavior of the integrand as a function of q when that variable is large. We take care of that by replacing $q \sin qr$ by $-\mathrm{d}/\mathrm{d}r \cos qr$. We then accomplish the integration by closing around that pole. The pole lies above the real q axis when $x < 0$ and below when $x > 0$. Looking carefully at the complex exponentials that enter the integration, we see that we obtain a non-zero result only when $v|x|t < r$. Otherwise, the contour can be closed in such a way that no singularity in q is enclosed. After a bit more work, we find that the expression of interest reduces to the following single integration

$$
\begin{aligned}
&\mathrm{e}^{-\gamma t}\,\mathrm{Re}\left\{\frac{1}{r}\int_0^1 \mathrm{d}x\left(\frac{\gamma}{v}\right)\left[\frac{\pi}{2}-\frac{\mathrm{i}}{2}\ln\left(\frac{1+x}{1-x}\right)\right]^2\right.\\
&\quad\left.\times\left(\mathrm{i}\frac{\partial}{\partial r}\right)\exp\left[-\mathrm{i}(r-vxt)\frac{\gamma}{v}\left(\frac{\pi}{2}-\frac{\mathrm{i}}{2}\ln\left(\frac{1+x}{1-x}\right)\right)\right]\Theta(r-vxt)\right\}\\
&=\mathrm{e}^{-\gamma t}\,\mathrm{Re}\left\{\frac{1}{r}\int_0^1 \mathrm{d}x\left(\frac{\gamma}{v}\right)^2\left[\frac{\pi}{2}-\frac{\mathrm{i}}{2}\ln\left(\frac{1+x}{1-x}\right)\right]^3\right.\\
&\quad\left.\times\exp\left[-\mathrm{i}(r-vxt)\frac{\gamma}{v}\left(\frac{\pi}{2}-\frac{\mathrm{i}}{2}\ln\left(\frac{1+x}{1-x}\right)\right)\right]\Theta(r-vxt)\right\}\\
&\quad+\mathrm{e}^{-\gamma t}\frac{\gamma}{v}\frac{1}{rvt}\frac{\pi}{2}\ln\left[\frac{1+r/vt}{1-r/vt}\right]\Theta(vt-r)
\end{aligned}
\tag{5.103}
$$

Here, Re means "real part of." The two terms on the right hand side of (5.103) arise from the action of the r-derivative on the exponent and on the step function.

It can be shown that when $r > vt$, the expression in (5.103) cancels (5.101). Here is how that is done. We note that when $r > vt$ the upper limit on the x-integration in (5.103) is 1. Then, we replace x in the integration by $\tanh y$, where the range of integration over y is from 0 to ∞. The integral is now

$$
\mathrm{e}^{-\gamma t}\,\mathrm{Re}\left\{\frac{1}{r}\int_0^\infty \frac{\mathrm{d}y}{\cosh^2 y}\left(\frac{\gamma}{v}\right)^2\left[\frac{\pi}{2}-\mathrm{i}y\right]^3\exp\left[-\mathrm{i}(r-vt\tanh y)\frac{\gamma}{v}\left(\frac{\pi}{2}-\mathrm{i}y\right)\right]\right\}
\tag{5.104}
$$

Now, we deform the contour of the y-integration. First, we integrate from $y = 0$ to $-\mathrm{i}\pi/2$ and then from $y = -\mathrm{i}\pi/2$ to $-\mathrm{i}\pi/2 + \infty$. The integrand along the second contour is readily shown to be pure imaginary. As for the integral from 0 to $-\mathrm{i}\pi/2$, we replace y in the integral by $-\mathrm{i}w$. The integral is then

$$
\mathrm{e}^{-\gamma t}\,\mathrm{Re}\left\{\int_0^{\pi/2}\left(\frac{\gamma}{v}\right)^2\frac{(-\mathrm{i}\mathrm{d}w)}{\cos^2 w}\left(\frac{\pi}{2}-w\right)^3\exp\left[-\mathrm{i}(r+\mathrm{i}vt\tan w)\frac{\gamma}{v}\left(\frac{\pi}{2}-w\right)\right]\right\}
\tag{5.105}
$$

We then replace w by $\pi/2 - u$, extract the real part – a straightforward procedure now – and end up with the following expression

$$
-\mathrm{e}^{-\gamma t}\int_0^{\pi/2}\frac{\mathrm{d}u}{\sin^2 u}\left(\frac{\gamma}{v}\right)^2 u^3\frac{\sin(\gamma ur/v)}{r}\mathrm{e}^{-\gamma ut\cot u}
\tag{5.106}
$$

which exactly cancels (5.101).

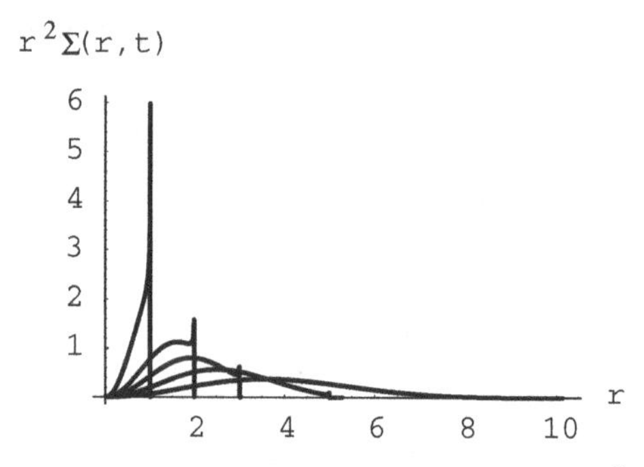

Fig. 5.6. Distributions of the walkers, $\Sigma(r, t)$, multiplied by r^2, for various values of t. The parameters v and γ have both been set equal to one. The functions $\Sigma(r, t)$ plotted in this graph are given by the sum of the expressions in (5.98), (5.101), and (5.103).

There appears to be no way to further reduce the integrals above. However, numerical integration generates results for the total density of persistent three-dimensional walks. Figure 5.6 shows a set of curves, without the delta function contributions for the distribution of such walkers, integrated over angles. Note that the distribution diverges logarithmically at the leading edge. This is because of the second term on the right hand side of (5.103). We have verified by numerical integration that the total number of walkers is conserved. In fact, if we combine all three terms, (5.98), (5.101), and (5.103), we end up with a density as a function of r that, when multiplied by r^2, integrates to $\pi/2$.

5.3 The continuous time random walk

We've looked at walks that take steps in a regular, predictable progression. All of the randomness appears in the choice of directions. We have also looked at walks that propagate continuously, making random changes of direction at random times. As yet another variation, we are going to imagine a walk that sits still between steps, which it takes after a variable time interval. Furthermore, the likelihood that the walk takes a step at a given point of time may possess a non-trivial time-dependence. The analysis again makes use of the Laplace transformation. In our study of persistent walks, we've seen that the Laplace transform is especially useful in the case of a process involving time delays.

We'll start by posing questions about the likelihood of a walker taking a certain number of steps in a given time interval. That is, we ask: what is the likelihood that at time t the walker has taken N steps? We will define the likelihood that the walker takes a step at some time t after it has arrived at a point in terms of the probability density $q(t)$ as follows. The probability that the walker takes a step in the time interval between t and $t + \mathrm{d}t$ after arriving at the point at which it now sits is equal to $q(t)\,\mathrm{d}t$, assuming that it has not taken a step before the time t. Then, the probability that it has not taken a step by the time τ after it has arrived at the point at which it now sits is equal to the product of the probability that it has not taken a step during any of the time intervals leading up to τ. This probability is equal to

$$\prod_{i=1}^{m} (1 - q(t_i)\,\mathrm{d}t_i) \tag{5.107}$$

where we have divided the time between $t = 0$ and $t = \tau$ into m intervals. The product will ultimately be over an infinite set of terms, each corresponding to an infinitesimal time interval. We can then reformulate the product as follows:

$$\begin{aligned}
\prod_{i=1}^{m} (1 - q(t_i)\,\mathrm{d}t_t) &= \exp\left[\sum_{i=1}^{m} \ln(1 - q(t_i)\,\mathrm{d}t_i)\right] \\
&= \exp\left[-\sum_{i=1}^{m} \left(q(t_i)\,\mathrm{d}t_i + O\left(\mathrm{d}t_i^2\right)\right)\right] \\
&\to \exp\left[-\int_0^{\tau} q(t)\,\mathrm{d}t\right] \\
&\equiv a(\tau)
\end{aligned} \tag{5.108}$$

If the integral $\int_0^{\infty} q(t)\,\mathrm{d}t$ is infinite, then the probability of not taking a step goes to zero as $\tau \to \infty$. On the other hand, if the integral does not diverge, there is a finite probability that the walker will never take a step after arriving at its current position.

Now, let's calculate the probability that the walker takes no step until $t = t_1$, and then takes a step between $t = t_1$ and $t = t_1 + \mathrm{d}t$. This probability is $a(t_1)q(t_1)\,\mathrm{d}t$. The probability that this will happen at any time is the sum over all times t_1, which is the integral

$$\begin{aligned}
\int_0^{\infty} a(t_1)q(t_1)\,\mathrm{d}t_1 &= \int_0^{\infty} \exp\left[-\int_0^{t_1} q(t)\,\mathrm{d}t\right] q(t_1)\,\mathrm{d}t_1 \\
&= \int_0^{\infty} \exp\left[-\int_0^{t_1} q(t)\,\mathrm{d}t\right] \left(\frac{\mathrm{d}}{\mathrm{d}t_1} \int_0^{t_1} q(t)\,\mathrm{d}t\right) \mathrm{d}t_1 \\
&= -\exp\left[-\int_0^{t_1} q(t)\,\mathrm{d}t\right]\Bigg|_0^{\infty} \\
&= a(0) - a(\infty) \\
&= 1 - a(\infty)
\end{aligned} \tag{5.109}$$

This result makes sense. The likelihood that the walker takes one step ever is equal to one minus the probability that it never takes a step.

Now, let's figure out the probability that the walker takes a single step between the time $t = 0$ and the time $t = \tau$. This is equal to the probability that the walker takes no steps until the time t_1, where $t_1 < \tau$, then takes a step in the time interval between t_1 and $t_1 + \mathrm{d}t$, then takes no more steps until $t = \tau$, integrated over all possible times t_1 between $t_1 = 0$ and $t_1 = \tau$. This combination is equal to

$$\int_0^{\tau} a(t_1)q(t_1)a(\tau - t_1)\,\mathrm{d}t_1 \tag{5.110}$$

We can clearly go further, and find out the likelihood that the walker will take m steps between a time $t = 0$ and $t = \tau$. This is the multiple integral

$$\int_0^{t_1} \mathrm{d}t_0 \int_0^{t_2} \mathrm{d}t_1 \cdots \int_0^{\tau} \mathrm{d}t_{m-1} a(t_0)q(t_0)a(t_1 - t_0)q(t_1 - t_0) \cdots a(\tau - t_{m-1}) \tag{5.111}$$

This multiple integral is in the form of a convolution. Its evaluation is not at all trivial. However, there is a way to simplify the task of determining its value, and that is to make use of the Laplace trasform. Recall, the Laplace transform, $F(\lambda)$, of a function of t, $f(t)$, which is defined for $t \geq 0$, is given by

$$F(\lambda) = \int_0^{\infty} \mathrm{e}^{-\lambda t} f(t)\,\mathrm{d}t \tag{5.112}$$

For example, the Laplace transform of the probability that the walker takes a step at exactly the time t after it has arrived at a certain position is given by

$$\int_0^{\infty} \mathrm{e}^{-\lambda t} a(t)q(t)\,\mathrm{d}t \tag{5.113}$$

If we define the function $Q(t)$ as follows:

$$Q(t) = \int_0^{t} q(t')\,\mathrm{d}t' \tag{5.114}$$

then,

$$a(t) = \mathrm{e}^{-Q(t)} \tag{5.115}$$

and the Laplace transform in (5.113) is equal to

$$\int_0^{\infty} \mathrm{e}^{-Q(t)} \frac{\mathrm{d}Q(t)}{\mathrm{d}t} \mathrm{e}^{-\lambda t}\,\mathrm{d}t \tag{5.116}$$

Let's define the following Laplace transforms

$$A(\lambda) = \int_0^\infty a(t)q(t)\mathrm{e}^{-\lambda t}\,\mathrm{d}t$$
$$= \int_0^\infty \mathrm{e}^{-Q(t)}\mathrm{e}^{-\lambda t}\frac{\mathrm{d}Q(t)}{\mathrm{d}t}\,\mathrm{d}t \tag{5.117}$$
$$B(\lambda) = \int_0^\infty a(t)\mathrm{e}^{-\lambda t}\,\mathrm{d}t \tag{5.118}$$

The relationship between $A(\lambda)$ and $B(\lambda)$ is readily established with the use of (5.117) and integration by parts.

$$A(\lambda) = \int_0^\infty \mathrm{e}^{-\lambda t}\mathrm{e}^{-Q(t)}\frac{\mathrm{d}Q(t)}{\mathrm{d}t}\,\mathrm{d}t$$
$$= -\mathrm{e}^{-Q(t)}\mathrm{e}^{-\lambda t}\Big|_0^\infty - \lambda\int_0^\infty \mathrm{e}^{-Q(t)}\mathrm{e}^{-\lambda t}\,\mathrm{d}t$$
$$= 1 - \lambda B(\lambda) \tag{5.119}$$

The probability displayed in (5.111) has a Laplace transform that is given by

$$\int_0^{t_1}\mathrm{d}t_0\int_0^{t_2}\mathrm{d}t_1\cdots\int_0^{\tau}\mathrm{d}t_{m-1}\int_0^\infty \mathrm{d}\tau\, a(t_0)q(t_0)a(t_1-t_0)q(t_1-t_0)\cdots a(\tau-t_{m-1})\,\mathrm{e}^{-\lambda\tau}$$
$$= \int_0^{t_1}\mathrm{d}t_0\int_0^{t_2}\mathrm{d}t_1\cdots\int_0^{\tau}\mathrm{d}t_{m-1}\int_0^\infty \mathrm{d}\tau\, a(t_0)q(t_0)\,\mathrm{e}^{-\lambda t_0}a(t_1-t_0)q(t_1-t_0)$$
$$\times\,\mathrm{e}^{-\lambda(t_1-t_0)}\cdots a(\tau-t_{m-1})\,\mathrm{e}^{-\lambda(\tau-t_{m-1})} \tag{5.120}$$

Further analysis of this multiple integral follows from a reordering of the integrations. We note that the function $a(t)$ is equal to zero when $t < 0$. Then, if the order is reversed, the integral over τ is from t_{m-1} to ∞. Then, the integral over t_{m-1} is from t_{m-2} to ∞, and so on. The expression is reduced to

$$\int_0^\infty a(\tau_1)q(\tau_1)\,\mathrm{e}^{-\lambda\tau_1}\int_0^\infty a(\tau_2)q(\tau_2)\,\mathrm{e}^{-\lambda\tau_2}\cdots\int_0^\infty a(\tau_{m+1})\,\mathrm{e}^{-\lambda\tau_{m+1}}\,\mathrm{d}\tau_{m+1}$$
$$= A(\lambda)^m B(\lambda) \tag{5.121}$$

What, then, is the Laplace transform of the probability that the walker will have taken no steps in the time interval τ, or that it will have taken one step in that time interval, or two, or three, or any number of steps in that time interval? This Laplace transform is given by the geometrical sum

$$\sum_{m=0}^{\infty} A(\lambda)^m B(\lambda) = \frac{B(\lambda)}{1 - A(\lambda)} \tag{5.122a}$$

$$= \frac{B(\lambda)}{1 - (1 - \lambda B(\lambda))} \tag{5.122b}$$

$$= \frac{1}{\lambda} \tag{5.122c}$$

In (5.122b), we made use of the relationship between $A(\lambda)$ and $B(\lambda)$ displayed in (5.119).

We have already seen how to perform the inverse Laplace transform (See (5.55) in Section 5.2.2). In the case at hand, we need only note that $1/\lambda$ is the Laplace transform of the constant 1:

$$\int_0^\infty e^{-\lambda t}\, dt = \frac{1}{\lambda} \tag{5.123}$$

Thus, we have recovered the pretty-much-obvious result that the probability that the walker will take any number of steps (including none at all) in a given time interval is equal to one, regardless of the length of the time interval.

Let's now take into account the fact that the walker moves some distance at each step. We'll do this by making use of the transform of the probability that the step takes it to a position a displacement vector $\vec{r}$ away from where it was standing. Recall that the probability that the walker has traversed a displacement vector $\vec{R}$ in two steps is given by

$$P\left(\vec{R}\right) = \int d^d r_1 \int d^d r_2 p\left(\vec{r}_1\right) p\left(\vec{r}_2\right) \delta\left(\vec{r}_1 + \vec{r}_2 - \vec{R}\right) \tag{5.124}$$

where $P(\vec{R})$ is the probability density that the walker steps a distance $\vec{R}$. If we take the spatial Fourier transform of this probability, we obtain

$$\begin{aligned}\int e^{-i\vec{q}\cdot\vec{R}} P\left(\vec{R}\right) d^d R &= \int d^d R \left[\int d^d r_1 \int d^d r_2 p\left(\vec{r}_1\right) p\left(\vec{r}_2\right) \delta\left(\vec{r}_1 + \vec{r}_2 - \vec{R}\right) e^{-i\vec{q}\cdot\vec{R}}\right] \\ &= \int d^d r_1 e^{-i\vec{q}\cdot\vec{r}_1} p\left(\vec{r}_1\right) \int d^d r_2 e^{-i\vec{q}\cdot\vec{r}_2} p\left(\vec{r}_2\right) \\ &\equiv \chi\left(\vec{q}\right)^2 \end{aligned} \tag{5.125}$$

where, as previously,

$$\chi\left(\vec{q}\right) = \int e^{-i\vec{q}\cdot\vec{r}} p\left(\vec{r}\right) d^d r \tag{5.126}$$

The spatial Fourier transform of the probability that the walker has traversed a net displacement vector in n steps is, by an extension of the above, equal to $\chi(\vec{q})^n$.

This means that the Laplace and spatial Fourier transform of the probability that the walker has taken n steps in an interval and has traversed a certain displacement vector as a result of those n steps is given by the formula

$$\left(A(\lambda)\chi\left(\vec{q}\right)\right)^n B(\lambda) \tag{5.127}$$

Summing over n, the Laplace and Fourier transform of the probability that the walk has arrived at a given location at a particular point in time (taking into account the possibility that any number of steps may have been taken) is equal to

$$\frac{B(\lambda)}{1-\chi\left(\vec{q}\right)A(\lambda)} \tag{5.128}$$

The likelihood that the walker has ended up a displacement vector $\vec{r}$ away from its point of origin at a time t after it started out is given by the inverse Laplace and spatial Fourier transform of (5.128). Let's see how the familiar Gaussian limit asserts itself. We'll do this by expanding to low order in both λ and $\vec{q}$. We know that if the walker takes steps that are not too long, we can write

$$\chi(\vec{q}) = 1 - Cq^2 \tag{5.129}$$

We'll assume that we can perform a similar expansion of $A(\lambda)$. Making use of (5.119):

$$\begin{aligned} A(\lambda) &= 1 - \lambda B(\lambda) \\ &\approx 1 - \lambda B(0) \end{aligned} \tag{5.130}$$

Then, the combined Laplace and Fourier transform of the desired probability is, approximately,

$$\frac{B(0)}{1-\left(1-Cq^2\right)\left(1-\lambda B(0)\right)} \approx \frac{B(0)}{\lambda B(0) + Cq^2} \tag{5.131a}$$

$$= \frac{1}{\lambda + (c/B(0))q^2} \tag{5.131b}$$

Now, we note that the Laplace transform of the function e^{-at} is given by

$$\int_0^\infty \mathrm{e}^{-at}\mathrm{e}^{-\lambda t}\,\mathrm{d}t = \frac{1}{\lambda + a} \tag{5.132}$$

which tells us that (5.131b) is the Laplace transform of $\mathrm{e}^{-Ctq^2/B(0)}$. The inverse

Fourier transform of this quantity is

$$\int d^d q e^{i\vec{q}\cdot\vec{r}} e^{-Ctq^2/B(0)} = \left(\frac{\pi B(0)}{Ct}\right)^{d/2} e^{-B(0)r^2/4Ct} \tag{5.133}$$

The probability distribution is again a Gaussian, with a width that is proportional to the time interval, t.

Exercise 5.8

Suppose that the function $q(t)$ is equal to a constant. Let this constant be p_0. Find the combined Laplace and Fourier transform of the probability that the walk has gotten to a given location at a specific time. That is, find the explicit form of the function (5.128).

Exercise 5.9

Another wrinkle on the continuous time random walk problem is to ask what happens when there is also the finite probability that a walker will disappear. Suppose that in a given small time interval Δt, the walker takes a step with a probability equal to $p_0 \Delta t$, and with a probability $p_1 \Delta t$ the walker disappears. We now define $a(t)$ as the number of remaining walkers that have not yet taken a step. Then, $a(t)$ can be shown to be given by

$$a(t) = e^{-(p_0+p_1)t}$$

and

$$a(t)q(t) = p_0 e^{-(p_0+p_1)t}$$

Show that this tells us that

$$A(\lambda) = \frac{p_0}{p_0 + p_1 + \lambda}$$

and that

$$B(\lambda) = \frac{1}{p_0 + p_1 + \lambda}$$

Use these results and (5.128) to find the Laplace and Fourier transform of the probability that a walker finds itself at a certain location at a particular time. Finally, make use of the result you have just obtained to calculate the probability as a function of space and time. You can use (5.129). Show that the total number of walks that remain at a time t decays exponentially. What is the rate of decay?

A discussion of the non-Gaussian case will be delayed until later (see Chapter 8), where our approach will be slightly different. We leave this chapter here, as our focus in the remaining chapters will be on other, less traditional, treatments of the random walk problem. However, we commend to the reader further study of the various "non-ideal" random walks introduced in this chapter; they are rich in theoretical insights and practical applications (Barber and Ninham, 1970; Hughes, 1995; Weiss, 1994).

6

The shape of a random walk

6.1 The notion and quantification of shape

Shape is an intuitively accessible notion. We organize visual information in terms of shapes, and the shape of an object represents one of the first of its qualities referred to in an informal descriptive rendering of it. While our language presents us with a wide repertoire of verbal images for the approximate portrayal of the shape of a physical entity ("round," "oblong," "crescent," "stellate" . . .) the precise characterization of a shape, in terms of a number, or set of numbers, has remained elusive. This is with good reason. It is well-known to mathematicians that the class consisting of the set of all curves is a higher order of infinity than the set of all real numbers. This means that there can be no one-to-one correspondence between curves and real numbers. As shapes, intuitively at least, bear a conceptual relationship to curves, it is plausible that the set of all shapes dwarfs in magnitude the set of real numbers, or of finite sets of real numbers.

On the other hand, if one is willing to content oneself with a general paradigm for the measurement of shape, there are ways of quantifying it in terms of numbers that have a certain descriptive and predictive utility. In fact, the numerical specification of shapes has acquired a certain urgency of late, in light of the widespread use of computer imaging and the concomitant focus on the development of codes for the creation and manipulation of pictorial quantities.

In this chapter, we will look at different ways of characterizing and measuring the shape of a random walk. We will focus on one particular method, based on calculations of the width of the distribution of steps about the "center of mass" of the walk. The particular quantity studied is the radius of gyration tensor, and the shape of the polymer is quantified in terms of the eigenvalues of this tensor, termed the principal radii of gyration of the walk. We will look at a particular combination of the principal radii of gyration that provides information with respect to the deviation from spherical symmetry of the shape of the walk. It will turn out that

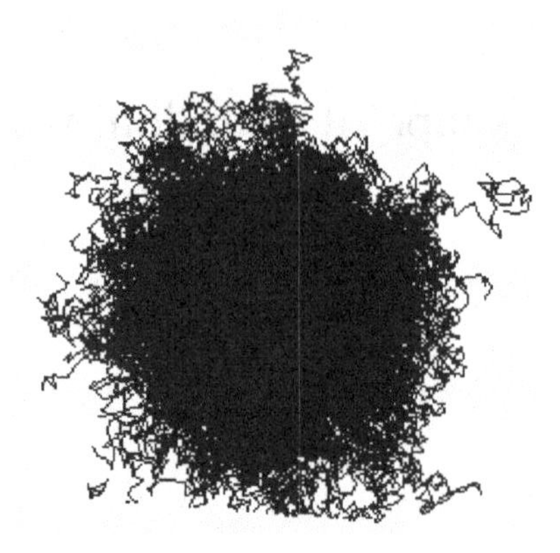

Fig. 6.1. A "cloud" consisting of the paths of 1000 random walkers, each of which has taken 100 steps from a common point of origin.

shape as a concept is, as one might expect, a bit elusive. For one thing, there is no generic "shape" for a random walk. However, statistical statements can be made, with regard to the probability that a walk takes on a particular shape, at least as characterized by the principal radii of gyration. In addition, there is one limit in which the shape of the trail left by a walker is fixed and predictable. That is the limit of a walker in an infinite dimensional space ($d = \infty$). We will discuss the construction of an expansion about that limit, the $1/d$-expansion. This expansion yields the shape distribution of a random walker's trail when the walker wanders in a high spatial dimension environment. As we will see, this expansion is – at least for some puroposes – respectably accurate in three dimensions.

6.1.1 Anisotropy of a random walk

When we talk about the distribution of points visited by a random walker, we generally do so in the context of ensemble averages. That is, we ask *on average* how many walks visit a given point. Looking at things this way can obscure the detailed structure of a given random walk. For example, if we are interested in how many times a given point at location $\vec{r}_1$ is visited by a walker that starts out at location $\vec{r}_0$, we find, after suitable averaging, that the answer depends only on the distance between those points in space, $|\vec{r}_1 - \vec{r}_0|$. This is true because for every walker that tends to go off in one direction there will be another walker that ends up going in the opposite direction. The statistical distribution of places visited is rotationally symmetric about the point of origin. In other words, the totality of walkers in the ensemble creates a "cloud" that is spherically symmetric. Figure 6.1 shows just such a cloud, which consists of the paths of 1000 random walkers each of whom has taken 100 steps from a common point of origin. The near-spherical symmetry of the cloud is evident from the figure.

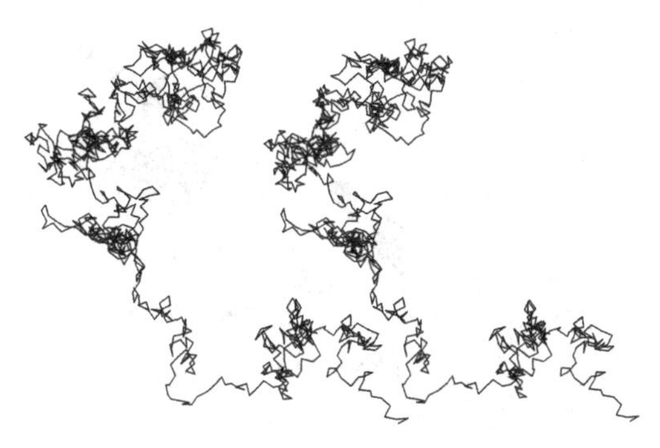

Fig. 6.2. Stereographic pair of images of a 1000-step three-dimensional random walk.

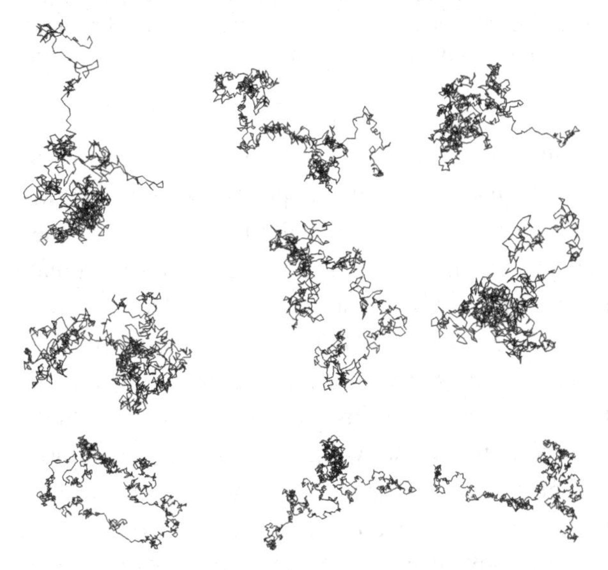

Fig. 6.3. Several examples of a 1000-step random walk.

This result of averaging obscures the fact that a given random walk can be quite anisotropic spatially. Figure 6.2 shows a stereographic pair of images of a single 1000-step random walk. The elongated nature of the walk shown in this figure is not a statistical anomaly. Figure 6.3 shows several examples of 1000-step walks. Note that not one of those walks is reminiscent of the cloud of walkers shown in Figure 6.1.

On the other hand, as Figure 6.3 makes abundantly clear, no typical, or definitive, shape can be assigned to a random walk. How, then, to quantify the shape of a walk?

6.1.2 Measures of the shape of a walk

The literature presents a number of algorithms for the characterization of the shape of an object (Bookstein, 1978; Costa and Cesar, 2001). Here, we choose one that is

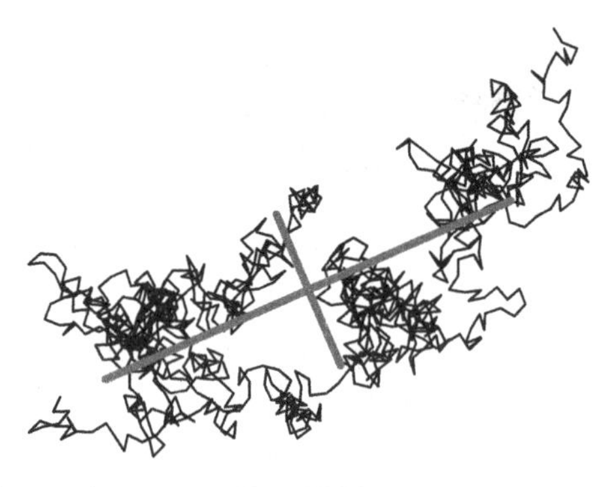

Fig. 6.4. The anisotropic nature of a 1000-step two-dimensional walker. The shaded lines indicate the directions in which its linear extent is the greatest and the smallest. The lines also run parallel to the eigenvectors of the matrix defined in (6.1)–(6.3). The point of intersection of those two lines is the "center of gravity" of the walk.

particularly well-suited to our needs. We construct a moment-of-inertia-like tensor (See Supplement 1 at the end of this chapter). By diagonalizing this tensor we are able to extract numbers that quantify the linear "size" of the walk in various directions, particularly in the directions in which it has the greatest linear extent and the direction in which it is most compact. For a visualization of this, see Figure 6.4, in which the extensions in both directions of a two-dimensional random walk are illustrated. The thick lines indicate the directions in which random walk has the greatest and the smallest extension, as determined by the tensor that we are about to introduce. Note that these two lines provide a quantitative representation of both the overall orientation of the walk and of its spatial anisotropy –that is, the degree to which the shape of the walker's path differs from that of a sphere. As we will see, the amount of anisotropy exhibited by the walks in Figures 6.2, 6.3, and 6.4 is not at all atypical.[1]

6.1.3 The radius of gyration tensor

The tensor that we are about to define is also used to determine the rotational inertia of a three-dimensional object. Supplement 1 at the end of this chapter reviews its use in that context. What this means is that the results to be derived here are relevant to the rotational motion of an object that mimics the form of the path followed by a random walker, assuming that the constituents of this object have an inertial mass, that they are uniformly distributed along the path it imitates, and that the object is, itself, rigid.

[1] For a characterization of the anistropy of a random walker using the notion of spans, see (Weiss and Rubin, 1976). Here, our approach will be somewhat different.

Here is how the tensor is constructed (Solc and Stockmayer, 1971). Given the location, $\vec{r}_i$ $(1 \leq i \leq N)$, of each step of a walker in d dimensions that has left N footprints, we construct a d-dimensional tensor, $\overleftrightarrow{T}$ with entries

$$T_{kl} = \frac{1}{N} \sum_{j=1}^{N} \left(r_{jk} - \langle r_k \rangle \right) \left(r_{jl} - \langle r_l \rangle \right) \tag{6.1}$$

Here, r_{jk} is the kth component of the position vector of the jth step, and $\langle r_k \rangle$ is the average of the kth component of the locations of the steps of the walker:

$$\langle r_k \rangle = \frac{1}{N} \sum_{j=1}^{N} r_{jk} \tag{6.2}$$

For example, a walker in two dimensions has *radius of gyration tensor* $\overleftrightarrow{T}$ with the following form

$$\overleftrightarrow{T} = \begin{pmatrix} \frac{1}{N} \sum_{j=1}^{N} \left(x_j - \langle x \rangle \right)^2 & \frac{1}{N} \sum_{j=1}^{N} \left(x_j - \langle x \rangle \right) \left(y_j - \langle y \rangle \right) \\ \frac{1}{N} \sum_{j=1}^{N} \left(x_j - \langle x \rangle \right) \left(y_j - \langle y \rangle \right) & \frac{1}{N} \sum_{j=1}^{N} \left(y_j - \langle y \rangle \right)^2 \end{pmatrix} \tag{6.3}$$

The eigenvectors and eigenvalues of this tensor quantify the linear dimensions of the walker – its girth – in various directions. The eigenvectors point in the direction in which this span is maximized, and the direction in which it is minimized. The eigenvalues tell us how extended the walk is in those extremal directions. In fact, the lines in Figure 6.4 lie along the directions in which those two eigenvectors point. The lengths of those lines are directly proportional to the eigenvalues of the matrix $\overleftrightarrow{T}$ appropriate to the walk in that figure.

For a discussion of the relationship between the eigenvectors and eigenvalues of $\overleftrightarrow{T}$ and the maximal and minimal spans of a walk see Section 6.4.1 in Supplement 1 at the end of this chapter.

6.1.4 Eigenvalues of the matrix $\overleftrightarrow{T}$: the asphericity of a random walk

The eigenvalues of the matrix $\overleftrightarrow{T}$ are the squares of the *principal radii of gyration*, R_i, of the object in question. They are essentially the mean square deviations of the steps of the walker from the "center of gravity" of the walk. In Figure 6.4 the walk's center of gravity lies at the point of intersection of the two thick lines, each of which lie in the direction of the eigenvectors of the matrix $\overleftrightarrow{T}$ for that walk. This means that diagonalized, the matrix $\overleftrightarrow{T}$ takes the form

$$\overleftrightarrow{T} = \begin{pmatrix} \lambda_1 & 0 & 0 \\ 0 & \lambda_2 & 0 \\ 0 & 0 & \lambda_3 \end{pmatrix} \equiv \begin{pmatrix} R_1^2 & 0 & 0 \\ 0 & R_2^2 & 0 \\ 0 & 0 & R_3^2 \end{pmatrix} \tag{6.4}$$

The relative magnitudes of the eigenvalues of the radius of gyration tensor $\overleftrightarrow{T}$ then tell us to what extent the object in question has a shape that differs from that of a sphere. Clearly if all R_i^2's in (6.4) are equal, then the linear span of the object will be the same in all directions, and it is reasonable to attribute a kind of spherical symmetry to it. However, if, for example, $R_1^2 \gg R_2^2, R_3^2$, which means that R_1 is significantly larger than R_2 and R_3, then the object can be thought of as greatly elongated, and not at all spherical.

The eigenvalues of an object's radius of gyration tensor are invariant with respect to the overall orientation of the object. That is, a rotation of the object will not change those eigenvalues. On the other hand, the tensor itself does change as the object is rotated. If the brackets $\langle \cdots \rangle_r$ stand for averaging with respect to overall orientation, then the average $\langle R_i^2 \rangle_r$ is just the same as R_i^2. On the other hand, performing the same average over $\overleftrightarrow{T}$ produces a matrix altered by the averaging process. In fact, it is pretty straightforward to argue that $\langle (x - \langle x \rangle)(y - \langle y \rangle) \rangle_r$ will average to zero, while $\langle (x - \langle x \rangle)^2 \rangle_r = \langle (y - \langle y \rangle)^2 \rangle_r = \langle (z - \langle z \rangle)^2 \rangle_r$ This means that

$$\langle \overleftrightarrow{T} \rangle = \begin{pmatrix} \bar{T} & 0 & 0 \\ 0 & \bar{T} & 0 \\ 0 & 0 & \bar{T} \end{pmatrix} \tag{6.5}$$

The eigenvalues of this matrix are clearly all equal to $\bar{T}$. In averaging the radius of gyration tensor, we are performing the kind of ensemble average that destroys information regarding the non-spherical shape of the object in question. This clearly means an informative characterization of the shape of the random walk is not contained in the averaged radius of gyration tensor.

We can, nevertheless extract useful shape information by averaging quantities that are directly derivable from the radius of gyration matrix. What we need to do is use quantities that are invariant with respect to rotations and reflections in space (the matrix is automatically invariant with respect to translations). All of these quantities are directly related to the eigenvalues of the matrix. In the case of a three-dimensional matrix there are three independent invariants. One choice of those three is

$$\begin{aligned}\mathrm{Tr}\,\overleftrightarrow{T} &= T_{11} + T_{22} + T_{33} \\ &= R_1^2 + R_2^2 + R_3^2 \end{aligned} \tag{6.6}$$

$$\mathrm{Tr}\,\overleftrightarrow{T}^2 = (R_1^2)^2 + (R_2^2)^2 + (R_3^2)^2 \tag{6.7}$$

$$\mathrm{Tr}\,\overleftrightarrow{T}^3 = (R_1^2)^3 + (R_2^2)^3 + (R_3^2)^3 \tag{6.8}$$

Another well-known invariant of the tensor, its determinant, is obtained as follows

$$\begin{aligned}\mathrm{Det}\,\overleftrightarrow{T} &= R_1^2 R_2^2 R_3^2 \\ &= \frac{1}{6}\left[\left(R_1^2 + R_2^2 + R_3^2\right)^3 + 2\left((R_1^2)^3 + (R_2^2)^3 + (R_3^2)^3\right)\right. \\ &\quad \left. - 3\left(R_1^2 + R_2^2 + R_3^2\right)\left((R_1^2)^2 + (R_2^2)^2 + (R_3^2)^2\right)\right] \\ &= \frac{1}{6}\left[\left(\mathrm{Tr}\,\overleftrightarrow{T}\right)^3 + 2\,\mathrm{Tr}\,\overleftrightarrow{T}^3 - 3\,\mathrm{Tr}\,\overleftrightarrow{T}\;\mathrm{Tr}\,\overleftrightarrow{T}^2\right]\end{aligned} \tag{6.9}$$

Now, it is possible to average the three invariants defined in (6.6)–(6.8). These averages retain important information regarding the devation from spherical symmetry of the shape of the "average" random walk. Consider, for example, the following combination of eigenvalues

$$\begin{aligned}&\left(R_1^2 - R_2^2\right)^2 + \left(R_1^2 - R_3^2\right)^2 + \left(R_2^2 - R_3^2\right)^2 \\ &\quad = 3\left((R_1^2)^2 + (R_2^2)^2 + (R_3^2)^2\right) - \left(R_1^2 + R_2^2 + R_3^2\right)^2 \\ &\quad = 3\,\mathrm{Tr}\,\overleftrightarrow{T}^2 - \left(\mathrm{Tr}\,\overleftrightarrow{T}\right)^2\end{aligned} \tag{6.10}$$

Both sides of this equation can be averaged over all orientations of an object, and, given the fact that they are invariants with respect to translation, rotation and reflection, they will remain unchanged. In the case of the random walk, this means that if we average the last line of (6.10) over an ensemble of walkers we are left with a quantity that tells us something about the differences between the various principal radii of gyration. That is, we find out how different the shape a random walk is, on the average, from that of a sphere.

To construct a quantity that interpolates between zero when all principal radii of gyration are equal and unity when one of the R_i's is much greater than the others we will divide by

$$2\left\langle\left(\sum_{i=1}^{3} R_i^2\right)^2\right\rangle = 2\langle\mathrm{Tr}\,\overleftrightarrow{T}\rangle^2 \tag{6.11}$$

We can, in fact, generalize this quantity and define the mean *asphericity*, A_d, of

d-dimensional random walks as follows (Aronovitz and Nelson, 1986; Rudnick and Gaspari, 1986a; Theodorou and Suter, 1985):

$$A_d = \frac{\sum_{i>j}^{d} \langle (R_i^2 - R_j^2)^2 \rangle}{(d-1)\langle (\sum_{i=1}^{d} R_i^2)^2 \rangle} \tag{6.12}$$

The numerator of (6.12) can be rewritten as follows:

$$\begin{aligned} &\left\langle \left(R_1^2 - R_2^2\right)^2 + \left(R_1^2 - R_3^2\right)^2 + \cdots \left(R_{d-1}^2 - R_d^2\right)^2 \right\rangle \\ &= d\,\mathrm{Tr}\langle \overleftrightarrow{T}^2 \rangle - \left\langle \left(\mathrm{Tr}\, \overleftrightarrow{T}\right)^2 \right\rangle \\ &= d(d-1)\left(\langle T_{11}^2 \rangle - \langle T_{11} T_{22} \rangle\right) + d^2(d-1)\langle T_{12}^2 \rangle \end{aligned} \tag{6.13}$$

The last line of (6.13) follows from the equations for the trace of a tensor and of its square. It also follows from the fact that $\langle T_{11}^2 \rangle = \langle T_{22}^2 \rangle = \cdots \langle T_{dd}^2 \rangle$, and similar equalities for $\langle T_{ii} T_{jj} \rangle$ and $\langle T_{ij}^2 \rangle$. The denominator of the last line of (6.12) can be reduced in the same way, leading to the following expression for the asphericity:

$$\begin{aligned} A_d &= \frac{d(d-1)\left(\langle T_{11}^2 \rangle - \langle T_{11} T_{22} \rangle\right) + d^2(d-1)\langle T_{12}^2 \rangle}{d(d-1)\langle T_{11}^2 \rangle + d(d-1)^2 \langle T_{11} T_{22} \rangle} \\ &= \frac{\left(\langle T_{11}^2 \rangle - \langle T_{11} T_{22} \rangle\right) + d\langle T_{12}^2 \rangle}{\langle T_{11}^2 \rangle + (d-1)\langle T_{11} T_{22} \rangle} \end{aligned} \tag{6.14}$$

The calculation of the asphericity reduces to the problem of determining the average values of powers of the entries in the radius of gyration tensor. The details, which are a bit involved, are in Supplement 2 at the end of this chapter. The end result of the calculation is the following general expression for the mean asphericity of a d-dimensional random walk:

$$A_d = \frac{4 + 2d}{4 + 5d} \tag{6.15}$$

The three-dimensional walk has a mean asphericity of 10/19, or a little more than a half, so in this sense the three-dimensional walk is, on the average, somewhere between an isotropic object and a highly elongated one.

Of course the notion of the mean asphericity of a random walk does not necessarily imply that there is a characteristic shape for three-dimensional walks. Given the examples depicted in Figure 6.3, it seems much more likely that random walks come in a wide variety of shapes and that a quantity such as the mean asphericity provides a very broad-brush characterization of that property of random walks. Figure 6.5 illustrates this point. It is a histogram of the distribution of the individual asphericities of 20000 three-dimensional walks, each comprising 100 steps.

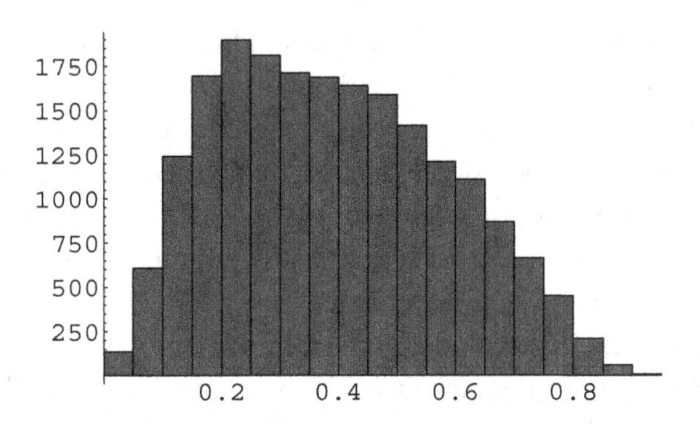

Fig. 6.5. The distribution of the individual asphericities of 20 000 three-dimensional 100-step walks.

Note that the distribution spans the range from 0 to 1, and that no narrow region dominates.

It is also important to note that what is presented in (6.15) is not, strictly speaking, the average of the individual asphericities of the walks, which is given by

$$A'_d = \left\langle \frac{\sum_{i<j} \left(R_i^2 - R_j^2\right)^2}{(d-1) \sum_{k=1}^{N} R_k^2} \right\rangle \tag{6.16}$$

This quantity can also be found exactly in the case of the ordinary d-dimensional walk. The analytical result for this quantity is (Diehl and Eisenriegler, 1989)

$$A'_d = \frac{d}{4}\left[3 + \frac{4}{d} - \frac{d}{2} M_{d/2}\right] \tag{6.17}$$

where

$$M_p = \int_0^\infty x^{p+1} \sinh^{-p} x \, dx \tag{6.18}$$

In three dimensions, $A'_d = 0.394274\ldots$. The average of the individual asphericities is somewhat smaller than the mean asphericity.

Exercise 6.1

Show that (6:S2-2b) in Supplement 2 is correct.

6.2 Walks in $d \gg 3$ dimensions

Often, mathematical models for physical processes are especially amenable to analysis if the system under consideration is imagined to occupy a fictitious universe consisting of a large number of spatial dimensions. We will encounter this when we discuss the statistics of self-avoiding walks in later chapters. In that case, if the dimensionality of the space exceeds four, the statistical properties of a self-avoiding walk are identical in all important respects to those of an ordinary walk. There are similar simplifications in other useful physical models. Given this, a question worth asking is what shape a random walk takes if the walker is free to explore a large number of spatial dimensions. Do shape statistics simplify? If so, is there any way of extrapolating from those statistics to three dimensions?

The mean asphericity, A_d, provides a clue to the behavior of the high-dimensional random walk. As $d \to \infty$, the mean asphericity approaches 2/5. This tells us that random walks in high dimensions are not spherical, nor are they highly elongated. They retain, statistically at least, a shape intermediate between those two extremes. What more can we say about the shape, or range of shapes, of the walk in $d \gg 1$ dimensions?

As it turns out, the shape of a walk in high spatial dimensionality is not only well-defined, it is, in a sense, *universal*, in that essentially every random walk has the same basic shape. We can see this by placing the walker on a d-dimensional cubic lattice, with d a very large number. In such a lattice each vertex is connected to $2d$ nearest neighbors by links that are parallel to the d Cartesian coordinate axes. A walker will set out from its point of origin along one of those links. Upon arriving at the nearest neighbor vertex joined to the point of origin by that link, the walker chooses another link along which to proceed. The choice is purely random, so it is equally likely to take its next step along any one of the $2d$ links attached to that vertex. With overwhelming likelihood, the walker will take its next step in a direction perpendicular to the one it traveled in its previous step. This is because there are $2(d-1)$ choices that cause it to do so and only two that have it moving parallel, or antiparallel, to its first step. At the next vertex, the walker will, with a probability essentially equal to one, take a step in a direction that is perpendicular to those of the first two. Continuing in this way, the walker executes an N-step walk that consists of a set of displacements, all of which are mutually perpendicular.[2]

Exercise 6.2

Consider a walker on a d-dimensional hypercubic lattice. For this walker, show that the number of N-step walks, in which the walker takes each step in a direction orthogonal to all previous steps, is approximately equal to

$$(2d)^N e^{-N^2/2d} \qquad \left(N \ll \sqrt{d}\right)$$

---

[2] This argument, and the results described immediately below, can be found in expanded form in (Rudnick *et al.*, 1987)

Taking into account the possibility of rotating the walk, and of forming mirror images, we see that every N-step walk in $d \gg 1$ dimensions will have the same shape. This, of course, holds true in the strict sense only when $N \ll d$. The construction of the radius of gyration tensor $\overleftrightarrow{T}$ for such a walk is relatively straightforward (Rudnick *et al.*, 1987). The willing reader is led through the essential steps in Exercise 6.3, and the results are obtained using an alternative approach in the next section. This tensor can be diagonalized, and one finds for the eigenvalues

$$R_n^2 = \frac{(N+1)}{\pi^2 n^2} \qquad n \geq 1 \tag{6.19}$$

The principal radii of gyration of the random walk in infinite spatial dimensions are, with perfect certainty, equal to the above set of values. The asphericity is given by

$$A_\infty = \frac{\sum_n \left(R_n^2\right)^2}{\left(\sum_n R_n^2\right)^2} \tag{6.20a}$$

$$= \frac{\sum_{n=1}^{\infty} n^{-4}}{\left(\sum_{n=1}^{\infty} n^{-2}\right)^2} \tag{6.20b}$$

$$= \frac{\pi^4/90}{\left(\pi^2/6\right)^2} \tag{6.20c}$$

$$= \frac{2}{5} \tag{6.20d}$$

Equation (6.20c) follows from known results for the infinite sums in the numerator and denominator in (6.20b) (Gradshteyn *et al.*, 2000).

The simplicity of the shape distribution when $d = \infty$ allows us to hope that we can extrapolate to finite dimensions and obtain information about the shapes of walkers in dimensions of direct interest to us, for instance $d = 3$.

Exercise 6.3

Here, we are going to go through the derivation of the eigenvalues of the radius of gyration tensor, $\overleftrightarrow{T}$, in the case of a walker that wanders on a hypercubic lattice in d dimensions in the limit $d = \infty$. This exercise asks that you fill in the steps in the calculation.

(a) The walker is assumed to start out from a site at the origin ($x_i = 0$ for all i). At the first of N steps, the walker moves in the direction of x_1, and takes a step with a length of l. At the second step the walker moves in the positive direction along the x_2 axis, at the third step in the positive direction along the x_3 axis, and so on. Note that we can always arrange our notation so that this is so. If the walker takes N steps, so that it visits $N + 1$ sites, show that $\langle x_n \rangle = l(N + 1 - n)/(N + 1)$, where the average is over points visited by the walker.

(b) Now, show that at the jth step the following is true

$$x_{j,n} - \langle x_n \rangle = \begin{cases} -l\left(\dfrac{N+1-n}{N+1}\right) & j < n \\ l\left(1 - \dfrac{N+1-n}{N+1}\right) & j \geq n \end{cases}$$

(c) Use the above results to show that, for this walk,

$$\sum_{j=1}^{N+1} \left(x_{j,n} - \langle x_n \rangle\right)\left(x_{j,m} - \langle x_m \rangle\right) = l^2 (N+1) \frac{n}{N+1} \left(1 - \frac{m}{N+1}\right) \qquad n \leq m$$

If $m < n$, the result is the same as the above with m and n interchanged.

(d) Define $x \equiv n/(N + 1)$, $y \equiv m/(N + 1)$. Note that the radius of gyration tensor becomes an operator expressed in terms of x and y. The range of this operator is the interval between 0 and 1. The operator is equal to zero whenever either x or y is on the boundary of the interval. Show by direct substitution that $\sin k\pi y$ is an eigenvector of this tensor, with k an integer. What is the eigenvalue?

(e) Use all the above to construct the eigenvalue spectrum of the radius of gyration tensor, $\overleftrightarrow{T}$.

Exercise 6.4

Show that the operator

$$L(x_1, x_2) = x_<(1 - x_>)$$

where $x_{<(>)}$ is the smaller (larger) of the two x_i's, is the inverse of the operator $-\mathrm{d}^2/\mathrm{d}x^2$. That is, show that

$$-\frac{\mathrm{d}^2}{\mathrm{d}x^2} L(x, y) = \delta(x - y)$$

Furthermore, show that this operator, when its range is restricted to the interval $0 < x < 1$, maps any bounded function onto a function that goes to zero at the end-points of the interval.

6.2.1 A key identity

An analysis of walks in high dimensions is greatly facilitated with the use of a key identity relating entries in the radius of gyration tensor $\overleftrightarrow{T}$ to the displacements corresponding to the steps in a walk. Suppose we denote by $\eta_{\alpha,i}$ the displacement in the direction of the ith cartesian coordinate that occurs in the αth step of the walk. It is then straightforward to show that the entries in the radius of gyration tensor can be expressed in terms of the displacement vectors as follows (Kramers, 1946)[3]

$$T_{ij} = \sum_{\alpha,\beta=1}^{N} a_{\alpha\beta}\eta_{\alpha,i}\eta_{\beta,j} \tag{6.21}$$

where $a_{\alpha\beta}$ are the entries of an N-by-N matrix:

$$a_{\alpha\beta} = \begin{cases} \frac{1}{(N+1)^2}\alpha\,(N+1-\beta)\,,\ \alpha < \beta \\ \frac{1}{(N+1)^2}\beta\,(N+1-\alpha)\,,\ \alpha > \beta \end{cases} \tag{6.22}$$

These relationships are established in Supplement 3 at the end of this chapter.

The operator $\overleftrightarrow{a}$ has, in the limit $N \gg 1$, a straightforward set of eigenvectors and eigenvalues. We can find them by inspection. Begin by rewriting the operator $\overleftrightarrow{a}$ as follows:

$$a_{x_1,x_2} = x_<(N+1-x_>) \tag{6.23}$$

where $x_{<(>)}$ is the smaller (greater) of the variables x_1 and x_2. Then, defining $x = \alpha/(N+1)$, $y = \beta/(N+1)$, we can rewrite this operator when $N \gg 1$ as

$$a_{x,y} = \begin{cases} x(1-y),\ x < y \\ y(1-x),\ x > y \end{cases} \tag{6.24}$$

The eigenvectors of this operator are

$$\psi_n(w) = \sin(n\pi w) \qquad 0 < w < 1 \tag{6.25}$$

where $n > 0$ is an integer. Direct substitution verifies that

$$\sum_{\beta=1}^{N+1} a_{\alpha,\beta}\psi_n(\beta/(N+1)) = (N+1)\int_0^1 a_{\alpha,y}\psi_n(y)\,\mathrm{d}y \tag{6.26a}$$

$$= (N+1)\frac{1}{(n\pi)^2}\psi_n(\alpha/(N+1)) \tag{6.26b}$$

The right hand side of (6.26a) follows from substitution of the variable y for $\beta/(N+1)$ and replacement of a sum over β by an integration over y. This replacement is well-justified in the limit of very large N. The final equality follows from a direct

[3] For an expanded version of the development that follows, see Gaspari *et al.* (1987).

evaluation of the resulting integration. This all means that the eigenvalues of the operator $\overleftrightarrow{a}$ are given by

$$\lambda_k = \frac{N+1}{\pi^2 k^2} \tag{6.27}$$

where the k's are integers.

Our calculations are greatly simplified – and the results are still valid – if we assume that the displacements $\eta_{\alpha,i}$, are subject to a Gaussian probability distribution:

$$P\left(\eta_{\alpha,i}\right) = \left(\frac{d}{2\pi}\right)^{1/2} \exp(-d\eta_{\alpha,i}^2/2) \tag{6.28}$$

Then, the ensemble average of the total length of a single link is unity:

$$\sum_{i=1}^{d} \langle \eta_{\alpha,i}^2 \rangle = 1 \tag{6.29}$$

The displacements corresponding to different steps are statistically independent:

$$\langle \eta_{\alpha,i}\eta_{\beta,j} \rangle = \frac{1}{d}\delta_{\alpha,\beta}\delta_{i,j} \tag{6.30}$$

Exercise 6.5

If the random walk is closed, so that the end-to-end distance, $\vec{R}$, is equal to zero, a correction to (6.30) results. Show that this correction takes the following form:

$$\langle \eta_{\alpha,i}\eta_{\beta,j} \rangle = -\frac{1}{Nd}\delta_{i,j} \quad (\alpha \neq \beta) \tag{6.31}$$

Statistical averages of the entries in the radius of gyration tensor, $\overleftrightarrow{T}$, are readily evaluated. For example

$$\langle T_{11}T_{11} \rangle = \sum_{\alpha,\beta,\gamma,\delta} a_{\alpha,\beta}a_{\gamma,\delta}\langle \eta_{\alpha,1}\eta_{\beta,1}\eta_{\gamma,1}\eta_{\delta,1} \rangle \tag{6.32}$$

As it turns out, the most important terms in the average on the right hand side of (6.32) have the indices paired off, i.e. $\alpha = \beta$, $\gamma = \delta$ or $\alpha = \gamma$, $\beta = \delta$ or $\alpha = \delta$, $\beta = \gamma$. Another possibility yielding a non-zero result on ensemble averaging is $\alpha = \beta = \gamma = \delta$. However, as it turns out, we do not need to pay attention to this special case, as the correction to the result that we will obtain will be smaller than that result by the factor $1/N$. Given (6.30) and the possibilities for pairing described above, the result of the averaging yields

$$\langle T_{11}T_{11}\rangle = \sum_{\alpha,\gamma} a_{\alpha,\alpha}a_{\delta,\delta}\frac{1}{d^2} + 2\sum_{\alpha,\delta} a_{\alpha,\delta}a_{\alpha,\delta}\frac{1}{d^2} \tag{6.33a}$$

$$= \frac{1}{d^2}\left[\left(\mathrm{Tr}\,\overleftrightarrow{a}\right)^2 + 2\,\mathrm{Tr}\,\overleftrightarrow{a}^2\right] \tag{6.33b}$$

$$= \frac{1}{d^2}\left[\left(\sum_k \lambda_k\right)^2 + 2\sum_k \lambda_k^2\right] \tag{6.33c}$$

In (6.33c), the quantities λ_k are the eigenvalues of the matrix $\overleftrightarrow{a}$. Given (6.27), we know that

$$\lambda_k = \frac{N+1}{\pi^2 k^2} \tag{6.34}$$

Then,

$$\langle T_{11}T_{11}\rangle = \frac{1}{d^2}\frac{(N+1)^2}{\pi^4 d^2}\left[\left(\sum_n \frac{1}{n^2}\right) + 2\sum_n \frac{1}{n^4}\right] \tag{6.35a}$$

$$= \frac{N^2}{\pi^4 d^2}\left[\frac{\pi^4}{36} + \frac{\pi^4}{90}\right] \tag{6.35b}$$

$$= \frac{N^2}{20d^2} \tag{6.35c}$$

In going to (6.35c), we have ignored the difference between $(N+1)^2$ and N^2, which is permissible for our purposes in the large-N limit. Furthermore, we have made use of the results for sums over $1/n^2$ and $1/n^4$ that were also utilized in (6.20d).

A similar set of calculations yields the following results

$$\langle T_{11}T_{22}\rangle = \frac{1}{d^2}\left(\mathrm{Tr}\,\overleftrightarrow{a}\right)^2 = \frac{N^2}{36d^2} \tag{6.36}$$

$$\langle T_{12}T_{12}\rangle = \frac{1}{d^2}\,\mathrm{Tr}\,\overleftrightarrow{a}^2 = \frac{N^2}{90d^2} \tag{6.37}$$

These averages can be inserted into (6.14) to yield the formula (6.15) for A_d, the asphericity in d dimensions.

Exercise 6.6
Verify (6.36) and (6.37).

6.2.2 The eigenvalue structure of $\overleftrightarrow{T}$ in high dimensionality

We have already established that the radius of gyration tensor $\overleftrightarrow{T}$ has a well-determined structure and a set of definite eigenvalues in the limit of very high dimensionality. We will now see in greater detail how the eigenvalue structure of this tensor is determined in the limit $d \to \infty$, and how one can construct an expansion in $1/d$ for the eigenvalues, and their distributions. The key to this investigation is the identity (6.21), and a quantity known as the *resolvent* of the tensor $\overleftrightarrow{T}$. This quantity, $\mathcal{R}(\lambda)$, is given by

$$\mathcal{R}(\lambda) = \mathrm{Tr}\left(\frac{1}{\lambda \overleftrightarrow{I} - \overleftrightarrow{T}}\right) \tag{6.38}$$

where the quantity λ is assumed to take on complex values. The matrix $\overleftrightarrow{I}$ is the d-dimensional identity tensor:

$$\overleftrightarrow{I} = \begin{pmatrix} 1 & 0 & 0 & \cdots \\ 0 & 1 & 0 & \cdots \\ 0 & 0 & 1 & \cdots \\ \vdots & \vdots & \vdots & \ddots \end{pmatrix} \tag{6.39}$$

If the eigenvalues of the $d \times d$ tensor $\overleftrightarrow{T}$ are λ_j $(1 \leq j \leq d)$, then

$$\mathcal{R}(\lambda) = \sum_{j=1}^{d} \frac{1}{\lambda - \lambda_j} \tag{6.40}$$

The function $\mathcal{R}(\lambda)$ thus has poles at the eigenvalues of $\overleftrightarrow{T}$. Given that $\overleftrightarrow{T}$ is a real, symmetric, matrix, which means that its eigenvalues are all real (and are, in fact, all real and positive), the poles of the resolvent all lie on the real λ axis. We now make use of the identity

$$\frac{1}{x - \mathrm{i}\epsilon} = \mathrm{P}\left(\frac{1}{x}\right) + \mathrm{i}\pi\delta(x) \tag{6.41}$$

which holds for real x and real and infinitesimal $\epsilon > 0$, and where P stands for principal part (Morse and Feshbach, 1953). This allows us to extract the eigenvalue distribution from the resolvent, via the relationship

$$\text{Im}\left[\mathcal{R}(\lambda - i\epsilon)\right] = \pi \sum_{j=1}^{d} \delta\left(\lambda - \lambda_j\right) \tag{6.42}$$

The relation assumes that λ is real, and, of course, that ϵ is both real and infinitesimal. If we perform ensemble averages over both sides of (6.42), we obtain the distributions of eigenvalues of the radius of gyration tensor $\overleftrightarrow{T}$.

Consider, now, the formal expansion of $\mathcal{R}(\lambda)$ in terms of the tensor $\overleftrightarrow{T}$:

$$\begin{aligned}\mathcal{R}(\lambda) &= \text{Tr}\,\frac{1}{\lambda \overleftrightarrow{I} - \overleftrightarrow{T}} \\ &= \text{Tr}\,\frac{1}{\lambda}\left[1 + \left(\frac{\overleftrightarrow{T}}{\lambda}\right) + \left(\frac{\overleftrightarrow{T}}{\lambda}\right)^2 + \cdots\right] \\ &= \frac{1}{\lambda}\sum_{n=0}^{\infty} \text{Tr}\left(\frac{\overleftrightarrow{T}}{\lambda}\right)^n \end{aligned} \tag{6.43}$$

This means that

$$\langle \mathcal{R}(\lambda) \rangle = \frac{1}{\lambda}\sum_{n=0}^{\infty}\left\langle \text{Tr}\left(\frac{\overleftrightarrow{T}}{\lambda}\right)^n \right\rangle \tag{6.44}$$

We can find the ensemble average of the resolvent if we can manage to find the ensemble average of the radius of gyration tensor raised to an arbitrarily high power. Making use of the identity (6.21), we have

$$\text{Tr}\,\overleftrightarrow{T}^{n} = \sum_{\alpha_1,\beta_1,\ldots\alpha_n,\beta_n} \sum_{i_1,\ldots i_n} \langle \eta_{\alpha_1,i_1} a_{\alpha_1,\beta_1} \eta_{\beta_1,i_2} \eta_{\alpha_2,i_2} a_{\alpha_2,\beta_2} \eta_{\beta_2,i_3} \cdots \eta_{\alpha_n,i_n} a_{\alpha_n,\beta_n} \eta_{\beta_n,i_1} \rangle \tag{6.45}$$

The matrices a_{α_k,β_k} are the same for every N-step random walk. The averages are all over the η's. Given the fact that the probability density governing the distribution of the displacements is Gaussian, we can generalize (6.30). For instance, if we have a product of six η's, then the average is given by

$$\begin{aligned}\langle \eta_{\alpha_1,i_1}\eta_{\alpha_2,i_2}\eta_{\alpha_3,i_3}\eta_{\alpha_4,i_4}\rangle &= \langle \eta_{\alpha_1,i_1}\eta_{\alpha_2,i_2}\rangle\langle \eta_{\alpha_3,i_3}\eta_{\alpha_4,i_4}\rangle \\ &\quad + \langle \eta_{\alpha_1,i_1}\eta_{\alpha_3,i_3}\rangle\langle \eta_{\alpha_2,i_2}\eta_{\alpha_4,i_4}\rangle + \langle \eta_{\alpha_1,i_1}\eta_{\alpha_4,i_4}\rangle\langle \eta_{\alpha_2,i_2}\eta_{\alpha_3,i_3}\rangle \end{aligned} \tag{6.46}$$

Fig. 6.6. Diagrammatic representation of the right hand side of (6.47). The text immediately below the equation contains a detailed discussion of the meaning of the various elements in the diagram.

Fig. 6.7 Diagrammatic representation of $(\overleftrightarrow{T}/\lambda)^n$.

Exercise 6.7

In general, the average of a product of $2m$ η's will be equal to a sum of products of the averages of m pairs of the η's. This is a classical version of what is known in field theory as Wick's theorem. Derive an expression for the number of ways of forming those m pairs.

6.2.3 Diagrammatic expansion

We can use what we now have to formulate a diagrammatic method that keeps track of the pairings of the η's. This method forms the basis of an expansion in $1/d$ of the eigenvalue distribution of the radius of gyration tensor $\overleftrightarrow{T}$. It has other uses as well, including the construction of an approximate analytical expression for the distributions of the individual eigenvalues of $\overleftrightarrow{T}$ directly from the average $\langle R(\lambda)\rangle$. We accomplish this by summing a class of terms in the $1/d$ expansion.

To start, consider the quantity

$$\left(\frac{\overleftrightarrow{T}}{\lambda}\right)^2_{i,j} = \sum_{\alpha_1,\alpha_2,\beta_1,\beta_2} \sum_{i_2=1}^{d} \eta_{\alpha_1,i} \frac{a_{\alpha_1,\beta_1}}{\lambda} \eta_{\beta_1,i_2} \eta_{\alpha_2,i_2} \frac{a_{\alpha_2,\beta_2}}{\lambda} \eta_{\beta_2,j} \tag{6.47}$$

The right-hand side of (6.47) is represented diagrammatically in Figure 6.6.

The crosses at the ends of the horizontal lines represent the displacement η, and the lines themselves stand for the element $a_{\alpha,\beta}$. The dot between the two adjacent crosses in the center of the figure is for accounting purposes only. The nth order term $(\overleftrightarrow{T}/\lambda)^n$ is represented by a string of lines with crosses at both ends, as shown in Figure 6.7.

We now introduce a new element into the diagrammatic method: the representation of the Gaussian pairing of two η's. This sort of pairing is symbolized by

Fig. 6.8. Representation of the pairing of two adjacent η's. See the text above (6.48), for an explanation.

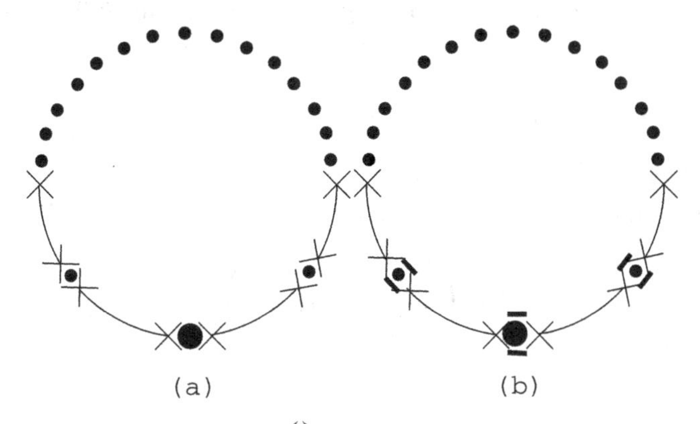

Fig. 6.9. (a) Representation of $\mathrm{Tr}(\overleftrightarrow{T}/\lambda)^n$. (b) Representation of the lowest order contribution to $\langle \mathrm{Tr}(\overleftrightarrow{T}/\lambda)^n \rangle$. The large dot at the bottom of each of the two figures separates the η's at the beginning and the end of the expansion of $\overleftrightarrow{T}^n$, and is for accounting puroposes only.

drawing two lines between them, as shown in Figure 6.8. This diagram represents

$$
\begin{aligned}
&\sum_{\alpha_1,\beta_1,\alpha_2,\beta_2} \sum_{i_2=1}^{d} \eta_{\alpha_1,i} \frac{a_{\alpha_1,\beta_1}}{\lambda} \langle \eta_{\beta_1,i_2} \eta_{\alpha_2,i_2} \rangle \frac{a_{\alpha_2,\beta_2}}{\lambda} \eta_{\beta_5,j} \\
&\quad = \sum_{\alpha_1,\beta_1,\alpha_2,\beta_2} \sum_{i_2=1}^{d} \eta_{\alpha_1,i} \frac{a_{\alpha_1,\beta_1}}{\lambda} \left(\frac{1}{d} \delta_{\beta_1,\alpha_2} \delta_{i_2,i_2} \right) \frac{a_{\alpha_2,\beta_2}}{\lambda} \eta_{\beta_2,j} \\
&\quad = \sum_{\alpha_1,\beta_2,\alpha_2} \eta_{\alpha_1,i} \frac{a_{\alpha_1,\alpha_2}}{\lambda} \frac{a_{\alpha_2,\beta_2}}{\lambda} \eta_{\beta_2,j} \\
&\quad = \sum_{\alpha_1,\beta_2} \eta_{\alpha_1,i} \left(\frac{\overleftrightarrow{a}}{\lambda} \right)^2_{\alpha_1,\beta_2} \eta_{\beta_2,j}
\end{aligned}
\tag{6.48}
$$

An important point to note is that the delta function, δ_{i_2,i_2}, the delta function for the components of two η's, is automatically satisfied for two adjacent η's. This guaranteed satisfaction of the delta function for components occurs whenever two adjacent η's are paired. There is *no* such guarantee when the pairing is between two non-adjacent η's. This turns out to be the basis of the $1/d$ expansion. An expansion up to nth order will result when account is taken of up to $2n$ pairings of non-adjacent η's.

We will start with the lowest order term in the expansion – the term of zeroth order in $1/d$. Consider $\mathrm{Tr}(\overleftrightarrow{T}/\lambda)^n$, the nth order term in the summation in (6.43). This term is represented as the ring diagram in Figure 6.9(a). The large dot that separates the crosses representing the η's at the two ends of the right hand side of (6.45) is, like the other dots in the diagram, for accounting purposes only. The zeroth order contribution to $\langle \mathrm{Tr}(\overleftrightarrow{T}/\lambda)^n \rangle$ is obtained by pairing off adjacent η's only. The diagram representing this pairing is shown in Figure 6.9(b). This diagram represents

$$\sum_{\alpha_1,\beta_1,\ldots,\alpha_n,\beta_n} \frac{a_{\alpha_1,\beta_1}}{\lambda}\delta_{\beta_1,\alpha_2}\frac{a_{\alpha_2,\beta_2}}{\lambda}\cdots\frac{a_{\alpha_n,\beta_n}}{\lambda}\delta_{\beta_n,\alpha_1} = \mathrm{Tr}\left(\frac{\overleftrightarrow{a}}{\lambda}\right)^n \tag{6.49}$$

Thus, to zeroth order in $1/d$,

$$\left\langle \mathrm{Tr}\left(\frac{\overleftrightarrow{T}}{\lambda}\right)^n \right\rangle = \mathrm{Tr}\left(\frac{\overleftrightarrow{a}}{\lambda}\right)^n \tag{6.50}$$

and

$$\left\langle \mathrm{Tr}\frac{1}{\lambda\overleftrightarrow{I} - \overleftrightarrow{T}} \right\rangle = \mathrm{Tr}\frac{1}{\lambda\overleftrightarrow{I} - \overleftrightarrow{a}} + O(1/d) \tag{6.51}$$

The ensemble average of the eigenvalue distribution of the radius of gyration tensor $\overleftrightarrow{T}$ is then given by

$$\mathrm{Im}\,(\mathcal{R}(\lambda - \mathrm{i}\epsilon)) = \sum_i \delta(\lambda - \lambda_i) + O(1/d) \tag{6.52}$$

where λ_i is the ith eigenvalue of the operator $\overleftrightarrow{a}$. As noted earlier (see (6.27)), the eigenvalues of this operator have the form $\lambda_k = (N+1)/\pi^2 k^2$.

To recapitulate, the principal radii of gyration of a walk in very high spatial dimensionality are well-defined, in that their distributions are essentially delta-function-like, and they are given by the formula (6.27). It is instructive to compare the ratio of the three largest eigenvalues in this series with the ratios of the principal radii of gyration of an ordinary walk in three dimensions, as determined in numerical simulations. In asymptotically large dimensionality, (6.27) tells us that

$$\langle R_1^2 \rangle : \langle R_2^2 \rangle : \langle R_3^2 \rangle = 9 : 2.25 : 1 \tag{6.53}$$

On the other hand, numerical studies yield (Bishop and Michels, 1986)

$$\langle R_1^2 \rangle : \langle R_2^2 \rangle : \langle R_3^2 \rangle = 13.76 : 3.03 : 1 \tag{6.54}$$

The agreement between the ratios in (6.53) and those in (6.54) is quite good, indicating that one might be able to learn something about the shapes of three-dimensional walks by loooking at walks in much higher spatial dimensionality.

Again, as in the case of asphericities, the ratios of principal radii of gyrations are not perfectly well-defined for an ensemble of walks, but rather occupy a reasonably

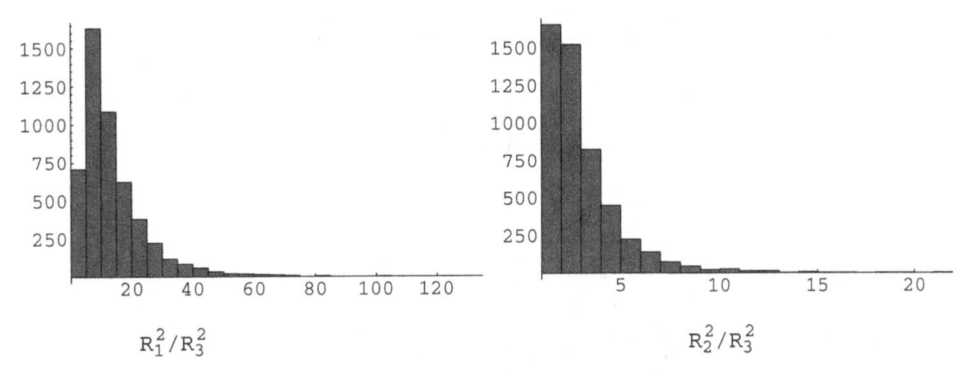

Fig. 6.10 The distributions of the ratios R_1^2/R_3^2 and R_2^2/R_3^2 for an ensemble of 5 000 three-dimensional random walks. Note the difference between the horizontal axes of the two historgams.

broad distribution. Figure 6.10 shows the distributions of the ratios R_1^2/R_3^2 and R_2^2/R_3^2 for an ensemble of 5000 random walks in three dimensions. In other words, the distribution of ratios of principal radii of gyration is also consistent with the notion that random walks come in a variety of shapes, rather than a single characteristic one. As we look more closely at random walks in high dimensionality, we expect to see how a distribution of finite width emerges from the perfectly localized distributions in infinite dimensionality.

6.2.4 The $1/d$ *expansion*

The next term in the $1/d$ expansion of the eigenvalue structure of the radius of gyration tensor follows from a pairing of non-adjacent $\eta_{\alpha,i}$'s. Two sets of non-adjacent displacements are shown encircled in Figure 6.11(a). The two η's are brought together in Figure 6.11(b). The three ways of pairing these η's are shown in Figures 6.11(c)–(e). The pairing shown in Figure 6.11(c) is just the pairing of adjacent η's that leads to the zeroth order result. The other two pairings are new. If the remaining pairings are all between adjacent η's, the first-order-in-$1/d$ contribution to $\langle \mathrm{Tr}(\overleftrightarrow{T}/\lambda)^n \rangle$ is obtained. Details of the analysis, which is straightforward, but fairly tedious, can be found in (Gaspari *et al.*, 1987). The end result for the average of the resolvent, $\langle \mathcal{R}(\lambda) \rangle$, is

$$\langle \mathcal{R}(\lambda) \rangle = \sum_i \left[\frac{1}{\lambda - \lambda_i} + \frac{1}{d} \sum_{\substack{j \\ j \neq i}} \frac{\lambda_i \lambda_j}{(\lambda - \lambda_i)^2 (\lambda - \lambda_j)} + \frac{2}{d} \frac{\lambda_i^2}{(\lambda - \lambda_i)^3} \right] + O\left[\left(\frac{1}{d} \right)^2 \right] \tag{6.55}$$

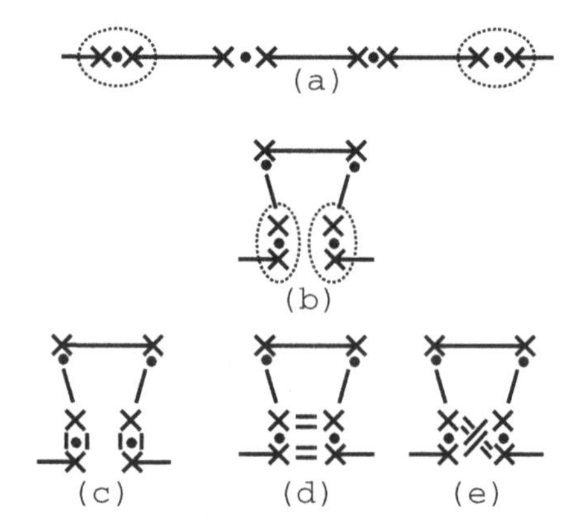

Fig. 6.11. (a) Two sets of η's in the diagrammatic representation of $(\overleftrightarrow{T}/\lambda)^n$, shown encircled. (b) The two sets are pulled together as a preliminary step in the pairing of those non-adjacent η's. (c)–(e) The three ways of pairing the four η's that have been brought together.

The quantities λ_i in (6.55) are the eigenvalues of the tensor $\overleftrightarrow{a}$. This expression does not match the zeroth order result for the averaged resolvent of $\overleftrightarrow{T}$, (6.51). If it did, then the eigenvalues of the moment of intertia tensor would be perfectly well-defined to first order in $1/d$. The correction to the zeroth order result for the resolvent can be interpreted in terms of an $O(1/d)$ shift in each ensemble-averaged eigenvalue and an $O(1/d)$ contribution to the width of the distribution of each eigenvalue about its average. That is, the eigenvalue distribution is no longer a set of delta functions, but rather a set of peaks of finite width.

To see how this interpretation works, imagine that each eigenvalue, Λ_i, of a given realization of the radius of gyration tensor $\overleftrightarrow{T}$ can be written as

$$\Lambda_i = \lambda_i(1 + \varepsilon_i) \tag{6.56}$$

where λ_i is the corresponding eigenvalue of the operator $\overleftrightarrow{a}$ and ε_i is the small relative shift in Λ_i from its limiting $d = \infty$ value. Then

$$\begin{aligned}
\left\langle \frac{1}{\lambda - \Lambda_i} \right\rangle &= \left\langle \frac{1}{\lambda - \lambda(1 + \varepsilon_i)} \right\rangle \\
&= \left\langle \frac{1}{\lambda - \lambda_i} + \frac{\lambda_i \varepsilon_i}{(\lambda - \lambda_i)^2} + \frac{\lambda_i^2 \varepsilon_i^2}{(\lambda - \lambda_i)^3} + \dots \right\rangle \\
&= \frac{1}{\lambda - \lambda_i} + \frac{\lambda_i}{(\lambda - \lambda_i)^2} \langle \varepsilon_i \rangle + \frac{\lambda_i^2}{(\lambda - \lambda_i)^3} \langle \varepsilon^2 \rangle + \dots
\end{aligned} \tag{6.57}$$

Referring back to (6.55), we are able to assert that this implies a fractional shift in the eigenvalues of the radius of gyration tensor in for a d-dimensional random walk given by

$$\begin{aligned}\langle \varepsilon_i \rangle &= \frac{1}{d} \sum_{j \neq i} \frac{\lambda_j}{\lambda - \lambda_j} \\ &\approx \frac{1}{d} \sum_{\substack{j \\ j \neq i}} \frac{\lambda_j}{\lambda_i - \lambda_j} \end{aligned} \tag{6.58}$$

The last approximate equality is asymptotically accurate as $d \to \infty$. Given the results (6.27) for the eigenvalues λ_i, we have for the order $1/d$ fractional shift in the eigenvalues

$$\begin{aligned}\langle \varepsilon \rangle &= \frac{1}{d} \sum_{\substack{j \\ j \neq i}} \frac{j^{-2}}{i^{-2} - j^{-2}} \\ &= \frac{1}{d} \sum_{\substack{j \\ j \neq i}} \frac{i^2}{j^2 - i^2} \\ &= \frac{3}{4d} \end{aligned} \tag{6.59}$$

The last line of (6.59) follows from the mathematical result (Gradshteyn *et al.*, 2000)

$$\sum_{\substack{m=1 \\ m \neq n}}^{\infty} \frac{n^2}{m^2 - n^2} = \frac{3}{4} \tag{6.60}$$

Thus, to order $1/d$,

$$\langle \Lambda_k \rangle = \frac{N+1}{\pi^2 k^2} \left(1 + \frac{3}{4d} \right) \tag{6.61}$$

Making use of a similar line of reasoning, we can also infer, from (6.55) and (6.57), that

$$\left\langle (\Lambda_k - \langle \Lambda_k \rangle)^2 \right\rangle = \frac{2}{d} \left(\frac{N+1}{\pi^2 k^2} \right)^2 \tag{6.62}$$

This tells us that both the shift of the mean principal radii of gyration and the widths of their distribution about that mean are of the same general magnitude as their value in the limit $d \to \infty$. The shifts and the widths are smaller than the asymptotic means by a factor that goes as $1/d$.

6.2.5 *Accuracy of the* $1/d$ *expansion*

We've already seen that the ratios of the mean pricipal radii of gyration are remarkably accurate, even at lowest order in $1/d$ (see (6.53) and (6.54)). We will now look

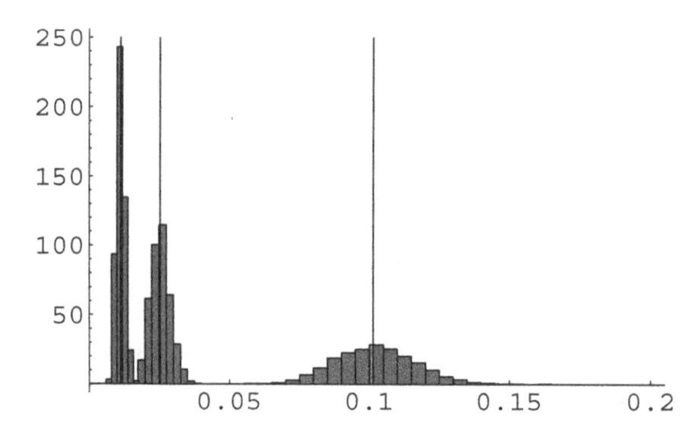

Fig. 6.12. Distribution of the three largest eigenvalues of the radius of gyration tensor for a walk in 100 dimensions. The vertical lines indicate the locations of the eigenvalues as given by the leading-order prediction of the $1/d$ expansion, (6.19).

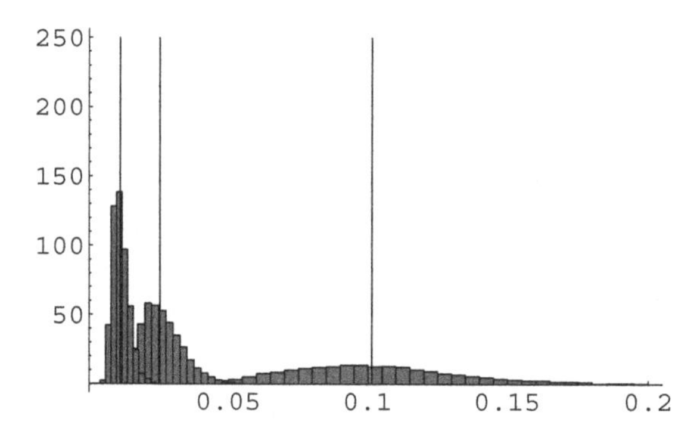

Fig. 6.13. Distribution of the three largest eigenvalues of the radius of gyration tensor for a walk in 20 dimensions. The vertical lines indicate the locations of the eigenvalues as given by the leading-order prediction of the $1/d$ expansion, (6.19).

further into the accuracy of the predictions for walks in high dimensions, as they apply to walks in three dimensions, and to walks taking place in what we would generally agree are a large number of spatial dimensions. For that purpose, displayed in Figures 6.12, 6.13, and 6.14 are the distributions of the three largest eigenvalues of the radius of gyration tensor $\overleftrightarrow{T}$ for walks in, respectively, 100 dimensions, 20 dimensions and three dimensions. Three points are worth considering here. The first is qualitative – are the distributions well-separated and at all well-represented by narrow, nearly delta-function-like, peaks? The second and third are how accurately the means and widths of the distributions are predicted by the $1/d$ expansion, carried out to first order in the expansion parameter. It is clear from the figures that, even in 100 dimensions, the delta-function limit is not achieved. However, the formula (6.62) tells us that the width of the distribution of an eigenvalue of $\overleftrightarrow{T}$ will not be all that close to zero unless the dimensionality is truly enormous. In fact, if

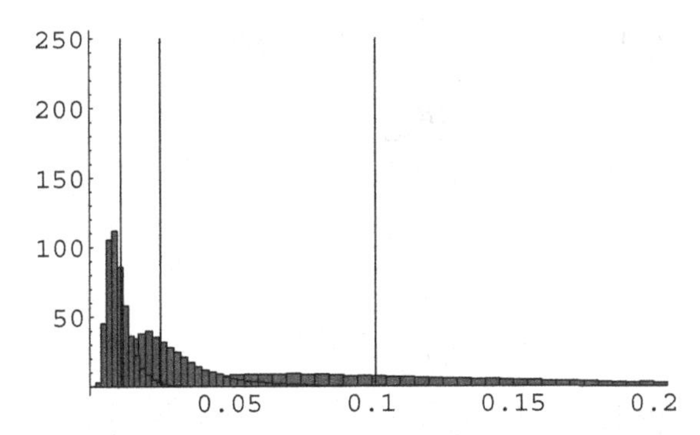

Fig. 6.14. Distribution of the three largest eigenvalues of the radius of gyration tensor for a walk in three dimensions. The vertical lines indicate the locations of the eigenvalues as given by the leading-order prediction of the $1/d$ expansion, (6.19).

we denote by v_i the variance of the distribution of the ith eigenvalue R_i^2, the ratios of the actual variances and the predicted ones are, in 100 dimensions, are

$$\frac{v_1}{2(N+1)^2/(100 \times \pi^4)} = 0.97 \tag{6.63}$$

$$\frac{v_2}{2(N+1)^2/(100 \times (4\pi^2)^2)} = 0.93 \tag{6.64}$$

$$\frac{v_3}{2(N+1)^2/(100 \times (9\pi^2)^2)} = 0.95 \tag{6.65}$$

As might be expected, the $1/d$ expansion works reasonably well when $d = 100$. Furthermore, even though the distributions are far from delta-function-like, they are relatively well separated.

When $d = 20$, the eigenvalue distributions are not quite as well distinguished, and in three dimensions, the distributions clearly run into each other. Nevertheless, some (but not all) of the predictions of the $1/d$ expansion hold up reasonably well. For example, if we compare the variances of the distributions of individual eigenvalues with those predicted, we find in three dimensions

$$\frac{v_1}{2(N+1)^2/(3 \times \pi^4)} = 0.98 \tag{6.66}$$

$$\frac{v_2}{2(N+1)^2/(3 \times (4\pi^2)^2)} = 0.48 \tag{6.67}$$

$$\frac{v_3}{2(N+1)^2/(3 \times (9\pi^2)^2)} = 0.23 \tag{6.68}$$

The width of the distribution is accurately given by (6.62) in the case of the largest eigenvalue, if not for the smaller ones. Comparing to the mean values, $\langle R_k^2 \rangle$ with

$\langle \Lambda_k \rangle$ as given by (6.61), we find in three dimensions

$$\frac{\langle R_1^2 \rangle}{\langle \Lambda_1 \rangle} = 1.01 \tag{6.69}$$

$$\frac{\langle R_2^2 \rangle}{\langle \Lambda_2 \rangle} = 0.91 \tag{6.70}$$

$$\frac{\langle R_3^2 \rangle}{\langle \Lambda_3 \rangle} = 0.75 \tag{6.71}$$

Again, the largest eigenvalue is best described by the $1/d$ expansion. However, in the worst case, the $1/d$ expansion is accurate to within 25%. Given that the comparison is with a formula that is correct to first order in $1/d$, it is reasonable to anticipate an error of order $1/d^2 \sim 0.1$ in three dimensions. This expectation is borne out, or exceeded, in the case of the two largest principal radii of gyration. The smallest eigenvalue of $\overleftrightarrow{T}$ is the most resistant to analysis in the context of the $1/d$ expansion.

The expansion in inverse dimensionality, with its clear shortcomings, is of some utility in the analysis of random walk shape.

6.2.6 Prediction of the distribution of the principal radii of gyration

It is possible to peform a partial summation of the $1/d$ expansion that extends to all orders in $1/d$, yielding an approximation for the distribution of the square of the ith principal radius of gyration, R_i^2. One finds (Gaspari *et al.*, 1987)

$$P(R_i^2) = \frac{1}{2\pi} \int_{-\infty}^{\infty} \exp\left[-\frac{d}{2} \ln\left(1 + \frac{2ix\alpha_i}{d}\right) + ixR_i^2 \right] dx \tag{6.72}$$

Here, α_i is the lowest-order-in-$1/d$ prediction for the mean value, $\langle R_i^2 \rangle$, as given by (6.19). A steepest-descents evaluation of the integral in (6.72) yields

$$P(R_i^2) \propto (R_i^2)^{d/2-1} \exp\left(-dR_i^2/2\alpha_i\right) \tag{6.73}$$

Figure 6.15 shows how this prediction compares with the actual distribution of the largest eigenvalue of $\overleftrightarrow{T}$ in three dimensions. The quality of the fit, if not perfect, is encouraging. If we combine the formulas (6.72) as they apply to all the principal radii of gyration, we end up with the following result for their sum

$$P(R^2) = \frac{1}{2\pi} \int_{-\infty}^{\infty} e^{isR^2} K(s)\, ds \tag{6.74}$$

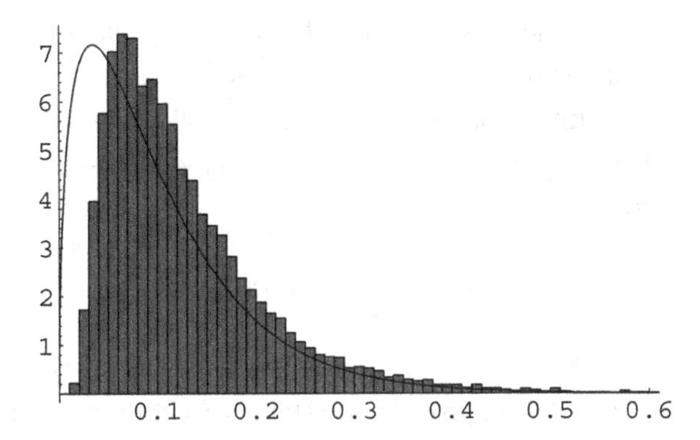

Fig. 6.15. Comparison of the approximate distribution of the largest eigenvalue, R_1^2, of the radius of gyration tensor $\overleftrightarrow{T}$ with the actual distribution of that quantity. Both distributions are normalized.

where

$$R^2 = \sum_{i=1}^{d} R_i^2 \tag{6.75}$$

and

$$K(R) = \prod_{n=1}^{N} \left(1 - \frac{2is\alpha_n}{d}\right)^{-d/2} \tag{6.76}$$

and $\alpha_n = (N+1)/\pi^2 n^2$. As it turns out, this is an exact result for the distribution of the sum of the principal radii of gyration, first derived by Fixman (Fixman, 1962). The sum is also known as the mean radius of gyration. It is the trace of the radius of gyration tensor, $\overleftrightarrow{T}$. Its mean value is calculated in the supplement 2 to this chapter. (See (6:S2-13)).

Exercise 6.8
Using (6.73), evaluate $\langle \left(R_i^2 - \langle R_i^2 \rangle\right)^2 \rangle$.

Exercise 6.9
Verify that (6.74) follows from (6.72).

6.2.7 Shape of a self-avoiding random walk

In subesequent chapters, we will turn our attention to walks that are self-avoiding, in that the walker refuses to take a path that intersects itself. This type of walk is important, in that it provides a model for the configurational statistics of a long chain polymer. Work on the shape of a self-avoiding walk has been performed (Aronovitz and Nelson, 1986). The calculation is based on an expansion in the difference between the dimensionality in which the walk takes place and an " upper critical dimensionality," equal to four. The quantity $\epsilon = 4 - d$ is the expansion parameter. To first order in ϵ

$$A_d = \frac{2d+4}{5d+4} + 0.008\epsilon \tag{6.77}$$

The main conclusion to be gleaned from this result is that self-avoidance plays a non-trivial, but far from decisive, role in the shape of a random walk.[4]

6.3 Final commentary

We've talked a bit on the issue of the shape of a random walk. Given the open-endedness of the notion of shape, it ought not be surprising that we have not come close to exhausting the subject in this chapter. The quantification of shape on which we have focused here has the virtue of amenability to analysis, and more than a little can be done with shape as measured in terms of the eigenvalue structure of the radius of gyration tensor. In light of the importance of the conformations taken by biological molecules that are topologically linear (RNA, DNA, and proteins in particular), it is clear that the subject of shape is a field that deserves continued attention. There will no doubt be important developments in the forseeable future. We look forward to their advent.

6.4 Supplement 1: principal radii of gyration and rotational motion

Recall Newton's second law:

$$\vec{F} = \frac{d\vec{p}}{dt} \tag{6:S1-1}$$

Here, $\vec{F}$ is the applied force and $\vec{p}$ is the linear momentum of the particle or collection of particles to which that force is applied. When the system undergoing the change in motion is a single particle, $\vec{p} = m\vec{v}$, where $\vec{v}$ is the particle's velocity. Now, assume that we have a system of particles. Newton's second law, as it refers to each point particle, is, then

$$\vec{F}_l = m_l \frac{d\vec{v}_l}{dt} \tag{6:S1-2}$$

[4] Notice that the first term on the right hand side of (6.77) has not been expanded about $d = 4$. This is a (minor) violation of the spirit of the expansion in $\epsilon = 4 - d$, which does not materially affect the conclusion stated above.

where l is the subscript that identifies the particle. If we sum up these equations we end up with (6:S1-1), where $\vec{F}$ is the net external force (internal forces cancel because of Newton's third law), and $\vec{p}$ is the total momentum:

$$\vec{p} = \sum_l m_l \vec{v}_l \tag{6:S1-3}$$

So far, so good. This is all pretty elementary. Now, let's focus on rotational motion. We derive Newton's second law, as it applies to rotational motion, by taking the cross product of the position vectors $\vec{r}_l$ with the corresponding equation in the set (6:S1-2). Defining the total torque $\vec{\tau}$ as the sum of the $\vec{r}_l \times \vec{F}_l$'s, we end up with the equation

$$\begin{aligned} \vec{\tau} &= \sum_l m_l \vec{r}_l \times \frac{\mathrm{d}\vec{v}_l}{\mathrm{d}t} \\ &= \frac{\mathrm{d}}{\mathrm{d}t} \sum_l m_l \vec{r}_l \times \vec{v}_l \\ &\equiv \frac{\mathrm{d}\vec{L}}{\mathrm{d}t} \end{aligned} \tag{6:S1-4}$$

The last two lines of (6:S1-4) constitute a definition of the angular momentum, $\vec{L}$ of a system of point particles. Note that the precise definition of angular momentum depends on the origin with respect to which the position of each particle is defined. It is often convenient to place the origin at the center of mass of the set of particles. If the internal force between each pair of particles is along the line joining them, then the internally generated torques cancel, and the total torque, $\vec{\tau}$, is entirely due to external forces.

Now, suppose that the motion of the system is entirely rotational, about some point $\vec{R}$. Then,

$$\vec{v}_l = \vec{\omega} \times \left(\vec{r}_l - \vec{R}\right) \tag{6:S1-5}$$

Here, $\vec{\omega}$ is the angular velocity of the system of particles (see Figure 6.16). Now, we can choose $\vec{R}$ as the center of our system of coordinates, so that $\vec{r}_l - \vec{R}$ is replaced by $\vec{r}_l$. In this case, the angular momentum becomes

$$\sum_l m_l \vec{r}_l \times (\vec{\omega} \times \vec{r}_l) \tag{6:S1-6}$$

We can rewrite the above relationship with the use of the standard identity for the triple product:

$$\vec{L} = \sum_l m_l \left(\vec{\omega}\left(\vec{r}_l \cdot \vec{r}_l\right) - \vec{r}_l\left(\vec{r}_l \cdot \vec{\omega}\right)\right) \tag{6:S1-7}$$

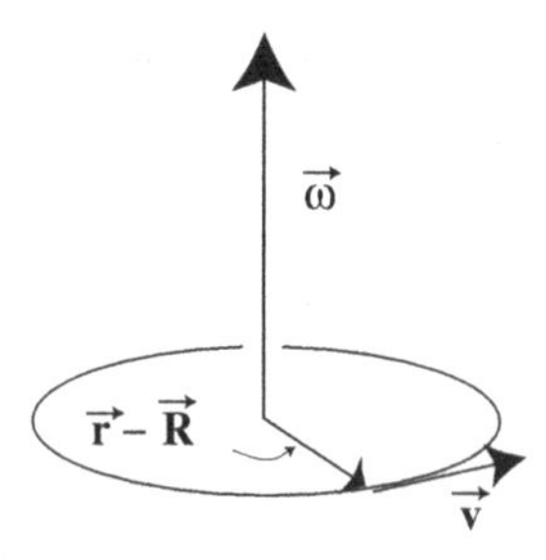

Fig. 6.16. The angular velocity, $\vec{\omega}$, and its relation to the velocity, $\vec{v}$, of a particle.

Suppose, now, we define the matrix $\overleftrightarrow{T}$ as follows:

$$T_{ij} = \sum_l m_l r_{l,i} r_{l,j} \tag{6:S1-8}$$

Then, the relationship between $\vec{L}$ and $\vec{\omega}$ is

$$\vec{L} = \mathrm{Tr}\, \overleftrightarrow{T}\, \vec{\omega} - \overleftrightarrow{T} \cdot \vec{\omega} \tag{6:S1-9}$$

Defining

$$\overleftrightarrow{C} = \overleftrightarrow{T} - \overleftrightarrow{I}\, \mathrm{Tr}\, \overleftrightarrow{T} \tag{6:S1-10}$$

where $\overleftrightarrow{I}$ is the identity operator we find that

$$\vec{L} = -\overleftrightarrow{C} \cdot \vec{\omega} \tag{6:S1-11}$$

Note that the trace of the operator $\overleftrightarrow{C}$ has a trace equal to twice the trace of $\overleftrightarrow{T}$. This is because the trace of the identity operator is three. The matrix $\overleftrightarrow{C}$ is the moment of inertia matrix. The fact that the angular velocity and the angular momentum are not parallel is just one of the complications of rotational motion.

Now, because the matrix $\overleftrightarrow{T}$ is real and symmetric, we know that it has real (in fact, positive) eigenvalues. Those eigenvalues have a name. They are known as the principal radii of gyration, R_i^2. If $\vec{\omega}$ points in the same direction as the eigenvector of one of them, say R_1^2, then the angular momentum and the angular velocity point in the same direction, and the relationship between the two becomes

$$\vec{L} = \left(R_2^2 + R_3^2\right) \vec{\omega} \tag{6:S1-12}$$

This is because $\mathrm{Tr}\, \overleftrightarrow{T} = R_1^2 + R_2^2 + R_3^2$, while $\overleftrightarrow{T} \cdot \vec{\omega} = R_1^2 \vec{\omega}$.

The eigenvalues of the tensor $\overleftrightarrow{T}$ can also be used as a measure of the extent to which the system of particles possesses spherical symmetry, at least in terms of rotational motion. If all the eigenvalues are equal, then the system has approximately the same weighted extent. In fact, in this case, $\vec{L}$ is always parallel to $\vec{\omega}$.

Now imagine that all the masses, m_l, are equal to one. Then, the matrix measures the extent to which the particles are in a spherically symmetric distribution. While equality of the principal radii of gyration is not equivalent to spherical symmetry, it provides a very useful quantitative measure of that property, and of departure from it.

We can construct a new tensor, $\overset{\leftrightarrow}{Q}$, as follows

$$\overset{\leftrightarrow}{Q}=\overset{\leftrightarrow}{T}-\frac{1}{3}\overset{\leftrightarrow}{I}\,\mathrm{Tr}\,\overset{\leftrightarrow}{T} \tag{6:S1-13}$$

The trace of this matrix is equal to zero, and, if all the principal radii of gyration are the same, then the diagonalized form of this matrix has all entries equal to zero.

Position vectors transform under rotations about the center of mass as follows

$$r'_k=\sum_l R_{kl}r_l \tag{6:S1-14}$$

where R_{kl} are the elements of the rotation matrix. This matrix has the property that its transpose is also its inverse. That is

$$\overset{\leftrightarrow}{R}\cdot\overset{\leftrightarrow}{R}^{\mathrm{T}}=\overset{\leftrightarrow}{I} \tag{6:S1-15}$$

A matrix whose transpose is also its inverse is known as an orthogonal matrix. Then, the matrix $\overset{\leftrightarrow}{T}$ with elements going as $r_l r_k$ transforms as follows

$$\begin{aligned}T'_{k_1k_2}&=\sum_{l_1,l_2}R_{k_1l_1}T_{l_1l_2}R_{k_2l_2}\\&=\sum_{l_1,l_2}R_{k_1l_1}T_{l_1l_2}R^{\mathrm{T}}_{l_2k_2}\\&=\sum_{l_1,l_2}R_{k_1l_1}T_{l_1l_2}R^{-1}_{l_2k_2}\end{aligned} \tag{6:S1-16}$$

or, in shorthand,

$$\overset{\leftrightarrow}{T}'=\overset{\leftrightarrow}{R}\cdot\overset{\leftrightarrow}{T}\cdot\overset{\leftrightarrow}{R}^{-1} \tag{6:S1-17}$$

The same relationship clearly holds for the tensor $\overset{\leftrightarrow}{Q}$. The demonstration that traces of powers of this tensor are invariant under rotations follows from this equation for the way in which rotations give rise to changes in $\overset{\leftrightarrow}{Q}$.

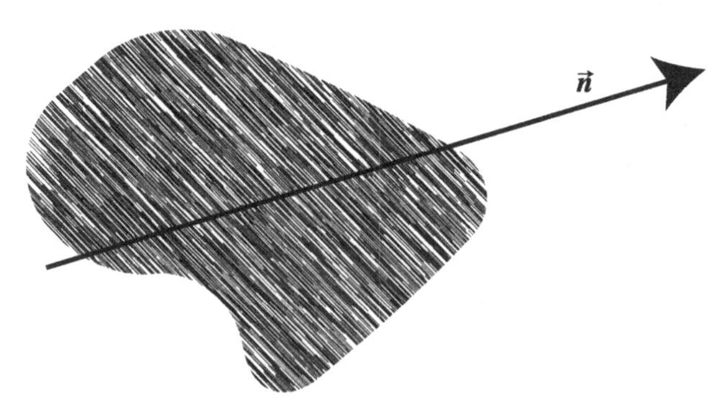

Fig. 6.17. Looking for the direction in which an object has the greatest spatial extent.

6.4.1 Just a little bit more on the meaning of the operator $\overleftrightarrow{T}$

Suppose we are interested in finding the direction in which an object has the greatest spatial extent. We start by assuming a vector $\vec{n}$, which points along the direction of interest. We will set the length of the vector at unity (see Figure 6.17). Then the extent of the object in the direction established by $\vec{n}$ is

$$\begin{aligned}\sum_l (\vec{r}_l \cdot \vec{n})^2 &= \sum_l \sum_{i,j} n_i r_{l,i} r_{l,j} n_j \\ &= \vec{n} \cdot \overleftrightarrow{T} \cdot \vec{n} \qquad (6\text{:S1-18})\end{aligned}$$

Remember that the position vectors are drawn with their tails at the center of mass of the object. If we wish to extremize the above quantity with respect to $\vec{n}$, subject to the condition that its length is held constant, we take the derivative with respect to each component of $\vec{n}$ of the expression below

$$\vec{n} \cdot \overleftrightarrow{T} \cdot \vec{n} - \lambda \vec{n} \cdot \vec{n} \qquad (6\text{:S1-19})$$

The quantity λ in (6:S1-19) is a Lagrange multiplier. The extremum equation easily reduces to

$$\overleftrightarrow{T} \cdot \vec{n} = \lambda \vec{n} \qquad (6\text{:S1-20})$$

That is, in order to find the direction of greatest (or least) extent of the object, we solve the eigenvalue equation of the operator $\overleftrightarrow{T}$. The largest eigenvalue is the greatest extent, as defined by (6:S1-18), with $\vec{n}$ chosen to extremize the quantity.[5] The smallest extent, similarly defined, is given by the smallest eigenvalue of $\overleftrightarrow{T}$.

[5] In other words, the extremizing choice for the vector $\vec{n}$ is the eigenvector of the operator $\overleftrightarrow{T}$ with the largest eigenvalue.

6.5 Supplement 2: calculations for the mean asphericity

6.5.1 The average $\langle T_{11}\rangle$

First, we slightly recast the general expression for the mean asphericity

$$A_d = \left[\left(\frac{\langle T_{11}^2\rangle}{\langle T_{12}^2\rangle} - \frac{\langle T_{11}T_{22}\rangle}{\langle T_{12}^2\rangle}\right) + d\right] \bigg/ \left(\frac{\langle T_{11}^2\rangle}{\langle T_{11}T_{22}\rangle} + d - 1\right) \tag{6:S2-1}$$

This means that to calculate the mean aspericity we need to find ratios of averages, rather than the averages themselves. This simplifies our task a bit.

Although the averages that we have to perform in order to arrive at a numerical result for the asphericity of the random walk involve squares of entries in the radius of gyration tensor, or products of two entries, it is useful to look at the average of a single element of that tensor, lying along the diagonal. Eventually, we will perform this calculation in another way when we develop an expansion in $1/d$ for the principal radii of gyration. However, we will start out by showing how the calculation can be done with the use of the generating functions that have proven so useful in the study of random walk statistics. As a first step, we recast the expression for the entries T_{kl}:[6]

$$T_{kl} = \frac{1}{N}\sum_{j=1}^{N}\left(r_{jk} - \langle r_k\rangle\right)\left(r_{jl} - \langle r_l\rangle\right) \tag{6:S2-2a}$$

$$= \frac{1}{2N^2}\sum_{i=1}^{N}\sum_{j=1}^{N}\left(r_{ik} - r_{jk}\right)\left(r_{il} - r_{jl}\right) \tag{6:S2-2b}$$

Equation 6:S2-2a is a recapitulation of (6.1); 6:S2-2b can be established by inspection. Consider, now the average

$$\langle T_{11}\rangle = \frac{1}{2N^2}\sum_{i=1}^{N}\sum_{j=1}^{N}\langle(x_i - x_j)^2\rangle \tag{6:S2-3}$$

We have reverted here to the notation appropriate to a three-dimensional walk, and replaced r_1 by x. The average in the sum on the right hand side of (6:S2-3) is directly proportional to a function that can be graphically represented as shown in Figure 6.18.

The dashed curve joins the ith and jth footprints on the walk. The solid lines stand for the walk that begins at the leftmost end of the three-line segment and ends at the rightmost point. There will, in general, be N_1 steps from the far left point to the leftmost vertex at which the dashed curve touches the line, N_2 steps in the central segment of the walk, and N_3 steps in the far right segment of the walk.

[6] Here, we make no distinction between N and $N + 1$.

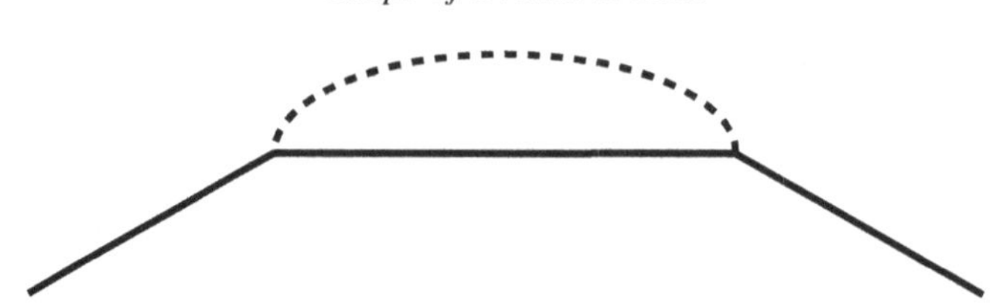

Fig. 6.18. Graphical representation of $\langle (x_i - x_j)^2 \rangle$.

Subject to the overall constraint that the N_i's add up to the total number of steps in the walk, we sum over all values of N_1, N_2, and N_3. The evaluation of the sum represented by this diagram is most conveniently carried out in the grand canonical ensemble, with the use of the generating function. We seek the coefficient of z^{N-1} in the direct product

$$\frac{1}{2N^2} \times 2 \int \mathrm{d}^d r_0 \int \mathrm{d}^d r_2 \int \mathrm{d}^d r_3 G(z; \vec{r}_1 - \vec{r}_0) G(z; \vec{r}_2 - \vec{r}_1) G(z; \vec{r}_3 - \vec{r}_2)(x_2 - x_1)^2 \tag{6:S2-4}$$

The "missing" integration in (6:S2-4), over $\vec{r}_1$, would yield a factor equal to the volume of the portion of space in which the random walk occurs. The factor of two multiplying the integral represent the two possible orderings of the indices i and j ($i > j$ and $i < j$). As the next step, we rewrite the generating functions in terms of their spatial Fourier transforms,

$$G(z; \vec{r}) = \frac{1}{(2\pi)^d} \int \mathrm{d}^d k g(z; \vec{k}) \mathrm{e}^{-\vec{k}\cdot\vec{r}} \tag{6:S2-5}$$

Making use of this representation, we find that the expression in (6:S2-4) reduces to

$$\frac{1}{N^2} g(z; \vec{k}) \left(-\frac{\partial^2}{\partial k_x^2} g(z; k) \right) g(z; \vec{k}) \Big|_{\vec{k}=0} \tag{6:S2-6}$$

The second derivative follows from the identity

$$(x_1 - x_2)^2 \mathrm{e}^{i\vec{k}\cdot(\vec{r}_1 - \vec{r}_2)} = -\frac{\partial^2}{\partial k_x^2} \mathrm{e}^{i\vec{k}\cdot(\vec{r}_1 - \vec{r}_2)} \tag{6:S2-7}$$

and an integration by parts. Once this identity and integration by parts have been implemented, the integrations over the $\vec{r}_i$'s produces Dirac delta functions in the k_i's, and we are immediately led to (6:S2-6)

Now, in the case of a random walk in d dimensions, we can write

$$g(z; \vec{k}) = \frac{1}{1 - z z_c^{-1} + k^2 l^2 / 2d} \tag{6:S2-8}$$

where l represents the mean distance covered by the walker in each step. This leaves us with the following result for the expression (6:S2-6):

$$\frac{1}{N^2}\frac{l^2}{d}\frac{1}{(1 - zz_c^{-1})^4} \tag{6:S2-9}$$

We now extract the coefficient of z^{N-1} in the power-series expansion of (6:S2-9).[7] This is straightforward given what we know about the process.[8] We find for this coefficient

$$\frac{1}{N^2}\frac{l^2}{d}z_c^{-(N-1)}\frac{(N+2)(N+1)N}{6} \approx \frac{1}{N^2}\frac{l^2}{d}z_c^{-(N-1)}\frac{N^3}{6} \tag{6:S2-10}$$

This result is the desired value of $\langle T_{11}\rangle$, multiplied by the total number of random walks with $N - 1$ steps. To obtain the average, we divide this by the total number of $N - 1$-step walks, which is equal to the coefficient of z^{N-1} in

$$\begin{aligned}\int \mathrm{d}^d r\, G(z;\vec{r}) &= g(z;\vec{k} = 0)\\ &= \frac{1}{1 - zz_c^{-1}}\end{aligned} \tag{6:S2-11}$$

The coefficient in question is $z_c^{-(N-1)}$. We thus find

$$\langle T_{11}\rangle = N\frac{l^2}{6d} \tag{6:S2-12}$$

The average of the trace of the tensor $\overleftrightarrow{T}$ is, by symmetry, equal to $d\langle T_{11}\rangle$. Given (6:S2-12) we have

$$\langle \mathrm{Tr}\, \overleftrightarrow{T}\rangle = Nl^2/6 \tag{6:S2-13}$$

The quantity above is also known as the mean radius of gyration.

6.5.2 Calculation of the asphericity

The quantities that contribute to the asphericity are $\langle T_{11}^2\rangle$, $\langle T_{11}T_{22}\rangle$ and $\langle T_{12}^2\rangle$. Again, reverting to standard cartesian notation we find

$$\langle T_{11}^2\rangle = \frac{1}{4N^4}\sum_{i=1}^{N}\sum_{j=1}^{N}\sum_{k=1}^{N}\sum_{l=1}^{N}\langle (x_i - x_j)^2 (x_k - x_l)^2\rangle \tag{6:S2-14}$$

Now, the graphical representation of the numerator in the expression leading to the desired average is a bit more complicated, in that there are three different

[7] The relevant power is $N - 1$ because there are $N - 1$ steps in a walk that leaves N footprints.
[8] See Supplement 3 in Chapter 2.

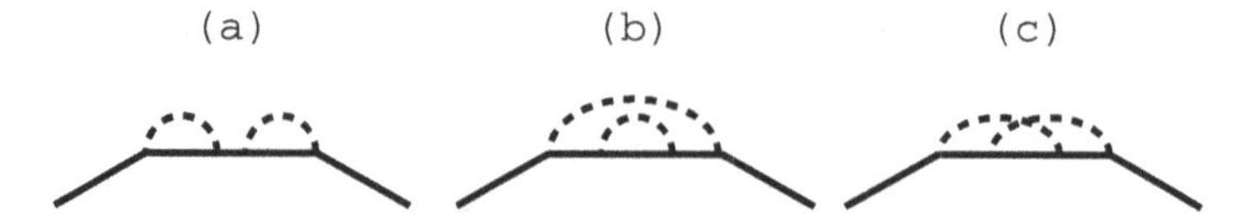

Fig. 6.19. The three different forms of the graph involved in the calculation of $\langle T_{11}^2 \rangle$.

forms, as shown in Figure 6.19. The dotted curves are a stand-in for the differences $(x_m - x_n)^2$, and the three types of graphical representations correspond to the three "topologically distinct" possibilities for the sequence of the indices i, j, k, l in (6:S2-14). The calculation proceeds along the same line as the one leading to a result for $\langle T_{11} \rangle$. As we can see from (6.14), the only information we need to extract from our calculation is the ratios $\langle T_{11}^2 \rangle / \langle T_{11} T_{22} \rangle$ and $\langle T_{11} \rangle / \langle T_{12} \rangle$. We will go over the determination of $\langle T_{11}^2 \rangle$ in detail. The other averages are determined in a similar way.

6.5.3 Determination of $\langle T_{11}^2 \rangle$

We proceed, following Figure 6.19 diagram-by-diagram.

Diagram (a)

Here, the calculation proceeds as it did in the evaluation of $\langle T_{11} \rangle$. Taking second derivatives and performing integrations by parts, we are left with the following expression

$$\frac{1}{4N^4} \times 8\, g(z;\vec{k}) \left(\frac{\partial^2}{\partial k_x^2} g(z;\vec{k}) \right) g(z;\vec{k}) \left(\frac{\partial^2}{\partial k_x^2} g(z;\vec{k}) \right) g(z;\vec{k}) \bigg|_{\vec{k}=0} \tag{6:S2-15}$$

The factor of 8 in the above expression counts the number of ways of constructing diagram (a), exchanging end-points of the two dashed curves, and permuting the two curves among themselves. The quantity of interest is, of course, the coefficient of z^{N-1} in (6:S2-15). We will defer the power-series expansion in the fugacity z. Making use of (6:S2-8) for the Fourier-transformed generating function, we end up with the result

$$\frac{2}{N^4} \left(\frac{l^2}{d} \right)^2 \frac{1}{\left(1 - z z_c^{-1}\right)^7} \tag{6:S2-16}$$

Diagram (b)

The evaluation of this diagram is a bit more involved. Utilizing the generating function in real space, we have the average of interest proportional to the following expression:

$$\int \mathrm{d}^d r_0 \int \mathrm{d}^d r_1 \int \mathrm{d}^d r_2 \int \mathrm{d}^d r_3 \int \mathrm{d}^d r_4 G\left(z;\vec{r}_1-\vec{r}_0\right) \times G\left(z;\vec{r}_2-\vec{r}_1\right) G\left(z;\vec{r}_3-\vec{r}_2\right) G\left(z;\vec{r}_4-\vec{r}_3\right) G\left(z;\vec{r}_5-\vec{r}_4\right) \times \left(x_4-x_1\right)^2 \left(x_3-x_2\right)^2 \tag{6:S2-17}$$

The next step is to express the generating functions in terms of their Fourier transforms. We end up with a product containg the factor

$$\left(x_4-x_1\right)^2\left(x_3-x_2\right)^2 \mathrm{e}^{\mathrm{i}\vec{k}_0\cdot(\vec{r}_1-\vec{r}_0)}\mathrm{e}^{\mathrm{i}\vec{k}_1\cdot(\vec{r}_2-\vec{r}_1)}\mathrm{e}^{\mathrm{i}\vec{k}_2\cdot(\vec{r}_3-\vec{r}_2)}\mathrm{e}^{\mathrm{i}\vec{k}_3\cdot(\vec{r}_4-\vec{r}_3)}\mathrm{e}^{\mathrm{i}\vec{k}_4\cdot(\vec{r}_5-\vec{r}_4)} = \left(\frac{\partial}{\partial k_{2x}}\right)^2\left(\frac{\partial}{\partial k_{1x}}+\frac{\partial}{\partial k_{2x}}+\frac{\partial}{\partial k_{3x}}\right)^2 \mathrm{e}^{\mathrm{i}\vec{k}_0\cdot(\vec{r}_1-\vec{r}_0)}\mathrm{e}^{\mathrm{i}\vec{k}_1\cdot(\vec{r}_2-\vec{r}_1)}\mathrm{e}^{\mathrm{i}\vec{k}_2\cdot(\vec{r}_3-\vec{r}_2)} \times \mathrm{e}^{\mathrm{i}\vec{k}_3\cdot(\vec{r}_4-\vec{r}_3)}\mathrm{e}^{\mathrm{i}\vec{k}_4\cdot(\vec{r}_5-\vec{r}_4)} \tag{6:S2-18}$$

After a series of integrations by parts in the variables $\vec{k}_i$ the derivatives above act on the Fourier transforms of the generating functions. The integrations over the $\vec{r}_l$'s produce delta functions, and we are left with the following result

$$\left(\frac{\partial}{\partial k_{2x}}\right)^2\left(\frac{\partial}{\partial k_{1x}}+\frac{\partial}{\partial k_{2x}}+\frac{\partial}{\partial k_{3x}}\right)^2 g\left(z;\vec{k}_0\right) g\left(z;\vec{k}_1\right) g\left(z;\vec{k}_2\right) g\left(z;\vec{k}_3\right) g\left(z;\vec{k}_4\right)\Big|_{\vec{k}_0=\vec{k}_1=\vec{k}_2=\vec{k}_3=\vec{k}_4=0} \tag{6:S2-19}$$

The polynomial expressions in the derivatives are now expanded, discarding in the process all terms that evaluate to zero when the $\vec{k}_i$'s are equal to zero.[9] This yields

$$\left(\frac{\partial^2}{\partial k_{1x}^2}\frac{\partial^2}{\partial k_{2x}^2}+\frac{\partial^2}{\partial k_{3x}^2}\frac{\partial^2}{\partial k_{2x}^2}+\frac{\partial^4}{\partial k_{2x}^4}\right) g\left(z;\vec{k}_0\right) g\left(z;\vec{k}_1\right) g\left(z;\vec{k}_2\right) g\left(z;\vec{k}_3\right) g\left(z;\vec{k}_4\right)\Big|_{\vec{k}_0=\vec{k}_1=\vec{k}_2=\vec{k}_3=\vec{k}_4=0} \tag{6:S2-20}$$

The remainder of the calculation is fairly straightforward. Inserting combinatorial factors noted above and taking the appropriate derivatives of the generating function, we end up with the contribution

$$\frac{16}{N^4}\left(\frac{l^2}{d}\right)^2 \frac{1}{\left(1-zz_{\mathrm{c}}^{-1}\right)^7} \tag{6:S2-21}$$

[9] The general operating principle is that any term that contains an odd-order derivative with respect to a k_{ix} will evaluate to zero.

Diagram (c)

In this case, the relevant identity is

$$(x_3 - x_1)^2 (x_4 - x_2)^2 \mathrm{e}^{\mathrm{i}\vec{k}_0\cdot(\vec{r}_1-\vec{r}_0)} \mathrm{e}^{\mathrm{i}\vec{k}_1\cdot(\vec{r}_2-\vec{r}_1)} \mathrm{e}^{\mathrm{i}\vec{k}_2\cdot(\vec{r}_3-\vec{r}_2)} \mathrm{e}^{\mathrm{i}\vec{k}_3\cdot(\vec{r}_4-\vec{r}_3)} \mathrm{e}^{\mathrm{i}\vec{k}_4\cdot(\vec{r}_5-\vec{r}_4)}$$
$$= \left(\frac{\partial}{\partial k_{1x}} + \frac{\partial}{\partial k_{2x}}\right)^2 \left(\frac{\partial}{\partial k_{2x}} + \frac{\partial}{\partial k_{3x}}\right)^2 \mathrm{e}^{\mathrm{i}\vec{k}_0\cdot(\vec{r}_1-\vec{r}_0)} \mathrm{e}^{\mathrm{i}\vec{k}_1\cdot(\vec{r}_2-\vec{r}_1)} \mathrm{e}^{\mathrm{i}\vec{k}_2\cdot(\vec{r}_3-\vec{r}_2)}$$
$$\times \mathrm{e}^{\mathrm{i}\vec{k}_3\cdot(\vec{r}_4-\vec{r}_3)} \mathrm{e}^{\mathrm{i}\vec{k}_4\cdot(\vec{r}_5-\vec{r}_4)} \qquad (6:S2\text{-}22)$$

The same set of steps as outined immediately above leads to the following non-vanishing contributions to the diagram, combinatorial factors having been left out,

$$\left(\frac{\partial^2}{\partial k_{1x}^2}\frac{\partial^2}{\partial k_{2x}^2} + \frac{\partial^2}{\partial k_{3x}^2}\frac{\partial^2}{\partial k_{2x}^2} + \frac{\partial^2}{\partial k_{1x}^2}\frac{\partial^2}{\partial k_{3x}^2} + \frac{\partial^4}{\partial k_{2x}^4}\right)$$
$$g\left(z;\vec{k}_0\right) g\left(z;\vec{k}_1\right) g\left(z;\vec{k}_2\right) g\left(z;\vec{k}_3\right) g\left(z;\vec{k}_4\right)\Big|_{\vec{k}_0=\vec{k}_1=\vec{k}_2=\vec{k}_3=\vec{k}_4=0} \qquad (6:S2\text{-}23)$$

Taking the derivatives indicated, evaluating the $\vec{k}_i = 0$ limits, and inserting the required combinatorial factors, we end up with

$$\frac{18}{N^4}\left(\frac{l^2}{d}\right)^2 \frac{1}{\left(1 - zz_c^{-1}\right)^7} \qquad (6:S2\text{-}24)$$

Summing all diagrams

Adding (6:S2-16), (6:S2-21), and (6:S2-23), we end up with the following total contribution to the generating function yielding $\langle T_{11}^2\rangle$

$$\frac{36}{N^4}\left(\frac{l^2}{d}\right)^2 \frac{1}{\left(1 - zz_c^{-1}\right)^7} \qquad (6:S2\text{-}25)$$

Actually, the generating function yields the numerator in a fraction. The denominator is the total number of $N-1$-step walks. However, as we are interested in ratios of averages, the common denominator is not important for our present purposes. For the same reason, we are not required to extract the coefficient of z^{N-1} in the power-series expansion of (6:S2-25), as that produces a common factor that cancels out when we evaluate ratios.

6.5.4 The ratios

The calculations of expressions contributing to $\langle T_{11}T_{22}\rangle$ and $\langle T_{12}^2\rangle$ proceed along the lines laid out above. For each of these averages, there are three contributions, corresponding to the three diagrams in Figure 6.19. Carrying out the required computations, we find

$$\frac{\langle T_{11}^2 \rangle}{\langle T_{11} T_{22} \rangle} = \frac{9}{5} \tag{6:S2-26}$$

$$\frac{\langle T_{11}^2 \rangle}{\langle T_{12}^2 \rangle} = \frac{9}{2} \tag{6:S2-27}$$

Inserting these results into the right hand side of (6:S2-1), we obtain (6.15) for the mean asphericity of a d-dimensional random walk.

6.6 Supplement 3: derivation of (6.21) for the radius of gyration tensor, $\overleftrightarrow{T}$, and the eigenvalues of the operator

The demonstration of (6.21) is straightforward. We start with (6:S2-2b) for the elements of $\overleftrightarrow{T}$. Now, we rewrite the difference $(r_{jl} - r_{kl})^2$ in terms of the displacements η_{il}:

$$\begin{aligned}(r_{jl} - r_{kl})^2 &= \left(\sum_{m=j+1}^{l} \eta_{ml} \right)^2 \\ &= \sum_{m=j+1}^{l} \sum_{n=j+1}^{l} \eta_{ml} \eta_{nl} \end{aligned} \tag{6:S3-1}$$

Now, if we look at the product $\eta_{ml}\eta_{nl}$, we can ask for the net contribution of this product to the entry T_{ll}. We do this by counting up the number of $(r_{jl} - r_{kl})^2$'s that generate, when expanded as above, in that product. Assume, for the time being, that $m \leq n$. Then, investigation leads to the result that there are $m(N + 1 - n)$ such terms. This is because there are m j's less than or equal to m and $N + 1 - n$ k's greater than or equal to n. We thus arrive at (6.21).

7

Path integrals and self-avoidance

The concept of a field dates back to Euler, who introduced the notion to describe fluid flows in his study of hydrodynamics. Methods and concepts based on field theory now pervade the physical sciences and engineering. Field-theoretical ideas exert a strong influence on physical intuition and shape modern nomenclature. In addition, some of the most powerful analytical tools available to the modern scientist are those developed to study the behavior of fields.

In the context of models designed to describe the physical world, a field is a quantity that varies continuously in space and time. Examples are the electric and magnetic fields, the velocity and density distributions of a liquid or vapor and the quantum-mechanical wavefunction of a microscopic particle. In some cases, such as the velocity and density fields introduced by Euler, the notion of continuity must be taken advisedly. Because of the atomic structure of matter, one cannot carry the notion of a smooth density distribution down to the length scales on which molecules can be distinguished. There, the classical description is necessarily in terms of particles. Quantum mechanically, wavefunctions replace the classical density and velocity fields as the appropriate mode of description. This proviso notwithstanding, in the regimes in which density and velocity fields accurately describe the state of a liquid or vapor, they form the basis of an extremely useful theoretical model that yields important physical properties of these systems.

It turns out that the random walk also lends itself to description in terms of a field. As in the case of a liquid or vapor, the field-based description maintains its validity in a restricted range of length scales. However, those restrictions allow for the investigation of random walks in most of the interesting regimes, and the field theoretical model of random walks has associated with it a wide range of very effective techniques.

These techniques will be especially useful when we confront the statistical consequences of self-avoidance, a restriction on the random walk that makes it relevant to the statistical mechanics of long chain polymers. We will discover that the self-avoiding random walk has many of the mathematical features of a self-interacting field, and that the statistics of such walks yield the conformational properties of long, flexible chain polymers.

A first step in the development of a field-theoretical description of the random walk is to express random walk statistics in terms of the various paths a walker can take as it makes its way from an initial point $\vec{x}$ to an ultimate destination $\vec{y}$. In appropriate limits, the generating function for this process takes the form of a path integral, of the sort utilized in the study of quantum-mechanical, and especially quantum field-theoretical, processes (des Cloiseaux and Jannink, 1990; Feynman and Hibbs, 1965; Freed, 1987; Itzykson and Drouffe, 1991; Kleinert, 1995). In subsequent chapters the path integral formulation will allow us to argue for the equivalence of the statistics of the self-avoiding random walker (to be defined below) and the statistical mechanics of a fictitious magnetic system. The spin system is an n-component vector model, in that one imagines the magnetic moments as vector-like objects, i.e. objects with a head and a tail, that are free to rotate in an n-dimensional "internal" space. The space is called internal to differentiate it from the actual, physical space in which the spins reside. This system is commonly referred to as the $O(n)$ model. The version of the $O(n)$ model that is relevant to the self-avoiding walker is the $n \to 0$ limit. This apparently non-physical (almost non-sensical) limit admits of the application of powerful field-theoretical methods. Among the methods that prove most useful are those associated with the renormalization group approach to critical phenomena. The development and exploitation of these matters will be the subject of the next several chapters.

7.1 The unrestricted random walk as a path integral

The first step in the development of a field-theoretical description of the random walk is to reformulate the process as a path integral. In a sense, this is just a matter of semantics. As we will see almost immediately, the random walk is nothing more than a discrete version of a path integral. However, the passage to a continuous process – as mathematically improbable as it may seem at the time we perform it – will lead us in some very interesting directions.

To begin, note that $C(N;\vec{x},\vec{y})$, the expression for the number of walks starting at $\vec{x}$ and ending at $\vec{y}$ is equal to the sum over all N-step paths that a random walker can take in going from $\vec{x}$ to $\vec{y}$. In this sense, we may describe $C(N;\vec{x},\vec{y})$ as a *path sum*. Formally, we can write

$$C(N;\vec{x},\vec{y}) = \sum_{\mathrm{P}} \mathcal{W}_{\mathrm{P}}(N;\vec{x},\vec{y}) \tag{7.1}$$

where $\mathcal{W}_{\mathrm{P}}(N;\vec{x},\vec{y})$ is the number of N-step walks from $\vec{x}$ to $\vec{y}$ that follow the path P. In the case of a walker on a lattice, the summand does not reduce to an expression in terms of other quantities. There is exactly one walk per allowable path. $\mathcal{W}_P(N;\vec{x},\vec{y})$ is equal to 1 for every possible path, and the mathematics of the random walk enters through the restriction on paths contributing to the sum. To develop an expression of greater use to us, we recast the problem slightly and recall results from the previous chapter.

Suppose we require that the walker pass through the point $\vec{r}$ on the way from $\vec{x}$ to $\vec{y}$. In particular, suppose that we restrict our considerations to walkers who go from $\vec{x}$ to $\vec{r}$ in M steps and from $\vec{r}$ to $\vec{y}$ in $N-M$ steps. The number of such walks is equal to

$$C(M;\vec{x},\vec{r})C(N-M;\vec{r},\vec{y}) \tag{7.2}$$

The total number of N-step paths from $\vec{x}$ to $\vec{y}$ is equal to the sum over intermediate points, $\vec{r}$, of the above expression. Thus,

$$C(N;\vec{x},\vec{y}) = \sum_{\vec{r}} C(M;\vec{x},\vec{r})C(N-M;\vec{r},\vec{y}) \tag{7.3}$$

When M is equal to $N-1$, the final portion of the walk consists of exactly one step, and (7.3) reduces to (2.7).

Let's carry the process further, and break the walk into several segments. Suppose we require that the walker pass through the point $\vec{r}_1$ after M_1 steps, through the point $\vec{r}_2$ after M_2 steps, and so on. If there are a total of k segments, then the number of N-step walks from $\vec{x}$ to $\vec{y}$ satisfying the above criteria is equal to the product

$$C(M_1;\vec{x},\vec{r}_1)C(M_2;\vec{r}_1,\vec{r}_2)\cdots C(M_{k-1};\vec{r}_{k-2},\vec{r}_{k-1})C(M_k;\vec{r}_{k-1},\vec{y}) \tag{7.4}$$

Summed over the intermediate points:

$$\begin{aligned} C(N;\vec{x},\vec{y}) = \sum_{\vec{r}_1,\ldots,\vec{r}_k} & C(M_1;\vec{x},\vec{r}_1)C(M_2;\vec{r}_1,\vec{r}_2) \\ & \cdots C(M_{k-1};\vec{r}_{k-2},\vec{r}_{k-1})C(M_k;\vec{r}_{k-1},\vec{y}) \end{aligned} \tag{7.5}$$

Now, if the numbers $M_1, M_2, \ldots, M_k$, where $\sum_{l=1}^{k} M_l = N$, are all reasonably large, then the individual contributions to the product in the sum can be replaced by the asymptotic Gaussian forms as given in (2.20). The sums over the intermediate points are also well-approximated by integrals, and we have

$$\begin{aligned} C(N;\vec{x},\vec{y}) = \int \cdots \int & \mathrm{d}\vec{r}_1 \cdots \mathrm{d}\vec{r}_k \left(2\pi M_1 a^2\right)^{-3/2} \mathrm{e}^{-3|\vec{x}-\vec{r}_1|^2/2M_1a^2} \\ & \times \left(2\pi M_2 a^2\right)^{-3/2} \mathrm{e}^{-3|\vec{r}_1-\vec{r}_2|^2/2M_2a^2} \times \\ & \cdots \times \left(2\pi M_k a^2\right)^{-3/2} \mathrm{e}^{-3|\vec{r}_k-\vec{y}|^2/2M_ka^2} \end{aligned} \tag{7.6}$$

Of course, we also know that the Gaussian form in (2.20) holds for the left hand side of (7.6). This means that it must be possible to show that the right hand side of (7.6) yields the appropriate Gaussian expression for $C(N, \vec{x}, \vec{y})$.

Exercise 7.1

Show that the right hand side of (7.6) yields the following:

$$C(N, \vec{x}, \vec{y}) = \left(2\pi N a^2\right)^{-3/2} \exp\left(-\frac{3|\vec{x} - \vec{y}|^2}{2Na^2}\right)$$

Now (7.6) tells us nothing new about the number of N-step walks between $\vec{x}$ and $\vec{y}$, beyond the somewhat interesting fact that it can be represented as a convolution of Gaussian integrals. The main utility of the formula lies in its connection to a form on which one can build the notion of the random walk as a path – or functional – integral. We arrive at the form as follows. First, we imagine that the differences $|\vec{r}_i - \vec{r}_{i+1}| \equiv |\Delta\vec{r}_i|$ can be considered to be small. If we replace M_i by ΔN_i – so that $N = \sum_{i=1}^{k} \Delta N_i$ – the exponents in (7.6) take the form

$$-\frac{3}{2a^2}\Delta N_i \left(\frac{|\vec{r}_i - \vec{r}_{i+1}|}{\Delta N_i}\right)^2 \sim -\frac{3}{2a^2}\Delta N_i \left(\frac{|\Delta\vec{r}_i|}{\Delta N_i}\right)^2 \qquad (7.7)$$

The term $\left(|\Delta\vec{r}_i|/\Delta N_i\right)^2$ looks like the square of a first derivative. As a brake on the temptation to denote it as such there is the fact that the denominator cannot be even close to an infinitesimal. Recall that the Gaussian forms for the number of random walks between the various locations hold *only* when the ΔN_i's are large compared to 1.

Nevertheless, we will, as needed, pretend that the limiting procedure

$$\left(\frac{|\Delta\vec{r}_i|}{\Delta N_i}\right)^2 \to \left(\frac{|\mathrm{d}\vec{r}_i|}{\mathrm{d}N_i}\right)^2$$

is not absurd. In that case, the right hand side of (7.6) has the form of a "true" functional integral, in that we can write

$$C(N; \vec{x}, \vec{y}) \propto \prod_s \int \mathrm{d}\vec{r}(s) \exp\left(-\frac{3}{2a^2}\int_0^N \left|\frac{\mathrm{d}\vec{r}(s)}{\mathrm{d}s}\right|^2 \mathrm{d}s\right). \qquad (7.8)$$

The above equation contains the newly introduced arc length parameter $\Delta s = \Delta N$, or, passing to the limit of infinitesimals, $\mathrm{d}s = \mathrm{d}N$. The path is now specified in terms of the continuous curve $\vec{r}(s)$, where the continuous variable s ranges from 0 to N, and the vector function $\vec{r}(s)$ satisfies the boundary conditions $\vec{r}(0) = \vec{x}$ and $\vec{r}(N) = \vec{y}$. The product in the above equation is clearly over an infinite number of factors, and the integration variables are the values of the function $\vec{r}(s)$ for all values of the variable s in the range $0 < s < N$. If we take the notion of s as a continuous variable seriously, there is an uncountable infinity of those factors.

It is, therefore, not surprising that the functional integral in (7.8) has mathematical peculiarities. Not the least of them is the fact that the multiplicative constant that turns the proportionality into an equality does not have a finite value. If placed on the right hand side of the equation it is infinite, and if the proportionality factor appears on the left hand side it is equal to zero. There are other, more subtle and vexing, complications that arise from the effectively pathological nature of the functional integral in (7.8). These complications can be collectively characterized as "ultraviolet divergences."

In practice, however, the path integral will always be over a finite number of discrete terms, and the ultraviolet problems that attend the functional integral as defined in (7.8) are not among the challenges facing whoever attempts evaluation of the path integral. Many of the principal difficulties arise when we modify the rules governing the random walk so as to make it self-avoiding.

7.2 Self-avoiding walks

The term "random walk" conjures up the image of a process that unfolds with time. Someone, or something, wanders along a path that meanders from a starting point to an eventual destination. In the cases we have considered up to now, the trail of the walker may intersect itself at any number of occasions. This particular kind of random walk is called *unrestricted*. The walker does not recall where it has been, and all steps are independent random variables. An interesting, and important, variation of the random walk endows the walker with just such a memory, which induces a correlation between steps. The *self-avoiding* walker remembers its past itinerary, and tries to avoid treading on the trail it has already left. It is thus clear the self-avoiding random walk as a sequence of random events is non-Markovian. In the most extreme case of self-avoidance, the walker stops cold when, by chance, it lands on a previously visited spot. In a "gentler" version of the model, the walk ceases with a probability that increases as the walker approaches, or comes into contact with, the trail it has left.

As another variation, the walker probes its immediate surroundings and utilizes information about whatever portion of the trail is nearby to bias the random process that controls the choice of the immediately subsequent steps. The altered process favors steps that avoid landing on, or crossing over, the walker's trail. No matter where the walker steps it keeps on going. This last kind of walker is called the "true" self-avoiding walker, as opposed to the walker described in the preceding pragraph, who executes what is known simply as the self-avoiding walk. There are other ways of altering the rules that govern the random walk, and in time we will explore their consequences. For the time being, our attention will be focused on the classical self-avoiding walk described in the first paragraph of this section.

7.2.1 Interactions

How does one alter the mathematical prescriptions developed in the last chapter so that they apply to the case of the self-avoiding walker? Recall (7.1), the formula for the number of walks starting at $\vec{x}$ and ending at $\vec{y}$. It consisted of a sum over all possible paths. If the walker avoids paths previously taken, then the sum suffers a modification. An appropriate modification is effected by the introduction of factors of the form $e^{-v(|\vec{w}_i-\vec{w}_j|)}$. In such a factor $\vec{w}_l$ is the position vector connecting the origin to the lth step of the walk. In fact, the sum over paths will be weighted by the factor

$$\prod_{i>j,\ i,j=1}^{N} e^{-v(|\vec{w}_i-\vec{w}_j|)} \tag{7.9}$$

If the function $v(x)$ becomes positive and infinite for any value of the argument x, then the weighting factor $e^{-v(x)}$ vanishes, and the path is deleted as a possible walk from $\vec{x}$ to $\vec{y}$. This means that the product in (7.9) can be made consistent with "strict" self-avoidance. We will assume for the present that $v(x)$ is always positive. The total number of N-step self-avoiding walks starting a $\vec{x}$ and terminating at $\vec{y}$ is then given by

$$C(N;\vec{x},\vec{y}) = \sum_{\mathrm{P}} \mathcal{W}_{\mathrm{P}}(N;\vec{x},\vec{y}) \left(\prod_{i=1}^{N} e^{-v(|\vec{w}_i-\vec{x}|)} e^{-v(|\vec{w}_i-\vec{y}|)} \right) \prod_{i>j,\ i,j=1}^{N} e^{-v(|\vec{w}_i-\vec{w}_j|)} \tag{7.10}$$

7.2.2 Path integral revisited: self-avoidance

The functional integral form for the unconstrained walk as prescribed in (7.8) is now easily modified to account for the fact that intersecting paths are to be avoided. When self-avoidance comes into play, extra terms enter into the integrand. These terms arise as a result of the recasting of the exponential in (7.8).

Following the development above, we can write for the number of non-intersecting walks beginning at $\vec{x}$ and ending at $\vec{y}$

$$\begin{aligned} C(N;\vec{x},\vec{y}) &= \int\cdots\int \mathrm{d}\vec{r}_1\cdots\mathrm{d}\vec{r}_k \left(2\pi M_1 a^2\right)^{-3/2} \mathrm{e}^{-3|\vec{x}-\vec{r}_1|^2/3M_1a^2} \times\cdots \\ &\times \left(2\pi M_k a^2\right)^{-3/2} \mathrm{e}^{-3|\vec{r}_k-\vec{y}|^2/3M_ka^2} \prod_l \mathrm{e}^{-v(|\vec{r}_l-\vec{x}|)-v(|\vec{r}_l-\vec{y}|)} \prod_{i>j} \mathrm{e}^{-v(|\vec{r}_i-\vec{r}_j|)} \end{aligned} \tag{7.11}$$

The product can be written as a sum:

$$\mathrm{e}^{-\sum_l v(|\vec{r}_l-\vec{x}|)-\sum_l v(|\vec{r}_l-\vec{y}|)}\mathrm{e}^{-\sum_{i>j} v(|\vec{r}_i-\vec{r}_j|)} \tag{7.12}$$

Under the same limiting procedures as applied in the case of the unrestricted walk, the factor above takes on the following continuous form

$$\exp\left[-v_0\int_0^N\int_0^{s_1}\mathrm{d}s_1\,\mathrm{d}s_2\, v\left(|\vec{r}(s_1)-\vec{r}(s_2)|\right)\right] \tag{7.13}$$

Equation (7.11) then becomes

$$\begin{aligned} C(N;\vec{x},\vec{y}) \propto \prod_s \int \mathrm{d}\vec{r}(s)\exp\Bigg[-\left(3/2a^2\right)\int_0^N\left|\frac{\mathrm{d}\vec{r}}{\mathrm{d}s}\right|^2\mathrm{d}s - v_0 \\ \times\int_0^N\int_0^{s_1}\mathrm{d}s_1\,\mathrm{d}s_2\, v\left(|\vec{r}(s_1)-\vec{r}(s_2)|\right)\Bigg] \end{aligned} \tag{7.14}$$

If the interaction leading to self-avoidance is extremely short-ranged, then the function $v(|\vec{r}_1-\vec{r}_2|)$ can be replaced by a Dirac delta function, and the weighting factor that guarantees self-avoidance takes the form

$$\exp\left[-v_0\int_0^N\int_0^{s_1}\mathrm{d}s_1\,\mathrm{d}s_2\,\delta\left(|\vec{r}(s_1)-\vec{r}(s_2)|\right)\right] \tag{7.15}$$

and the functional integral in (7.14) is replaced by

$$\begin{aligned} C(N;\vec{x},\vec{y}) \propto \prod_s \int \mathrm{d}\vec{r}(s)\exp\Bigg[-\left(3/2a^2\right)\int_0^N\left|\frac{\mathrm{d}\vec{r}}{\mathrm{d}s}\right|^2\mathrm{d}s - v_0 \\ \times\int_0^N\int_0^{s_1}\mathrm{d}s_1\,\mathrm{d}s_2\,\delta\left(|\vec{r}(s_1)-\vec{r}(s_2)|\right)\Bigg] \end{aligned} \tag{7.16}$$

In the next chapter we will return to this functional integral and show that it predicts interesting and useful scaling properties of the function $C(N;\vec{x},\vec{y})$.

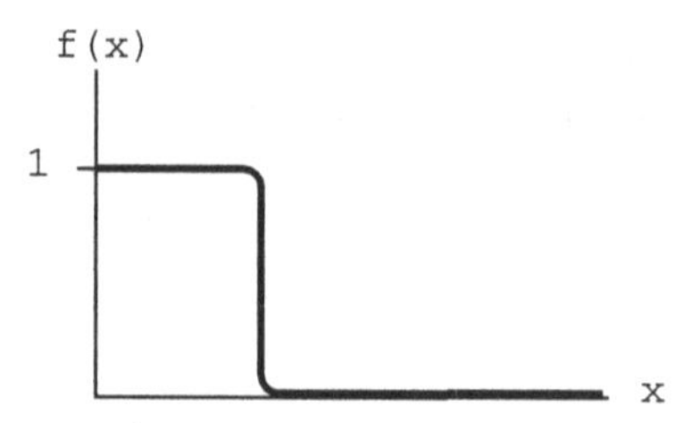

Fig. 7.1. The function $f(x)$ defined in (7.17). The function shown here is appropriate to the case of short-ranged, and very strong, self-avoidance of the random walk.

Expansion in self-avoidance

The product of factors that enforce self-avoidance seriously complicates the evaluation of random walk sums. In fact, there is no known exact expression for the number of self-avoiding walks in three dimensions that satisfy any reasonable set of self-avoidance requirements. However, approximate methods and some very clever analytical tricks have yielded non-trivial results of practical value. One such set of approaches is based on an expansion in the tendency to avoid previously visited portions of space. Recall an individual term in the product $e^{-v(x)}$. This term can be rewritten as follows

$$e^{-v(x)} = 1 + \left(e^{-v(x)} - 1\right) \equiv 1 - f(x) \tag{7.17}$$

The quantity $f(x)$ represents the modification of random walk statistics induced by self-avoidance. If a walker's tendency to avoid itself as expressed in the function $v(x)$ is short-ranged, then the function $f(x)$ will be equal to zero except in a small region in the vicinity of $x = 0$ (see Figure 7.1). On the other hand, when $f(x)$ differs from zero it never exceeds 1 in absolute value. Rewriting each term in the product that appears on the right hand side of (7.10) as above we obtain for the effect of self-avoidance

$$\begin{aligned} C(N;\vec{x},\vec{y}) &= \sum_{\mathrm{P}} \mathcal{W}_{\mathrm{P}}(N;\vec{x},\vec{y}) \left(\prod_{i=1}^{N} \left(1 - f\left(|\vec{w}_i - \vec{x}|\right)\right) \left(1 - f\left(|\vec{w}_i - \vec{y}|\right)\right) \right) \\ &\quad \times \prod_{i>j \;\; i,j=1}^{N} \left(1 - f\left(|\vec{w}_i - \vec{w}_j|\right)\right) \end{aligned} \tag{7.18}$$

The next step is to expand (7.18) in powers of the factors f. The zeroth order term in the expansion is just the number of ordinary, unrestricted walks. The first

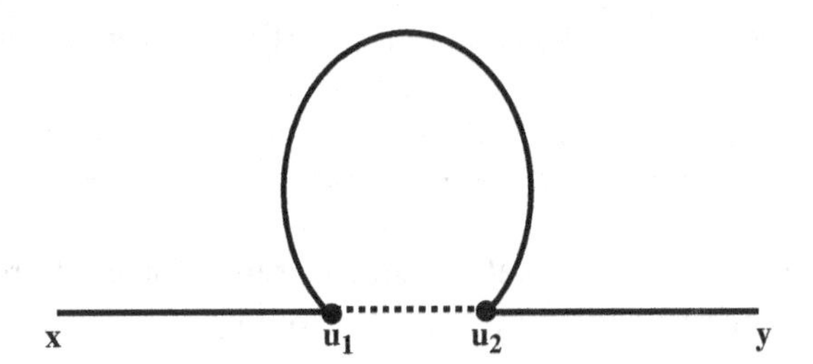

Fig. 7.2. The first order diagram.

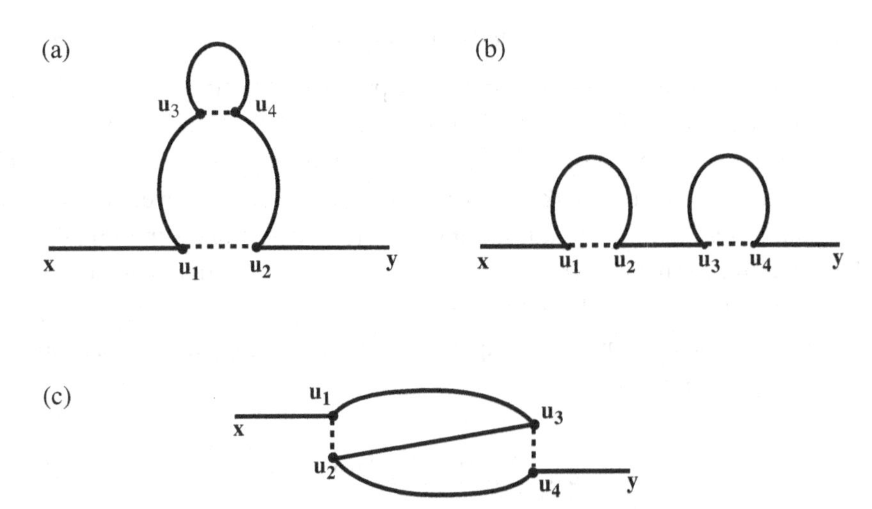

Fig. 7.3. The second order diagrams.

order term is

$$-\sum_{M_1,M_2,M_3}\sum_{\vec{u}_1,\vec{u}_2\,(\vec{u}_1>\vec{u}_2)} C(M_1;\vec{x},\vec{u}_1)C(M_2;\vec{u}_1,\vec{u}_2)C(M_3;\vec{u}_2,\vec{y})f\left(|\vec{u}_1-\vec{u}_2|\right) \quad (7.19)$$

where the position vectors $\vec{u}_1$ and $\vec{u}_2$ range from $\vec{x}$ to $\vec{y}$ through all the $\vec{w}_i$'s. The quantities M_i satisfy $M_1 + M_2 + M_3 = N$. Figure 7.2 gives a pictorial representation of the above first order correction to $C(N;\vec{x},\vec{y})$. The second order correction is a bit long to write out. The pictorial representation provides a useful shorthand. Figure 7.3 displays the three different expressions that are generated when one expands the right hand side of (7.18) to second order in the factor f. These pictures depict all the topologically distinct ways in which a walker can travel from $\vec{x}$ to $\vec{y}$ in $k+1$ segments that connect the starting and ending points to the intermediate locations $\vec{u}_1 \ldots \vec{u}_k$. The figures are more than a convenient pictorial representation. Given the appropriate interpretations, they can be constructed so as to contain all the information needed to reconstitute the expressions they represent.

The expansion we have developed here is a particular example of a *perturbation theoretical* expansion.

Exercise 7.2

Pictorially represent the topologically distinct graphs for the third-order-in-$f(x)$ correction to the generating function for self-avoiding random walks.

7.2.3 Perturbation theoretical expansion of the generating function

It is considerably more convenient to develop a perturbation series for the generating function of the number of walks than for the quantity itself. Here's why.

The series for the number of walks resulted from the expansion of the product in (7.18), which generated terms like (7.19). This series is readily converted to a series for the generating function by multiplying each term by z^N and summing over N. The key simplification arises from the fact that sums over intermediate numbers of steps, i.e. $M_1, M_2, \ldots M_k$, where $M_1 + M_2 + \cdots + M_k = N$ – that is, sums in the form of convolutions – convert to simple products. To see how this works consider the following convolution

$$\sum_{M_1=0}^{N} \sum_{M_2=0}^{N} H_1(M_1)H_2(M_2) \tag{7.20}$$

with $M_1 + M_2 = N$.

Now if the sum above is multiplied by z^N and N is summed from 0 to ∞, one obtains

$$\begin{aligned}
&\sum_{N=0}^{\infty} \sum_{M_1=0}^{N} z^N H_1(M_1)H_2(N - M_1) \\
&= \sum_{N=0}^{\infty} \sum_{M_1=0}^{N} z^{M_1} H_1(M_1) Z^{N-M_1} H_2(N - M_1) \\
&= \sum_{K_1=0}^{\infty} z^{K_1} H_1(K_1) \sum_{K_2=0}^{\infty} Z^{K_2} H_2(K_2) \\
&= h_1(z)h_2(z)
\end{aligned} \tag{7.21}$$

where

$$h_1(z) = \sum_{K=0}^{\infty} H_1(K)z^K \tag{7.22}$$

and similarly for $h_2(z)$. The reader can readily verify that the change in summation variables in (7.21) from N and M_1 to K_1 and K_2 is justified in that there is a one-to-one relation between the terms in the sums on the second and third lines of the equation.

Fourier transforms also have the effect of transforming convolutions into simple products. Consider, for example the following integral

$$\int_{-\infty}^{\infty} F_1(\vec{x} - \vec{w}) F_2(\vec{w} - \vec{y}) \, \mathrm{d}^d w \tag{7.23}$$

Multiplying the expression above by $\mathrm{e}^{\mathrm{i}\vec{q}\cdot(\vec{x}-\vec{y})}$ and integrating over $\vec{y}$ we obtain

$$\begin{aligned} &\int\int \mathrm{e}^{\mathrm{i}\vec{q}\cdot(\vec{x}-\vec{y})} F_1(\vec{x} - \vec{w}) F_2(\vec{w} - \vec{y}) \, \mathrm{d}^d w \, \mathrm{d}^d y \\ &= \int \mathrm{e}^{\mathrm{i}\vec{q}\cdot(\vec{x}-\vec{w})} F_1(\vec{x} - \vec{w}) \, \mathrm{d}^d w \int \mathrm{e}^{\mathrm{i}\vec{q}\cdot(\vec{w}-\vec{y})} F_2(\vec{w} - \vec{y}) \, \mathrm{d}^d y \\ &= \int F_1(-\vec{r}_1) \mathrm{e}^{-\mathrm{i}\vec{q}\cdot\vec{r}_1} \, \mathrm{d}^d r_1 \int F_2(-\vec{r}_2) \mathrm{e}^{-\mathrm{i}\vec{q}\cdot\vec{r}_2} \, \mathrm{d}^d r_2 \\ &\equiv f_1(\vec{q}) f_2(\vec{q}) \end{aligned} \tag{7.24}$$

where

$$f_1(\vec{q}) = \int F_1(\vec{r}) \mathrm{e}^{\mathrm{i}\vec{q}\cdot\vec{r}} \, \mathrm{d}^d r \tag{7.25}$$

and similarly for the function $f_2(\vec{q})$.

Because of this simplification of sums in the form of convolutions, the mathematics of the expansion is greatly simplified when the quantity under consideration is the Fourier transform of the generating function.

Diagrammatic rules

The generating function for unrestricted walks is as given by (2.13). The full generating function follows from the Fourier transform of (7.18), which is then multiplied by z^N and summed over N. The first order term in the expansion in powers of the function f, as illustrated in Figure 7.2, is given by

$$\begin{aligned} &\frac{1}{1 - z\chi(\vec{q})} \left(\sum_{\vec{Q}} F(\vec{q} - \vec{Q}) \frac{1}{1 - z\chi(\vec{Q})} \right) \frac{1}{1 - z\chi(\vec{q})} \\ &\equiv g_0(z; \vec{q}) \left(\sum_{\vec{Q}} F(\vec{q} - \vec{Q}) g_0(z; \vec{Q}) \right) g_0(z; \vec{q}) \end{aligned} \tag{7.26}$$

The function $g_0(z; \vec{q})$ is the "bare," or unrestricted generating function. Notice that the transformation from number of steps to fugacity as an independent variable

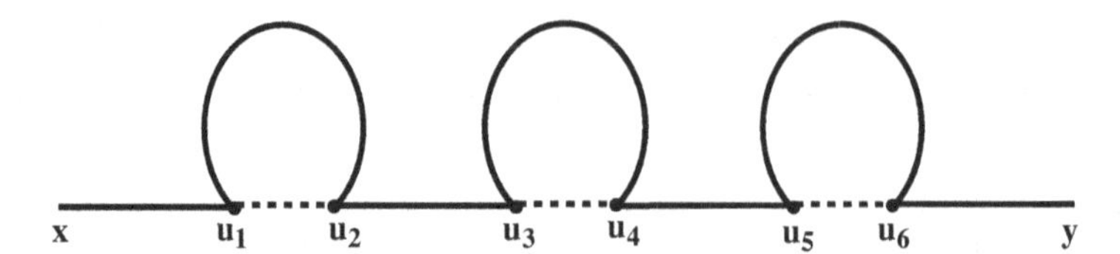

Fig. 7.4. The third order figure.

greatly simplifies the calculation of terms in the expansion, in that it is no longer necessary to perform convolutions in the number of steps. By the same token, the spatial Fourier transform allows for a simplification of important aspects of the calculation. Figure 7.3(a), for instance, corresponds to the following expression.

$$g_0(z;\vec{q})\left(\sum_{\vec{Q}} F(\vec{q}-\vec{Q})g_0(z;\vec{Q})\right) g_0(z;\vec{q})\left(\sum_{\vec{Q}} F(\vec{q}-\vec{Q})g_0(z;\vec{Q})\right) g_0(z;\vec{q})$$

$$= g_0(z;\vec{q})\left(\sum_{\vec{Q}} F(\vec{q}-\vec{Q})g_0(z;\vec{Q})g_0(z;\vec{q})\right)^2 \tag{7.27}$$

Exercise 7.3

Write down the explicit expressions for the contributions to the generating function corresponding to the second order graphs depicted in Figure 7.3.

This expression, and the first order correction in (7.26), look like terms in a geometric series. If we add them to the generating function of the unrestricted walk, we have the sum

$$\sum_{n=0}^{2} g_0(z;\vec{q})\left(\sum_{\vec{Q}} F(\vec{q}-\vec{Q})g_0(z;\vec{Q})g_0(z;\vec{q})\right)^n \tag{7.28}$$

Now, as the order of the expansion increases, additional terms appear that complete the sum above. For example, there is the third order contribution (shown in Figure 7.4) that corresponds to

$$g_0(z;\vec{q})\left(\sum_{\vec{Q}} F(\vec{q}-\vec{Q})g_0(z;\vec{Q})g_0(z;\vec{q})\right)^3 \tag{7.29}$$

Assuming, as must now seem to be eminently reasonable, that the expansion yields all the terms in the infinite geometric sum, we are left with a new, approximate result for $g(z;\vec{q})$

$$\begin{aligned}
g(z;\vec{q}) &\approx \sum_{n=0}^{\infty} g_0(z;\vec{q}) \left(\sum_{\vec{Q}} F(\vec{q}-\vec{Q}) g_0(z;\vec{Q}) g_0(z;\vec{q}) \right)^n \\
&= \frac{g_0(z;\vec{q})}{1 - \sum_{\vec{Q}} F(\vec{q}-\vec{Q}) g_0(z;\vec{Q}) g_0(z;\vec{q})} \\
&= \frac{1}{\left(g_0(z;\vec{q})\right)^{-1} - \sum_{\vec{Q}} F(\vec{q}-\vec{Q}) g_0(z;\vec{Q})} \\
&= \frac{1}{1 - z\chi(\vec{q}) - \sum_{\vec{Q}} F(\vec{q}-\vec{Q}) g_0(z;\vec{Q})} \\
&\equiv \frac{1}{1 - z\chi(\vec{q}) - \Sigma_1(z;\vec{q})}
\end{aligned} \tag{7.30}$$

The expression $\Sigma_1(z;\vec{q})$ is the first order term in the expansion in f of the *self-energy*. This function is also called the *mass operator* because of the role it plays in field-theoretical models of elementary particles. The sum we've performed is equivalent to solving the field-theoretical recursion relation known as *Dyson's equation*. This is all very nice, and there is much that can be done with a certain amount of cleverness (and hindsight). However, to make the most of the expansion to which we've just been introduced, it is useful to develop a set of rules that allow one in principle to set up the calculation of an arbitrary term in the expansion, and – as it turns out – to produce systematic groupings of those terms. This accounting procedure involves a diagrammatic representation of the terms, the diagrams being of the sort contained in Figures 7.2–7.4.

This diagrammatic representation of an expansion is standard in the physics of systems with a large number of degrees of freedom. In the case of quantum field theories the diagrams are known generically as Feynman diagrams. When the focus is a classical fluid or gas, the names associated with the expansion are Ursell and Mayer. The advantage of diagrammatic expansions is that a pictorial representation of various terms elucidates the underlying processes and lends itself quite naturally to various useful groupings and redefinitions. Explicit demonstrations of the utility of the diagrammatic expansion will be found in this and subsequent chapters.

The development of the rules to be listed below involves a relatively lengthy and tedious set of steps. Although there are rules that apply to the expansion as carried out above, the most straightforward set, and the one giving rise to terms that can be most easily evaluated, applies to an expansion for $g(z;\vec{q})$, the spatial Fourier transform of the random-walk-generating function. The rules are as follows.

1. The diagrams consist of lines and vertices. There are two kinds of lines. One is a *propagator line*, the other is an *interaction line*. The propagator lines, connected end-to-end at the vertices form a single segmented representation of the random walk.
2. All lines in the expansion of the Fourier-transformed generating function carry a wave-vector. Because of the connection between the diagrammatic expansion generated here and the expansion utilized in quantum field theory, the phrase "wave-vector" is replaced by the term "momentum." The "direction" of the momentum is indicated by an arrowhead on the line. The arrowhead on the propagator lines point from the starting point of the random walk to its ultimate destination.
3. A propagator line carrying a momentum $\vec{q}$ stands for the expression $g_0(z;\vec{q}) = 1/(1 - z\chi(\vec{q}))$. The interaction line carrying a momentum $\vec{Q}$ stands for the expression $-F(\vec{Q})$. The net expression is the product of all the above factors. Lines meet at vertices. Each vertex enforces "conservation of momentum," in that the total momentum carried into the vertex by the lines that meet there is equal to the total momentum carried out of that vertex. A line with momentum $\vec{q}$ whose arrowhead points towards the vertex carries a momentum $\vec{q}$ into the vertex. If the arrowhead points away from the vertex then the line carries a momentum equal to $\vec{q}$ out of the vertex.
   In this set of diagrams, the vertex is the meeting point of two propagator lines and one interaction line. A vertex such as this, at which three lines meet is called a *three-point* vertex.
4. The "external" momentum of the diagram is carried by the propagator line that emanates from the starting point. That momentum is held fixed.
5. With the exception of the external momentum, the ultimate expression is obtained by integrating over all momenta in the various factors, subject, of course, to the constraints imposed as the result of momentum conservation at the vertices.

This dizzying collection of rules merits a closer going over. Let's review them in a little more detail.

Rule 1

The propagator line is represented as a solid line. The interaction line is drawn dotted. The vertex is heavy dot. These elements are shown in detail in Figure 7.5(a).

Rule 2

Figure 7.5(b) shows the lines displayed in Figure 7.5(a) with arrowheads now attached.

Rules 3 and 4

The expression corresponding to the set of lines and vertices in Figure 7.6 is $g_0(z;\vec{q}_1)g_0(z;\vec{q}_2)(-F(\vec{q}_3))g_0(z;\vec{q}_4)$. Momentum conservation has not yet been invoked, and no integrations are as yet specified. Figure 7.7 illustrates the meeting of lines at a vertex. Momentum conservation is made explicit.

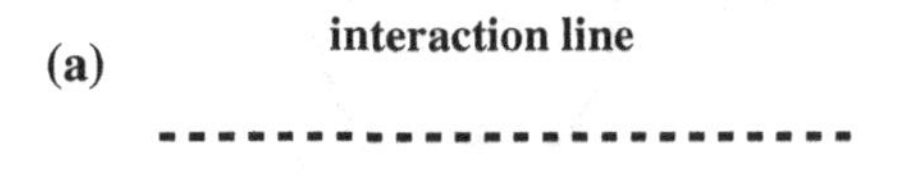

Fig. 7.5. Elements of the diagrammatic approach.

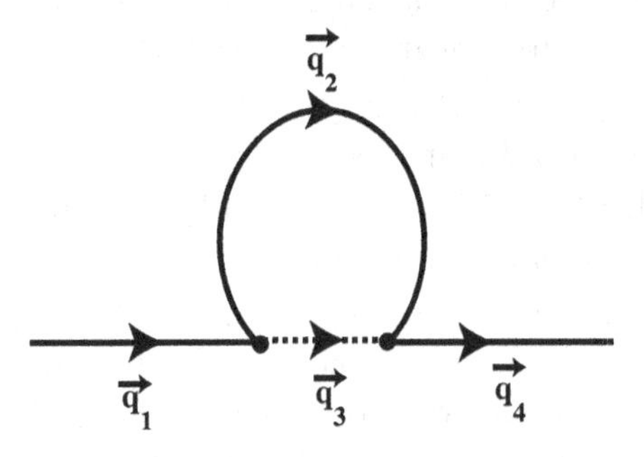

Fig. 7.6. The one loop-correction to the propagator line.

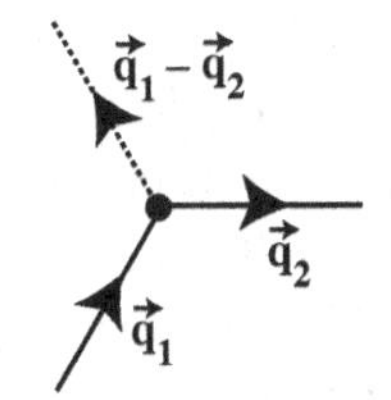

Fig. 7.7. The vertex.

Fig. 7.8. The one-loop self-energy.

Rules 5 and 6

The full expression corresponding to the diagram in Figure 7.6 is

$$g_0(z;\vec{q})\left(-\sum_{\vec{Q}} F(\vec{q}-\vec{Q})g_0(z;\vec{Q})\right)g_0(z;\vec{q})$$

Sums convert to integrals in infinite systems, as indicated in Supplement 2 of Chapter 2.

7.2.4 Dyson's equation

Given the rules above we are now in a position to take a closer look at the expansion in powers of self-avoidance. The fundamental structure of a term in the expansion is as follows. There is a central line that carries the external momentum, $\vec{q}$. This line reappears as segments joining self-energy terms. In the case of the sum of terms that produced the result in (7.30), the self-energy has the diagrammatic form shown in Figure 7.8. Note that these have the form of "amputated" diagrams, in that no propagator lines lead into or out of them.

The mathematical structure of a self-energy is as follows. There will be an even number of vertices, say $2m$. There will be m internal interaction lines and $2m-1$ propagator lines. With the exception of two of them, each vertex will have two propagator lines and one interaction line attached to it. The two exceptional vertices will have an interaction line and only *one* internal propagator line attached. The diagram will be "one particle irreducible" (1PI) in that it will not be possible to split the diagram that represents it into two by eliminating one propagator line. The last phrase is a residue of the quantum field theoretical development of diagrammatic techniques. Because of momentum conservation there are m integrations. Figure 7.9 contains an example of a self-energy diagram. The actual expression is

$$\int \frac{d^d q_1}{(2\pi)^d}\int \frac{d^d q_2}{(2\pi)^d} F(\vec{q}_1)g_0(z;\vec{q}_2)g_0(z;\vec{q}-\vec{q}_1)g_0(z;\vec{q}_2) \\ \times g_0(z;\vec{q}_1-\vec{q}_2)F(\vec{q}-\vec{q}_1-\vec{q}_2) \tag{7.31}$$

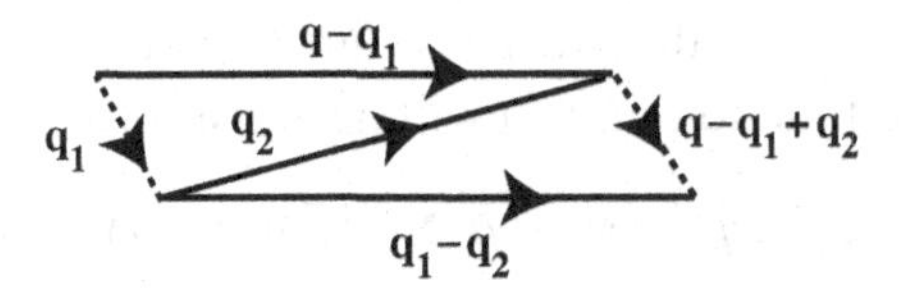

Fig. 7.9. The more general self-energy. The q_i's label the wave-vectors carried by each line.

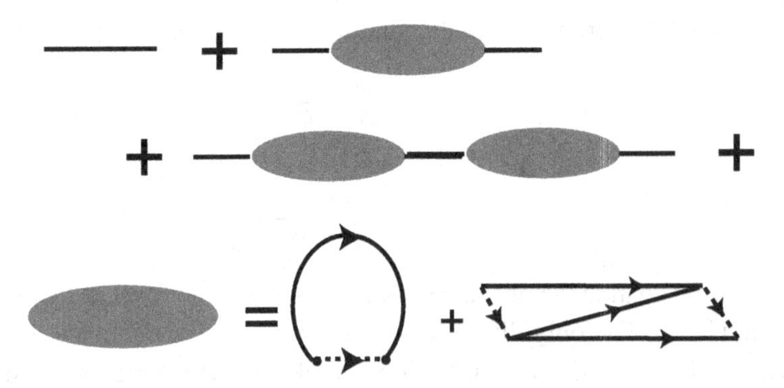

Fig. 7.10. The summation leading to Dyson's equation.

These contributions enter into the ultimate expression for the generating function according to the following diagrammatically based argument. Consider the following set of contributions to the generating function

$$g_0(z;\vec{q}) + g_0(z;\vec{q})\Sigma_1 g_0(z;\vec{q}) + g_0(z;\vec{q})\Sigma_1 g_0(z;\vec{q})\Sigma_1 g_0(z;\vec{q}) + \cdots \tag{7.32}$$

where $g_0(z;\vec{q}) = 1/(1 - z\chi(\vec{q}))$ and $\Sigma_1 = \sum_{\vec{Q}} F(\vec{q} - \vec{Q})g_0(z;\vec{Q})$, and the series was shown to sum to $1/(g_0^{-1}(z;\vec{q}) - \Sigma_1)$.

This argument can be generalized. Consider the contribution to $g(z;\vec{q})$ displayed in Figure 7.10 Mathematically, this contribution is

$$= g_0 + g_0\,[\Sigma_1 + \Sigma_2]\,g_0 + g_0\,[\Sigma_1 + \Sigma_2]^2\,g_0 + \cdots \tag{7.33}$$

with Σ_1 as given above and

$$\Sigma_2 = \sum_{\vec{Q}_1,\vec{Q}_2} F(\vec{q} - \vec{Q}_1)g_0(z;\vec{Q}_1)F(\vec{Q}_1 - \vec{Q}_2)g_0(z;\vec{Q}_2)g_0(z;\vec{Q}_1) \tag{7.34}$$

Again, the geometrical series is readily summed to yield

$$\frac{g_0(z;\vec{q})}{1 + (\Sigma_1 + \Sigma_2)\,g_0(z;\vec{q})} = \frac{1}{g_0^{-1} - (\Sigma_1 + \Sigma_2)} \tag{7.35}$$

The above generalizes even further. The ultimate perturbation theoretical expression for the generating function $g(z;\vec{q})$ will have the form

$$g(z;\vec{q}) = g_0(z;\vec{q}) + g_0(z;\vec{q})\Sigma(z;\vec{q})g_0(z;\vec{q}) + g_0(z;\vec{q})\Sigma(z;\vec{q})g_0(z;\vec{q})\Sigma(z;\vec{q})g_0(z;\vec{q}) + \cdots \tag{7.36}$$

where $\Sigma(z;\vec{q})$ is the sum of all *mass operator insertions*. Thus, a given contribution to $g(z;\vec{q})$ will consist of a string of self-energy diagrams connected by "bare" propagator lines. The series is formally summed, and the resulting solution has the form

$$g(z;\vec{q}) = \frac{g_0(z;\vec{q})}{1 - \Sigma(z;\vec{q})g_0(z;\vec{q})} = \frac{1}{g_0^{-1}(z;\vec{q}) - \Sigma(z;\vec{q})} \tag{7.37}$$

Notice that the left hand side of (7.37) has exactly the same form as the final result displayed in (7.30). The only difference is that in the latter, complete version, the self-energy is the sum of an infinite number of terms in the expansion in self-avoidance. The problem of determining the generating function reduces to the task of evaluating the members of the series that is summed to obtain the self-energy, and then of summing the series for that quantity.

At first sight the task may seem hopeless. In fact, it is. There is no known method of exactly performing the indicated sum. Not only is there an infinite number of terms, but the multiple integrals become increasingly complicated at higher and higher order. This accumulation of difficulties characterizes essentially all expansions of the type discussed here. Fortunately, a number of techniques now exists for the extraction of the key features of the quantity of interest. Some are relatively simple to implement, relying on clever, intuitively based, heuristic arguments. Others are more technically formidable, and involve substantial calculational effort. In the sections and chapters to follow, we will make use of some of the methods that have proven particularly successful when applied to the problem of self-avoiding random walks.

Before doing so, we will perform a quick but revealing analysis of the dependence on fugacity of contributions to the self-energy. This analysis tells us something very important: *that no low order truncation of the expansion will suffice to reveal the asymptotic statistics of a long self-avoiding walk.*

7.2.5 Breakdown of perturbation theory

Although the multiple integrals corresponding to a given high order contribution to the self-energy will, in all probability, defy attempts to produce a closed form expression, important aspects of the result of the integration can be inferred with the use of simple tricks. Here we will use one such stratagem to derive the dependence on the difference $z_c - z$ of a term in the expansion of $\Sigma(z; \vec{q})$ in powers of the factor f. Suppose the power is l. There will be $2l$ vertices in the diagram corresponding to the term and $2l - 1$ propagator lines. Each propagator line contributes a factor to the integrand of the form $1/(1 - z\chi(\vec{q}_i)) \sim 1/((z - z_c) + q_i^2)$. We will assume that the momentum dependence of the functions $f(\vec{q}_j)$ is unimportant, or uninteresting. In fact, it is generally a very good approximation to physical reality to assume that $f(\vec{q}_j)$ is independent of $\vec{q}_j$. Because there are l integrations over the $\vec{q}_i$'s, the integral will, in very general terms, look like

$$\int f^l \left(\frac{1}{(z_c - z) + q_i^2} \right)^{2l-1} \left(\mathrm{d}^d q_i \right)^l \tag{7.38}$$

We extract the effect of the combination $z_c - z$ on the integral by replacing the q_i's in the integral by $(z_c - z)^{1/2} q$. This has the effect of replacing the random walk propagators by $1/(1 + q_i^2)$ and generating the overall factor

$$(z_c - z)^{-(2l-1)+dl/2} = (z_c - z)^{1-l(4-d)/2} \tag{7.39}$$

All dependence on $z_c - z$ has been removed from the integrand with one exception. Because the external momentum $\vec{q}$, is not rescaled or integrated over, the integrand will depend on the rescaled external momentum $\vec{q}(z_c - z)^{-1/2}$. Aside from this, the entire effect of the combination $(z_c - z)$ can be extracted from the factor in (7.39). The overall task of performing the multiple integral has not, of course, been simplified in any real way. However, the general form of a term in the expansion for $\Sigma(z; \vec{q})$ is now apparent. We expect that the mass operator is representable as the following kind of sum

$$\Sigma(z; \vec{q}) = (z_c - z) \sum_n \left((z_c - z)^{-(4-d)/2} f \right)^n \mathcal{X}_n \left(\vec{q}(z_c - z)^{-1/2} \right) \tag{7.40}$$

The quantity f in (7.40) is the momentum-independent amplitude of the term enforcing self-avoidance.

Now, when f is small, the penalty for self-intersection is mild. The probability that a walker will disappear because it has trespassed on its own path is small. It is reasonable to expect that the modification of the self-energy resulting from self-avoidance will be correspondingly unimportant. However, high order terms in the sum in (7.40), as reduced in amplitude as they are because of the factor equal to a high power of the small amplitude f, also contain the difference $z_c - z$ raised to a large *negative* power. It is this latter term in the high order correction to the self-energy that ultimately controls the effect of the correction on the generating function. In order to see how perturbation theory breaks down, let's reconstitute from the generating function the expression for the total number of N-step walks that start out at $\vec{x}$ and end up anywhere. This is just

$$\int C(N; \vec{x}, \vec{y})\, \mathrm{d}y \tag{7.41}$$

The above is equal to the coefficient of z^N in the corresponding integral of the generating function $G(z; \vec{x}, \vec{y})$. However, given the definition of the spatial Fourier transform, it is clear that the integral gives rise to $g(z; \vec{q} = 0)$. Thus, the answer we seek is the coefficient of z^N in

$$\begin{aligned} g(z;0) &= \frac{1}{(g_0(z;0))^{-1} - \Sigma(z_c - z; 0)} \\ &\to \frac{1}{(z_c - z)\sum_n \mathcal{X}_n(0)\left((z_c - z)^{-(4-d)/2} f\right)^n} \end{aligned} \tag{7.42}$$

Now, as demonstrated in the third supplement to Chapter 2, the coefficient of z^N when N is large is determined by the location and order of the singularities in the complex z-plane. In particular, the leading contribution to the coefficient of z^N is controlled by the singularity of the form $(z_s - z)^{-p}$ for which z_s lies closest to the origin. If there is more than one such singularity, then the one with the largest value of the exponent p contributes the greatest part of the coefficient of interest. The expression (7.42) exhibits singular behavior at $z = z_c$. This is the dominating point of non-analyticity in the complex z-plane, in that it is the singularity that is closest to the origin. The difficulty lies in the nature of the singularity. The power p is equal to $1 - n_h(4 - d)/2$, where n_h is the order beyond which no terms in the perturbation series are retained.[1] The higher the order at which the series is truncated, the lower the value of the exponent. If all powers are retained, a naive interpretation of the consequences of the above argument is that the exponent is lower than any given value, or $-\infty$. However, the sum of the terms may easily lead to a modification of the analytic structure that is entirely different from this guess and from the result that follows from any truncation of the perturbation series. In all events, it is clear that there is no reason to trust the results of finite order perturbation theory.

7.2.6 Mean field theory

There are a number of approximate solutions to the problem of summing the perturbation theoretical series for the generating function. One of the most venerable

[1] This result tells us that $d = 4$ dimensions is special in some sense. In fact, we will see that the perturbation expansion derived in this chapter does, indeed, make perfect sense as an expansion in a small quantity in more than four dimensions. In fewer dimensions we can "rescue" perturbation theory by constructing an expansion in the difference between four and the dimensionality of interest to us.

is called the *mean field* solution. This nomenclature derives from the genesis of the technique in the context of the behavior of magnetic systems and, in particular, an approximation to the temperature dependence of the magnetism of a set of interacting moments that is based on the notion of an effective average (or mean) magnetic field. Here, the mean field solution follows from an approximation to the bare generating function based on an expansion of $G(z;\vec{x},\vec{y})$ in a complete set of states.

Recall that when translational invariance holds, the propagator can be expanded in plane waves. That is, one can write

$$G(z;\vec{x},\vec{y}) = \sum_{\vec{q}} \frac{e^{i\vec{q}\cdot\vec{x}}}{\sqrt{V}} g_0(z;\vec{q}) \frac{e^{-i\vec{q}\cdot\vec{y}}}{\sqrt{V}} \tag{7.43a}$$

$$\equiv \sum_{\vec{q}} \psi_{\vec{q}}(\vec{x}) g_0(z;\vec{q}) \psi^*_{\vec{q}}(\vec{y}) \tag{7.43b}$$

$$\approx \sum_{\vec{q}} \psi_{\vec{q}}(\vec{x}) \frac{z_c}{z_c - z + z_c z a^2 q^2} \psi^*_{\vec{q}}(\vec{y}) \tag{7.43c}$$

$$\equiv \sum_{\vec{q}} \psi_{\vec{q}}(\vec{x}) \frac{z_c}{z_{\vec{q}} - z} \psi^*_{\vec{q}}(\vec{y}) \tag{7.43d}$$

The quantity $z_{\vec{q}}$ in the last line of (7.43d) is equal to $z_c + z_c z a^2 q^2$. For our purposes we can write

$$z_{\vec{q}} = z_c + z_c^2 a^2 q^2 \tag{7.44}$$

(7.43c) holds to the extent that the sum is dominated by the terms for which $\vec{q}$ is small. Now, (7.43d) is a generally proper representation of the unrestricted walk generating function, even when the environment in which the walk takes place is not invariant with respect to spatial translations. The more general form of $G(z;\vec{x},\vec{y})$ is

$$G(z;\vec{x},\vec{y}) = \sum_{i} \psi_i(\vec{x}) \frac{z_c}{z_i - z} \psi^*_i(\vec{y}). \tag{7.45}$$

The functions ψ, ψ^* form a complete orthonormal set, in that

$$\int \psi_i(\vec{x}) \psi^*_j(\vec{x})\, d^d x = \delta_{i,j} \tag{7.46}$$

The expansion above follows from the mathematical equivalence between the generating function and a Green's function. In the continuum limit the quantity

$G(z;\vec{x},\vec{y})$ satisfies the partial differential equation

$$za^2\nabla^2 G(z;\vec{x},\vec{y}) - (z/z_c - 1)G(z;\vec{x},\vec{y}) = a^d\delta(\vec{x}-\vec{y}) \tag{7.47}$$

To derive (7.43d) one expresses the Dirac delta function in terms of the complete set of eigenfunctions of the operator ∇^2. These functions satisfy $\nabla^2\psi_i = -\lambda_i\psi_i$. In particular

$$\psi_i(\vec{r}) = e^{-\vec{q}\cdot\vec{r}} \tag{7.48}$$

and

$$\lambda_i = |\vec{q}_i|^2 \tag{7.49}$$

In terms of these eigenfunctions

$$\begin{aligned} G(z;\vec{x},\vec{y}) &= z_c a^d \sum_i \frac{\psi_i(\vec{x})\psi_i(\vec{y})}{(z_c - z) + z_c z a^2 q^2} \\ &= z_c a^d \sum_i \frac{\psi_i(\vec{x})\psi_i(\vec{y})}{z_i - z} \end{aligned} \tag{7.50}$$

where $z_i = z_c + z_c z a^2 q^2$.

The expansion in (7.45) is entirely equivalent to (7.43dd) if one makes the connections $z_i \leftrightarrow z_{\vec{q}} = z_c(1 + z_c a^2 q^2)$, $\psi_i(\vec{x}) \leftrightarrow \psi_{\vec{q}}(\vec{x}) = e^{i\vec{q}\cdot\vec{x}}/\sqrt{V}$ and similarly for ψ_i^*. Here V is the volume of the system.

Now, imagine that the sum in (7.45) is dominated by a single term. That is, imagine that we can approximate $G_0(z;\vec{x},\vec{y})$ as follows

$$\begin{aligned} G(z;\vec{x},\vec{y}) &\approx G_0(z;\vec{x},\vec{y}) \\ &= z_c \frac{\psi_0(\vec{x})\psi_0^*(\vec{y})}{z_0 - z} \end{aligned} \tag{7.51}$$

If the walk takes place in a large system that has no walls, or for which periodic boundary conditions hold, then an appropriate choice for the dominant mode ψ_0 is the plane wave with infinite wavelength. In this case $\psi_0(\vec{x}) = 1/\sqrt{V}$, and $z_0 = z_c$. The other plane wave modes, neglected in this approximation, represent, in the terminology that has developed around this approach, the effects of fluctuations about the mean field solution.

The results that follow from the approximation we have just made will yield useful insights into the consequences of self-avoidance. In particular, they will provide motivation for a powerful set of heuristic arguments first put forth by Flory (1953; 1969), leading to some very nice predictions for the behavior of polymers, where the effects of excluded volume due to the strong repulsive interactions between monomeric units comprising the polymer are modeled by self-avoiding walks.

Because of the simple form of the approximate expression for the propagator in (7.51) the perturbation series is most easily analyzed in real space.

Now, the insertion of a single interaction line, connected to the propagator line at two vertices is accomplished by acting on the bare propagator line with the following operator:

$$O(f;z) = -\int \left(\psi(\vec{w}_1)\psi^*(\vec{w}_1)\right)^2 \mathrm{d}^d w_1 \frac{f}{2}\frac{\mathrm{d}^2}{\mathrm{d}z^2} \tag{7.52}$$

The result is the following first order correction to the approximate unrestricted walk generating function

$$-\int \mathrm{d}^d w_1 \psi(\vec{x})\frac{z_\mathrm{c}}{z_0 - z}\psi^*(\vec{w}_1)\psi(\vec{w}_1)\frac{z_\mathrm{c}}{z_0 - z}\psi^*(\vec{w}_1)\psi(\vec{w}_1)\frac{z_\mathrm{c}}{z_0 - z}\psi^*(\vec{y})f \tag{7.53}$$

This expression has the same general form as the first order correction to the unrestricted random-walk-generating function exhibited in (7.26), with the exception that here the unrestricted random walk propagator has a specific, approximate form. Note that in this case we assume that the factor f is independent of momentum.

Further terms in the expansion in f of the generating function are generated with additional applications of the operator in (7.52). To eliminate overcounting, it is necessary to multiply the operator applied n times by the factor $1/n!$. All terms in the expansion of the propagator of a self-avoiding walk in terms of the self-avoidance factor f arise if one takes the following sum

$$\sum_n \frac{O(f;z)^n}{n!}\frac{\psi(\vec{x})\psi^*(\vec{y})}{z_0 - z}z_\mathrm{c} = \mathrm{e}^{O(f;z)}\frac{\psi(\vec{x})\psi^*(\vec{y})}{z_0 - z}z_\mathrm{c} \tag{7.54}$$

A few manipulations suffice to derive a complete expression for the full mean field self-avoiding walk propagator. First we rewrite $1/(z_0 - z)$ as

$$\int_0^\infty \mathrm{d}t\, \mathrm{e}^{-t(z_0 - z)/z_\mathrm{c}} \tag{7.55}$$

Then, we note that $\mathrm{d}^2/\mathrm{d}z^2 \mathrm{e}^{tz} = t^2 \mathrm{e}^{tz}$. This means that the right hand side of (7.54) is equal to

$$\psi(\vec{x})\psi^*(\vec{y})\int_0^\infty \exp\left[-ft^2/2\int \mathrm{d}^d w \left(\psi(\vec{w})\psi^*(\vec{w})\right)^2 - t\frac{(z_0 - z)}{z_\mathrm{c}}\right]\mathrm{d}t \tag{7.56}$$

The final steps leading to the mean field generating function follow from the identification of ψ_0 with $\sqrt{a^d/V} = 1/\sqrt{N}$, N being the number of lattice sites available to the walker and z_0 with z_c. This, we can write

$$
\begin{aligned}
G_0 &= G_{\mathrm{MF}}(z; \vec{x}, \vec{y}) \\
&= \frac{1}{N} \int_0^\infty \exp\left[-\frac{f t^2}{N} - t\,(1 - z/z_c) \right] \mathrm{d}t \\
&= \int_0^\infty \exp\left[N \left(-f\lambda^2 - \lambda\,(1 - z/z_c) \right) \right] \mathrm{d}\lambda
\end{aligned} \tag{7.57}
$$

The integral expression for G_{MF} is actually valid for all z, though our derivation assumes $z < z_c$. In evaluating the integral some care must be taken. The cases $z < z_c$ and $z > z_c$ must be treated separately. The relevant case for us is $z > z_c$. For this case, the expansion in z is obtained in a straightforward way by expanding the exponential $\mathrm{e}^{-Nz\lambda/z_c}$.

$$
\begin{aligned}
G_{\mathrm{MF}}(z; \vec{x}, \vec{y}) &= \sum_{m=0}^{\infty} z^m \frac{N^m}{m!} \int_0^\infty \mathrm{e}^{-Nf\lambda^2 - N\lambda} \lambda^n \, \mathrm{d}\lambda \\
&= \sum_m C(m; \vec{x}, \vec{y}) \left(\frac{z}{z_c} \right)^m
\end{aligned} \tag{7.58}
$$

We immediately find that the number of m-step *non-intersecting* walks that start out at $\vec{x}$ and end at $\vec{y}$ is, in the mean field approximation, given by

$$
C_{\mathrm{MF}}(m; \vec{x}, \vec{y}) = \frac{1}{z_c^m} \frac{N^m}{m!} \int_0^\infty \mathrm{e}^{-N(f\lambda^2 + N\lambda) + m \ln \lambda} \, \mathrm{d}\lambda \tag{7.59}
$$

When m is large, the leading contribution to the integral is

$$
C_{\mathrm{MF}}(m; \vec{x}, \vec{y}) = \frac{1}{N} z_c^{-m} \mathrm{e}^{-fm^2/N} \tag{7.60}
$$

Exercise 7.4

Show that for large N, (7.60) always follows from (7.59).

Exercise 7.5

Using (7.57) for the mean field generating function, show that the number of m-step walks starting at x and ending at y for the case $z < z_c$ is just z_c^{-m}/N, and hence self-avoidance has no effect. The interpretation of this result is fully explained in Chapter 10.

If the linear extent of the region to which the walker is confined is R, so that the volume is proportional to R^d (i.e. $N = \alpha R^d$), we can write

$$\begin{aligned} C_{\mathrm{MF}}(N; \vec{x}, \vec{y}) &= \frac{z_c^{-m}}{N} \mathrm{e}^{-f m^2/\alpha R^d} \\ &= \frac{(2d)^m}{N} \mathrm{e}^{-u m^2/R^d} \end{aligned} \tag{7.61}$$

The last equality in (7.61) follows from the definition of the quantity $u = f/\alpha$ and the replacement of z_c by its value for the case of a simple cubic lattice.

7.2.7 Flory's argument

The final result above is very revealing. Since N is just the total number of sites available to the walker, (7.61) tells us how the number of m-step unrestricted walks with beginning and end-points fixed, $(2d)^m/N$, is modified as a result of self-avoidance.

A bit of interpretation is in order at this point. The quantity $(2d)^m/N$ is the total number of paths available to an unrestricted walker divided by the number of available sites in the volume. Now, if the natural linear dimension of the walk were small enough so that it fits into the volume V comfortably, then the number of unrestricted walks between two points in the volume would be given by a Gaussian formula. In the case at hand, we have a path sum that is independent of the starting and ending point, so that requiring it to end at a given location simply reduces the total number of paths by the ratio of the number of sites at the given location (one) to the total number of sites at which the walk can terminate. This result follows from the presumption that the walker would naturally fill a much larger volume than the one available to it. By our choice of ψ_0 we assume that the boundary conditions are periodic, so the walker stays in the volume by reappearing at the far side whenever it executes an exit. If the walk is a very long one, then the probability that it ends up somewhere in the volume is independent of both its point of origin and its final resting place. This is because after many exits and re-entries it has lost memory of where it started. Therefore, our result is, strictly speaking, relevant only to highly "compressed" walks.

The Boltzmann-like weighting factor incorporates the *average* effects of the interactions that enforce self-avoidance. Indeed, such a factor plays a key role in arguments put forth by Flory for the dependence on the number of units of the spatial extent of a randomly coiled linear polymer. According to these arguments, this factor favors an extended polymer, in that it increases as the linear extent of the polymer coil, R, grows. On the other hand, an "entropic" factor, based on the expression for $C(m; \vec{x}, \vec{y})$ for the unrestricted walk favors a more tightly coiled polymer. This heuristic expression is

$$e^{-R^2/m} \tag{7.62}$$

Combining (7.62) with (7.61) one obtains as the weight for an N-step self-avoiding walk of net extension R

$$e^{-R^2/m-um^2/R^d} \tag{7.63}$$

This weight is maximized when the exponent is at a maximum, which occurs when

$$R \propto m^{3/(d+2)} \tag{7.64}$$

As the result of a patching together of mean field results for the interaction between monomers insuring self-avoidance and expressions applying to unrestricted walks, Flory was able to derive a simple, but extraordinarily accurate, scaling relationship between the size of a polymer and the number of monomeric units of which it is comprised (Flory, 1969).

8

Properties of the random walk: introduction to scaling

We now have had an introduction to the random walk, and there has been a discussion of some of the most useful methods that can be utilized in the analysis of the process. The unifying theme of this chapter is the introduction of scaling arguments to provide insights into the behavior of the walker under a variety of circumstances. First, we will address in more depth the notion of universality of ordinary random walk statistics. Then, we will discuss the (mathematical) sources of non-Gaussian statistics. Finally, we will develop a few simple but central scaling results by looking once more at the path integral formulation of the random walk. In particular, using simple scaling arguments, we will be able to provide heuristic arguments leading to Flory's formula for the influence of self-avoidance on the spatial extent of a random walker's path.

8.1 Universality

Notions of universality play an important role in discussions of the statistics of the random walk. We have already seen a version of universality in Chapter 2, in which it was shown that the statistics of a long walk are Gaussian, regardless of details of the rules governing the walk, as long as the individual steps taken by the walker do not carry it too far. In the remainder of this section, the idea of universality will be developed in a way that will allow us to apply it to the random walk when conditions leading to Gaussian behavior do not apply.

As it turns out, universality in random walk statistics has a close mathematical and conceptual connection to fundamental properties of a system undergoing a special type of phase transition. The justification of universality in a thermodynamic setting rests on one of the most important theoretical developments in late twentieth-century physics: the renormalization group. The recognition of the connection between random walk statistics and the statistical mechanics of continuous phase transitions has allowed the very powerful calculational machinery of the renormalization to be applied to random walks, especially self-avoiding walks. In this chapter and in chapters to follow, we will explore the way in which universality arises, how one understands its basis, and how it is possible to calculate properties arising from its influence on random walk statistics.

8.1.1 Non-Gaussian behavior

Recall the discussion in Chapter 2. In Section 2.1.4, we found that the leading order contribution to the distribution of paths, in terms of end-to-end distance, is Gaussian, when the paths are left by unrestricted walkers that have taken a sufficiently large number of steps. This follows from the fact that the asymptotic distribution is independent of higher order moments of the probability that individual steps are of a given length. The principal determinant of the properties of the end-to-end distribution is the second moment of the step length probability distribution, $p(|\vec{r}|)$. Higher order moments of this probability are not relevant to the asymptotic statistics of the long random walk.

However, it may not be possible to carry out the above moment expansion to higher than quadratic order. This is because integrals of the form $\int |\vec{r}|^n p(|\vec{r}|)\, d\vec{r}$ may or may not converge at the value of n in which we are interested. If the integrals do begin diverging at some value of n, then our analysis of the asymptotic behavior of $C(N; \vec{x} - \vec{y})$ can become more complicated. For example, if the integral above does not have a finite value when $n = 4$, the method we used to investigate the effect on the asymptotic form of $C(N; \vec{x} - \vec{y})$ of the quartic term in the expansion of $\chi(\vec{q})$ may not be useful, simply because there is no quartic term to consider.

Matters can be even more complicated. The moment expansion can fail at *quadratic* order. This will occur if the walker takes arbitrarily long steps with a probability that decays sufficiently slowly with the length of the step. Such a case was discussed as a worked-out example in Chapter 2. Suppose, for example, that

$$p(|\vec{r}|) \propto |\vec{r}|^{-\alpha} \tag{8.1}$$

And that the exponent α is less than 5. Then, the integral

$$\int p(|\vec{r}|)\, |\vec{r}|^2\, d\vec{r} \propto \int |\vec{r}|^{4-\alpha}\, d|\vec{r}| \tag{8.2}$$

does not converge, and we are no longer able to expand $\chi(\vec{q})$ as we did. In this case, the dependence of the function $\chi(\vec{q})$ on its argument is singular. For example, if

$$p(|\vec{r}|) \propto |\vec{r}|^{-5+\delta} \tag{8.3}$$

then, for small $|\vec{q}|$,

$$\chi(\vec{q}) \propto A - B|\vec{q}|^{2-\delta} \tag{8.4}$$

It is possible to perform the inverse Fourier transform of $\chi(\vec{q})^N$, and, thus find $C(N;\vec{x},\vec{y})$. The steps are similar to those taken in (2.17)–(2.19). Now, the integral needed, $C(N;\vec{x}-\vec{y})$, is easily demonstrated to behave asymptotically as

$$\int e^{i\vec{q}\cdot(\vec{x}-\vec{y})} A^N e^{-N(B/A)|\vec{q}|^{2-\delta}} \, d\vec{q} \propto |\vec{x}-\vec{y}|^{-5+\delta} \tag{8.5}$$

The Gaussian form no longer holds. We will have a good deal to say about the different forms that functions such as $C(N;\vec{x}-\vec{y})$ take on as the rules governing the random walker are changed.

The slightly frustrated reader will have noticed that we have not indicated how the results on the right hand sides of (8.4) and (8.5) were obtained. The general strategy for performing the sorts of integrals that lead to the results above will be outlined in the next section. By way of introduction to the overall approach, we note that one way to find the r-dependence of the integral

$$\int e^{ikr} k^y \, dk \tag{8.6}$$

is to replace the integration variable k by $w = k/r$. One is then left with the expression

$$r^{-(y+1)} \int e^{iw} w^y \, dw \tag{8.7}$$

We now replace the integral in (8.7) by a constant. Note that the limits of integration have been suppressed in both (8.6) and (8.7). The contributions associated with the limits of integration will undoubtedly have some dependence on r. However, we assume that this does not alter our conclusions. If this is, indeed, the case, the dominant dependence of (8.7) on r is contained in the factor $r^{-(y+1)}$.

Exercise 8.1

Using the strategy described above for working out the r-dependence from the Fourier transform, verify that (8.5) follows from (8.4).

8.1.2 Analytic structure, scaling, and the role of spatial dimensionality

In this section we learn a little more about the dependence of the generating function on the variable z. As we will soon see, the number of dimensions in which the random walk takes place has an important effect on the way in which it depends on

z, and, most decisively, on the way it behaves when z approaches a critical value, z_c, which turns out to be the coordination number of the lattice. This, in turn, exerts a profound effect on the statistics of long random walks.

How does the generating function depend on the dimensionality of the space explored by the walker? We start with the expression for the function $g(z;\vec{q})$

$$g(z;\vec{q}) = \frac{1}{1 - z\chi(\vec{q})}$$

where

$$\chi(\vec{q}) = a - b|\vec{q}|^2 + O\left(|\vec{q}|^4\right)$$

If we make use of the above form for $\chi(\vec{q})$, truncating beyond quadratic order in $\vec{q}$, we have

$$g(z;\vec{q}) \approx \frac{1}{1 - az + zb|\vec{q}|^2} \tag{8.8a}$$

$$= \frac{1}{a} \frac{1}{z_c - z + zb/a|\vec{q}|^2} \tag{8.8b}$$

$$\approx \frac{1}{a} \frac{1}{z_c - z + b/a^2|\vec{q}|^2} \tag{8.8c}$$

(8.8c) holds when $z \approx z_c = 1/a = 1/2d$, if the lattice is cubic. The quantity z_c can be termed the *critical fugacity*. The approximate equality above is sufficiently accurate for our purposes in the regimes of interest to us – in particular, when the walks consist of a large number of steps.

The result above tells us that the generating function $g(z;\vec{q})$ has a *scaling form*. That is, we can write

$$g(z;\vec{q}) = \frac{1}{a} \frac{1}{z_c - z + b/a^2|\vec{q}|^2}$$

$$= \frac{1}{a} \frac{1}{|\vec{q}|^2} \frac{1}{b/a^2 + (z_c - z)|\vec{q}|^{-2}}$$

$$\equiv |\vec{q}|^{-2} f\left(\vec{q}(z_c - z)^{-1/2}\right) \tag{8.9}$$

A generalization of the above scaling form is

$$g(z;\vec{q}) = |\vec{q}|^{-2+\eta} f\left(\vec{q}(z_c - z)^{-\nu}\right) \tag{8.10}$$

The introduction of the exponents ν and η, which, in the case at hand, are equal to $1/2$ and 0, respectively is not frivolous. As we will soon see, restrictions on the random walk, such as a prohibition against intersections in the trail left by the walker – the constraint that defines a self-avoiding walk – can have a substantial effect on the generating function, which is, in part, summarized by the generalization of the scaling form displayed in (8.10). The consequences of the scaling form above will be discussed at length in a subsequent chapter.

For the time being, let's see what the scaling form for $g(z;\vec{q})$ in (8.10) implies for the generating function in real space. Instead of undertaking an honest calculation we'll make use of some simple tricks to produce the general form. We start with the relationship between $G(z;\vec{r})$ and $g(z;\vec{q})$:

$$G(z;\vec{r}) = \int g(z;\vec{q}) e^{i\vec{q}\cdot\vec{r}}\, d^d q \tag{8.11}$$

We then have

$$G(z;\vec{r}) = \int |\vec{q}|^{-2+\eta} f\left(\vec{q}(z_c - z)^{-\nu}\right) e^{i\vec{q}\cdot\vec{r}}\, d^d q \tag{8.12}$$

Replacing $\vec{q}$ by $\vec{\kappa}/|\vec{r}|$ in the integral, we are left with

$$\begin{aligned} G(z;\vec{r}) &= |\vec{r}|^{-d+2-\eta} \int |\vec{\kappa}|^{-2+\eta} f\left[\vec{\kappa}/\left(|\vec{r}|(z_c - z)^{\nu}\right)\right] e^{i\vec{\kappa}\cdot\vec{r}/|\vec{r}|}\, d^d \kappa \\ &= |\vec{r}|^{-d+2-\eta} F\left(|\vec{r}|(z_c - z)^{\nu}\right) \end{aligned} \tag{8.13}$$

where

$$F(x) \equiv \int |\vec{\kappa}|^{-2+\eta} f\left(\vec{\kappa}/|\vec{x}|\right) e^{i\vec{\kappa}\cdot\vec{x}/|\vec{x}|}\, d^d \kappa \tag{8.14}$$

As in the previous section, we have neglected the effect limits have on the quantities defined in terms of them.

The generating function has the same kind of scaling form in real space as it has as a function of wave number. Because of the short cuts we have taken in getting to (8.14) we do not yet know much about what the generating function actually looks like in real space. For the time being, we will make do with assertions concerning the limiting behavior of $F(x)$. They are:

(1) as its argument goes to zero, $F(x)$ approaches a finite constant;
(2) the function $F(x)$ decays exponentially to zero as $|x|$ goes to infinity.

We will now make use of the properties of the generating function that we have just discussed to rederive the result that long, unrestricted random walks obey Gaussian statistics. This useful exercise allows us to test the consequences of the various scaling conjectures that have just been advanced, and it also offers us a chance to practice the application of the steepest-descents approximation as a method for extracting the coefficient of a high order term in a power series. As an added benefit, we will see that when the exponents in the scaling form differ from those that apply to the unrestricted walk – as is the case for self-avoiding random walks – then the statistics of long random walks are not Gaussian.

Steepest-descents evaluation of an integral

The coefficient of z^N in the power-series expansion of

$$G(z;\vec{r}) = |\vec{r}|^{-d+2-\eta} F\left(|\vec{r}|(z_c - z)^\nu\right)$$

is given by the Cauchy formula

$$\frac{1}{2\pi \mathrm{i}} \oint \frac{G(z;\vec{r})}{z^{N+1}} \, \mathrm{d}z \tag{8.15}$$

where the contour encircles the origin. By now the reader should be familiar with this formula or the method of steepest descents described below. If this is not the case, the supplemental notes at the end of Chapter 1 may be of some help. For large N the steepest-descents method derives that coefficient as the function

$$|\vec{r}|^{-d+2-\eta} F\left(|\vec{r}|(z_c - z)^\nu\right) z^{-(N+1)} \tag{8.16}$$

where the fugacity z is adjusted to extremize the above expression. To perform this mathematical operation, we will first exponentiate all z-dependent contributions to the expression and then find the extremum of the exponent. The task at hand is thus to find the maximum, or the minimum, value of

$$\ln\left[F\left(|\vec{r}|(z_c - z)^\nu\right)\right] - (N+1)\ln z \equiv A\left(|\vec{r}|(z_c - z)^\nu\right) - (N+1)\ln z \tag{8.17}$$

The limiting behavior of the function $F(x)$ as described above implies that the function $A(x) \equiv \ln(F(x))$ goes to a constant as $x \to 0$, and that, as $x \to \infty$, $A(x)$ decreases as a linear function of x, i.e. $A(x) = -Bx$ when $x \gg 1$. Then, the extremum equation is

$$\nu|\vec{r}|(z_c - z)^{\nu-1} A'\left(|\vec{r}|(z_c - z)^\nu\right) - (N+1)/z = 0 \tag{8.18}$$

To obtain the solution to this equation when $N \gg 1$ we anticipate that the argument $|\vec{r}|(z_c - z)^\nu$ will be very large, while, at the same time, $z \approx z_c$. Then the slope of $A\left(|\vec{r}|(z_c - z)^\nu\right)$ will be constant and negative ($\equiv -B$), and we can write $z = z_c - \delta$, where the deviation of z from z_c is small. The solution to (8.18) is

$$\delta \approx \left(\frac{N+1}{\nu B z_c |\vec{r}|}\right)^{1/(\nu-1)} \tag{8.19}$$

Substituting this solution for $\delta = z_c - z$ into the right hand side of (8.17) we have

$$\begin{aligned} C(N;\vec{r}) &\approx \exp\left(-B|\vec{r}|\delta^\nu - (N+1)/z_c\right) \\ &\propto \exp\left[-((\nu z_c)^\nu B)^{1/(1-\nu)}(N+1)^{-\nu/(1-\nu)}|\vec{r}|^{1/(1-\nu)}\right] \end{aligned} \tag{8.20}$$

Exercise 8.2
Complete the steps leading to (8.20).

Exercise 8.3
Use (8.10) and additional assumptions about the structure of the function f in that equation to show that the generating function for the total number of walks goes as $(z_c - z)^{-\gamma}$, where $\gamma = (2 - \eta)\nu$. Make clear all the assumptions that go into the derivation of this result.

Exercise 8.4
Make use of the result of Exercise 8.3 to show that the total number of N-step walks governed by the generating function in (8.10) is proportional to $x^N N^p$. What are the values of the number, x, and the power, p?

Exercise 8.5
Here, we are going to work backwards from known results for self-avoiding walks. It turns out that the number of N-step self-avoiding walks that end a distance r away from their starting point go as $z_c^{-N} N^{-d\nu}$, where ν is the critical exponent in the equations above. What does this tell us about the way in which the function F in (8.13) behaves? How does the number of such walks depend on the distance, r?

Since the exponent ν is always positive and less than one, it is relatively easy to verify that the two assumptions leading to (8.20) indeed hold, as long as $|\vec{r}| \sim N^\nu$. Setting $\nu = 1/2$, the value this exponent takes for the kind of random walks that we have studied so far, we regain a Gaussian form for $C(N; \vec{r})$. Note, however, that when the exponent ν has a value different from 1/2 the Gaussian form implied by the central limit theorem does *not* follow.

8.1.3 Scaling and dimensionality: rederivation of Flory's exponent

In Chapter 7 we demonstrated how the random walk process can be recast as a sum over all the allowed paths of the walker. If the walker is forbidden to cross the path it has already left, then an appropriate interaction term in the path integral formulation can be introduced to eliminate such paths.

Recall the functional integral version of the expression for the total number of N-step walks from $\vec{x}$ to $\vec{y}$

$$C(N; \vec{x}, \vec{y}) \propto \prod_s \int \mathrm{d}\vec{r}(s) \exp\left[-\frac{3}{2a^2}\int_0^N \left|\frac{\mathrm{d}\vec{r}}{\mathrm{d}s}\right|^2 \mathrm{d}s - \frac{v_0}{2}\int_0^N \int_0^N \mathrm{d}s_1\,\mathrm{d}s_2\,\delta\left(\vec{r}(s_1) - \vec{r}(s_2)\right)\right] \tag{8.21}$$

We employ the following scaling transformations.[1] Suppose we redefine the arc length variable s and the position vector $\vec{r}(s)$ by effecting two rescalings. The arc length goes to $s' = l_2 s$ while $\vec{r}(s) \to \vec{r}\,'(s') = l_1 \vec{r}(l_2 s)$. The path integral on the right hand side of (8.21) can be re-expressed in terms of these new variables. The new relationship between $C(N; \vec{x}, \vec{y})$ and a functional integral is

$$C(N; \vec{x}, \vec{y}) \propto \prod_{s'} \int \mathrm{d}\vec{r}\,'(s') \exp\left[-\frac{3}{2a^2} \frac{l_2}{l_1^2} \int_0^{l_2 N} \left| \frac{\mathrm{d}\vec{r}\,'}{\mathrm{d}s'} \right|^2 \mathrm{d}s' - \frac{v_0}{2} \frac{l_1^d}{l_2^2} \int_0^{l_2 N} \int_0^{l_2 N} \mathrm{d}s_1' \, \mathrm{d}s_2' \, \delta\left(\vec{r}\,'(s_1') - \vec{r}\,'(s_2')\right) \right] \tag{8.22}$$

The factor l_1^d multiplying the double integral in the exponent on the right hand side of (8.22) arises because $\delta(ax) = \frac{1}{a}\delta(x)$, and because $\delta(\vec{r}) = \delta(x_1)\delta(x_2)\ldots\delta(x_d)$, where $x_1 \ldots x_d$ are the components of the vector $\vec{r}$.

The right hand side of (8.22) looks like the functional integral representation of a random walk consisting of a different number of steps – and for which the length is measured with respect to a new scale. There are a few additional differences between the new and the old random walk. Each of the integrals in the exponent is now multiplied by an additional factor. The integral over $|\mathrm{d}\vec{r}/\mathrm{d}s|^2$ has the new factor l_2/l_1^2. In the case of the double integral the factor is l_1^d/l_2^2.

Exercise 8.6

By defining a new length $a' = a l_1/\sqrt{l_2}$, show that the self-avoidance term, represented by the double integral in (8.22), is rescaled by a factor l_1^d/l_2. Show that a redefinition of l_1 in terms of the factor l_2 that keeps the first term in the double integral constant replaces the factor multiplying the self-avoidance term by $l_2^{(d-4)/2}$

After rescaling the length as prescribed in Exercise 8.6, we are left with an expression for the number of N-step random walks that take the walker a distance $|\vec{R}| = |\vec{y} - \vec{x}|$ that looks like the expression for the number of steps of an N'-step walk that takes the walker a distance $|\vec{R}\,'|$. The relationship between the primed and the unprimed quantities is $N' = l_2 N$ and $\vec{R}\,' = l_2^{1/2} \vec{R}$. If the rescaling factor l_2 is less than one, we are able to reduce the problem of an N-step walk to the problem of a walk with fewer than N steps.

[1] First suggested by Kosmas and Freed (1978).

Exercise 8.7

In Exercise 8.6, take the extreme step of setting l_2 equal to $1/N$. In this case, the "reduced" walk consists of exactly one step. What is the length of that step? What does this tell us about the distance away from the origin that a walker travels in N steps, assuming that self-avoidance plays no role? Now, look at the self-avoidance contribution to the multiple integral in (8.22). How does the amplitude of this term change? In particular at what spatial dimensionality does the importance of self-avoidance remain unchanged as the rescaling described here is carried out? This dimensionality is known as the *upper critical dimensionality*. Show that when the dimensionality, d, is smaller than the upper critical dimensionality the importance of self-avoidance is greatly enhanced as rescaling is implemented, while above the upper critical dimensionality self-avoidance reduces in significance under this type of rescaling.

Exercise 8.8

Suppose the interaction leading to self-avoidance has a power-law form, so that the generating function for self-avoiding walks is given by

$$C(N;\vec{x},\vec{y})$$
$$\propto \prod_s \int \mathrm{d}\vec{r}(s) \exp\left[-\frac{3}{2a^2}\int_0^N \left|\frac{\mathrm{d}\vec{r}}{\mathrm{d}s}\right|^2 \mathrm{d}s - \frac{v_0}{2}\int_0^N\int_0^N \mathrm{d}s_1\,\mathrm{d}s_2\,|\vec{r}(s_1)-\vec{r}(s_2)|^{-q}\right]$$

where the power q is positive. Perform the same kind of scaling analysis as was done in the case of delta-function-like self-avoidance and find the upper critical dimension of this system. How does this dimensionality depend on the power q?

For the moment, we will leave rigor in the dust and construct another heuristic derivation of Flory's exponent that relates the average size of a self-avoiding walk to the number of steps in it (Flory, 1953). Recall that this exponent $\nu = 3/(d+2)$ enters into the relationship between walk size, R and walk length, N, as follows:

$$R \propto N^{\nu} \tag{8.23}$$

Now, if we look once again at the rescaled version of the functional integral, as displayed in (8.22), we notice that the relative influence of the two integrals in the exponent changes under rescaling by the factors l_1 and l_2. Previously, we adjusted the rescaling factor l_1 so as to maintain the amplitude of the first integral. Suppose, on the other hand, that we fixed l_1 so that the relative amplitude of the two integrals remains unchanged under rescaling. In order for this to be the case, the following equality must clearly hold:

$$\frac{l_2}{l_1^2} = \frac{l_1^d}{l_2^2} \tag{8.24}$$

or

$$l_1 = l_2^{3/(d+2)} \tag{8.25}$$

Given this relationship between l_1 and l_2 we can duplicate the discussion at the end of Section 8.1.2 and claim that the mean distance traveled by the N-step walk is equal to the unit distance traveled by the "reduced" walk multiplied by the factor $l_1^{-1} = l_2^{-3/(d+2)} \rightarrow N^{3/(d+2)}$. Flory's result is thus recovered.

There are, of course, difficulties with this set of arguments. Flory's result is *not* exact, so the development leading to it cannot be entirely correct. There's not all that much to be gained in addressing the subtleties of scaling at this point. Such a discussion is more profitably deferred until a firmer foundation has been laid, and we are now ready to embark on that project.

9

Scaling of walks and critical phenomena

9.1 Scaling and the random walk

The previous chapters contain a variety of arguments in support of the conclusion that the mean square end-to-end distance of an N-step random walk obeys the following power-law relationship

$$\langle R^2 \rangle \propto N^p \tag{9.1}$$

In the case of unrestricted walks, in which there is no penalty associated with a trail that crosses itself, the power p is equal to one. When self-avoidance in introduced, p deviates from this value. Flory used what amounted to an energy–entropy argument (restated in slightly altered form at the end of Chapter 7) to derive the dimensionality-dependent result

$$p = \frac{6}{d+2} \tag{9.2}$$

This formula turns out to be exact in four, two and one dimensions, and p as given by the above expression is within a percent or so of the correct exponent when the spatial dimensionality, d, is three.

Given the remarkable accuracy of Flory's formula, the conclusion that the theoretical arguments mustered to justify (9.2) contain the seeds of a rigorous theoretical model of the self-avoiding random walk seems inescapable. While Flory's expression for the exponent p, (9.2), is clearly not exact, given the small but non-zero discrepancy in $d = 3$ dimensions, the general form of the power-law relationship in (9.1) holds quite generally. In fact, there is a set of relations for the self-avoiding walk that mirror the power-laws characterizing the thermodynamic properties of systems in the vicinity of a critical point. Pierre-Giles de Gennes was the first to point out that if one writes $p = 2\nu$, so that $\langle R^2 \rangle \propto N^{2\nu}$, the quantity ν can be identified with one of the critical exponents of a fictitious magnetic system (de Gennes, 1972). In particular, he demonstrated that the statistics of the self-avoiding random walk derive from the equilibrium statistical mechanics of an n-component spin system with ferromagnetic interactions. The magnetic system is known as the $O(n)$ model, and correspondence with the random walk problem is achieved in the limit $n = 0$. This suggestion by de Gennes set the stage for a quantum leap in our ability to obtain information about self-avoiding random walks. Investigators were provided access to an arsenal of new and powerful analytical tools, which were brought to bear on the problem. A variety of quantities can be calculated by appealing to the connection between the self-avoiding walk and this spin system. A detailed demonstration of the connection between the statistics of the self-avoiding walk and the statistical mechanics of the n-component spin system in the limit $n = 0$ is contained in Section 9.5.

Before embarking on an exploration of the properties of the self-avoiding walk as revealed in the behavior of the $n = 0$ limit of the $O(n)$ model, it is useful to review the general subject of phase transitions and critical points, so as to establish the general framework in which the phenomena will be described.

9.2 Critical points, scaling, and broken symmetries

The notion of scaling is intimately connected with the behavior of systems in the vicinity of critical points. Because of this, many scaling concepts are framed in the language of critical phenomena. In order to efficiently exploit the theoretical advances that have led to a fuller understanding of the properties of random walks it is useful to establish a working knowledge of the vocabulary of phase transitions. This section consists of a review of phase transitions and critical phenomena, with special (essentially exclusive) reference to a set of magnetic systems in which the critical point also marks the onset of a symmetry-breaking phase transition.

As an example of the type of transition we will be talking about, consider what happens to a typical insulating ferromagnetic system. This system consists of a set of magnetic moments attached to atoms that are fixed in space. Imagine that the atoms are located at the sites of some lattice and they possess a net magnetic moment. The spins on neighboring sites interact in such a way that there is an energetic tendency of those moments to align. At $T = 0$ K, all moments will align. In the absence of the disruptive effects of thermal fluctuations, the system is a permanent magnet.

Suppose the magnetic system is placed in contact with a heat bath at some finite temperature T. If T is sufficiently high the moments fluctuate randomly, and there is no net magnetization. This state of affairs persists as the temperature is decreased until a critical temperature, T_c, is reached.[1] As T passes through T_c, the system

[1] In the case of ferromagnets, the critical temperature is also called the Curie temperature.

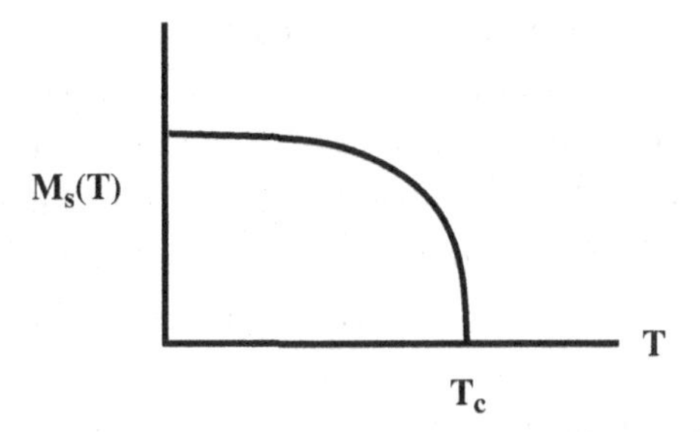

Fig. 9.1. The spontaneous magnetization of a ferromagnet as a function of temperature.

reacquires a net magnetization. This magnetization, called the *spontaneous magnetization* because no external fields are needed to maintain it, obeys the following power-law at temperatures sufficiently close to the critical temperature

$$M_s \propto (T_c - T)^{\beta} \tag{9.3}$$

with $\beta \approx 1/3$. A plot of the spontaneous magnetization, M_s, is displayed in Figure 9.1. The exponent β is known as a *critical exponent*. There are many others.

The noteworthy feature of the relationship governed by (9.3) is that it describes the temperature dependence of the spontaneous magnetization for a whole class of ferromagnets. Each member of this class will have a remanent magnetization described by (9.3), and for all the members of this class, the value of the exponent β is the same. This feature is one of the ways in which the transition to the ferromagnetic state exhibits *universal* behavior.

9.2.1 The $O(n)$ model

The system of interest here is a version of the insulating ferromagnet. The atoms carrying the magnetic moments are, as previously noted, fixed in space. The moments rotate in an n-dimensional "internal" space. Such a system is known as the $O(n)$ model. One standard realization of this system is the so-called *Heisenberg ferromagnet*, in which the moments are carried on atoms that are fixed in place on a crystalline lattice. In this case, the spins are three-component vectors, so the dimensionality of the internal space is the same as the spatial dimensionality inhabited by the atoms. The $O(3)$ model applies to this system. The set of moments may have an "easy plane," to which the moments are effectively confined. In that case $n = 2$. This model is also called the xy model, in reference to the two cartesian coordinates that span such a plane. There are magnetic realizations of this model, but the most widely studied system to which the $O(2)$ model applies is the superconductor or the superfluid. In the case of such a system, the two components of the "magnetization" are the real and imaginary parts of a macroscopic wavefunction.

Instead of an easy plane, there may be an "easy axis," which means the moments tend to point in one direction or another along a *one*-dimensional subspace. The version of the model that applies has $n = 1$. A number of magnetic materials are well-described in terms of the $O(1)$ model, but the most famous realization is the theoretical construct called the *Ising model.* Systems exist in which the internal subspace is more than three dimensional, and in that case the $O(n > 3)$ model may be invoked.

Finally, there are the instances in which the dimensionality of the internal space is less than one. For example, consider $n = 0$. While its analysis may seem to be an exercise in absurdity, the $O(n = 0)$ model will soon emerge as the target of our investigations.

9.2.2 Symmetries, respected and broken

There is more to the $O(n)$ model than the dimensionality of the internal space. A complete specification of its properties for the purposes of a study of its equilibrium thermal behavior must also include a description of its *symmetries*. The $O(n)$ model is symmetric under a simultaneous rotation of all the moments. This symmetry restricts the dependence of the energy of the model on the configurations of the moments. If we denote by the n-dimensional vector $\vec{S}_i$ the orientation and amplitude of the ith moment, the energy of interaction of the moments has the form

$$H_{\text{int}}\left[\left\{\vec{S}\right\}\right] = -J \sum_{i,j} \vec{S}_i \cdot \vec{S}_j \tag{9.4}$$

The coupling constant J quantifies the strength of the interaction between nearest neighboring atoms i and j. The negative sign on the right hand side of (9.4) indicates that the interaction is "ferromagnetic," in that the energy of a pair of interacting moments is lowest when they point in the same direction.

Recall that the scalar, or dot product, $\vec{A} \cdot \vec{B}$, of the two vectors $\vec{A}$ and $\vec{B}$ is equal to the magnitude of one of the vectors multiplied by the projection of the other one onto it, and that this combination depends *only* on their individual amplitudes and the angle between them. In particular, the scalar product does not change if both vectors rotate simultaneously through the same angle. Thus, the value of the expression on the right hand side of (9.4) remains the same if all the moments simultanenously rotate through the same angle. In this sense the set of moments exhibits a *rotational symmetry*.

Now, there is a perfectly plausible argument that leads inevitably to the conclusion that the set of moments will not, on the average point in any particular direction, at any non-zero temperature. This argument follows from one of the tenets of statistical mechanics, the *ergodic hypothesis*, and the rotational symmetry of the set of moments. The ergodic hypothesis, the (unproven) linchpin of equilibrium statistical mechanics, asserts that a system will sample, *in an absolutely unbiased fashion*, all the configurations accessible to it. Given the fact that no state of the set of moments in an $O(n)$ system is distinguishable, in terms of its interaction energy, from a state related to it by a simultaneous rotation of all of its moments, there cannot be an overall preference for a state in which the moments point in any particular direction. More quantitatively, if we define the net magnetization, $\vec{M}$, of the set of moments as

$$\vec{M} = \sum_i \vec{S}_i \tag{9.5}$$

then, for any state in which net magnetization has a particular value, there are states having exactly the same energy in which the value of $\vec{M}$ is related to the the net magnetization of the original state by a simple rotation. These states are sampled with the same frequency as the one mentioned above. This unbiased sampling of states with $\vec{M}$'s pointing in all possible directions leads to an average net magnetization that equals zero.

We thus arrive at the conclusion that, because of the $O(n)$ model's rotational symmetry, its average magnetization must be zero.

The above argument is compelling, but, as we've already noted, it cannot represent the last word on the subject. At sufficiently low temperatures, systems that possess the symmetry embodied in the interaction energy (9.4) may well violate the conclusion above. For example, the average net magnetization of the $O(n)$ model does not necessarily equal zero. When the temperature T is low enough, an argument based on the ergodic hypothesis ceases to hold and $\langle \vec{M} \rangle \neq 0$. There is, in fact a *symmetry-breaking phase transition* as the temperature decreases to a value lower than the *critical temperature*, T_c, of the system, from a high-temperature, symmetry-respecting, "paramagnetic" phase to a low-temperature, broken symmetry, "ferromagnetic" phase.

It is worthwhile to review in a bit more detail how the magnetic system defies the ergodic hypothesis in its low-temperature phase. In fact, the ergodic hypothesis holds at all temperatures. Left to itself the magnet will eventually sample all configurations. If it is allowed to carry out this sampling long enough, it will produce a time-averaged magnetization that is equal to zero. The real question is how long this averaging-out will take. As it turns out, a system in the ferromagnetic phase fluctuates through the set of states available to it at such a slow rate that no reasonable observation will yield the result $\langle \vec{M} \rangle = 0$ that follows from symmetry and ergodicity. The time scales involved are, literally, astronomical. Furthermore, they grow in magnitude with the size of the system. In the case of a magnet containing Avogadro's number of atoms, the projected lifetime of our universe is not long enough to wait. When the size of the system is infinite (the so-called thermodynamic limit), one has to wait literally forever to observe a restoration of ergodicity.

The full Hamiltonian

To facilitate a detailed investigation of the properties of the $O(n)$ model it proves convenient to introduce a term in the energy that explicitly breaks the rotational symmetry. This term looks like the "Zeeman" energy of interaction between a magnetic moment and the magnetic field in which it finds itself. Introducing the symmetry-breaking field $\vec{h}$, the new contribution to the energy of the collection of magnetic moments is, then

$$-\vec{h}\cdot\vec{M} = -\vec{h}\cdot\sum_i \vec{S}_i \tag{9.6}$$

and the total energy of the set of moments is

$$H\left[\left\{\vec{S}\right\}, J, \vec{h}\right] = -J\sum_{i,j}\vec{S}_i\cdot\vec{S}_j - \vec{h}\cdot\sum_i \vec{S}_i \tag{9.7}$$

The (canonical) partition function of this system, from which all thermodynamic properties derive, is given by

$$Z\left(T,\vec{h}\right) = \sum_{\text{Conf.}} \mathrm{e}^{-H\left[\left\{\vec{S}\right\},J,\vec{h}\right]/k_{\mathrm{B}}T} \tag{9.8}$$

The quantity k_{B} is Boltzmann's constant and T is the absolute temperature. The sum $\sum_{\text{Conf.}}$ in (9.8) is over all configurations of the moments, $\vec{S}_i$. The partition function yields the magnetic Gibbs free energy of the system, $F(T,\vec{h})$, through the relationship

$$F(T,\vec{h}) = -k_{\mathrm{B}}T\ln\left(Z\left(T,\vec{h}\right)\right) \tag{9.9}$$

The exponential in (9.8) is instrumental in the evaluation of thermal averages. Since $\mathrm{e}^{-H/k_{\mathrm{B}}T}/Z$ is the probability of realizing the configuration $\{\vec{S}\}$ for the system in contact with a heat bath at temperature T, it follows that the mean value of a quantity $X[\{\vec{S}\}]$ depending on the configuration of the moments is given by

$$\left\langle X\left[\left\{\vec{S}\right\}\right]\right\rangle = \frac{1}{Z}\sum_{\text{Conf.}} X\left[\left\{\vec{S}\right\}\right]\mathrm{e}^{-H\left[\left\{\vec{S}\right\},J,\vec{h}\right]/k_{\mathrm{B}}T} \tag{9.10}$$

Phase diagram and power law behavior of thermodynamic functions

The existence of the ferromagnetic phase is made manifest in the phase diagram depicted in Figure 9.2. The thick horizontal line running from the origin ($h = T = 0$) to the critical point at $h = 0, T = T_{\mathrm{c}}$ is a line of *first order phase transitions*, in that

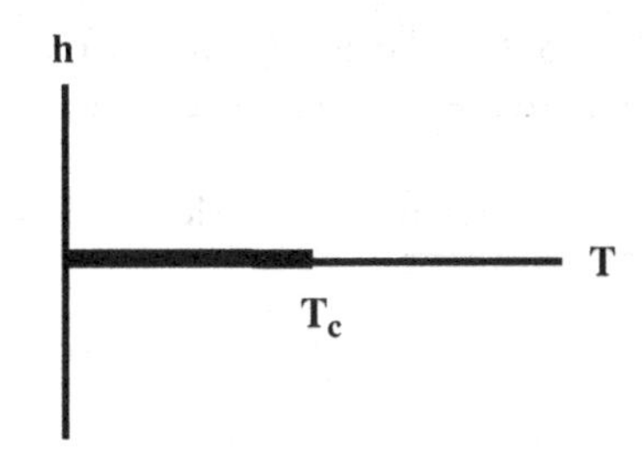

Fig. 9.2. The phase diagram of the ferromagnetic system. The thick line on the $h = 0$ axis extending from $T = T_c$ to the origin is a line of first order phase transitions.

the magnetization changes discontinously as the symmetry-breaking field traverses the abscissa when $T < T_c$. According to the classification of phase transitions by order, a first order phase transition is one in which first derivatives of the thermodynamic potential suffer discontinuities, and the average net magnetization is equal to the derivative of the magnetic Gibbs free energy, $F(T, \vec{h})$ with respect to $\vec{h}$ as follows:

$$\langle \vec{M} \rangle = -\left(\frac{\partial F(T, \vec{h})}{\partial \vec{h}} \right)_T \tag{9.11}$$

The derivative $\partial / \partial \vec{h}$ in (9.11) is shorthand for the gradient with respect to h. Another way to write it is as $\vec{\nabla}_h = \vec{i}\partial/\partial h_x + \vec{j}\partial/\partial h_y + \vec{k}\partial/\partial h_z$.

The critical point terminates the line of first order transitions. It is in the vicinity of this point that thermodynamic functions exhibit power-law behavior. For example, the isothermal susceptibility of the system, defined by

$$\chi_T = \left(\frac{\partial \langle \vec{M} \rangle}{\partial \vec{h}} \right)_T = -\left(\frac{\partial^2 F(T, \vec{h})}{\partial \vec{h}^2} \right)_T \tag{9.12}$$

has the following temperature dependence

$$\chi_T \propto |T_c - T|^{-\gamma} \quad \left(T \approx T_c,\ \vec{h} = 0 \right) \tag{9.13}$$

The two-point correlation function:

$$C(T, \vec{h}; \vec{\rho}) = \langle \vec{S}(\vec{R}) \vec{S}(\vec{R} + \vec{\rho}) \rangle \tag{9.14}$$

also exhibits scaling behavior:

$$\begin{aligned} C(T, \vec{h}; \vec{\rho}) &\propto \frac{1}{|\vec{\rho}|^{d-2+\eta}} f\left(|\vec{\rho}|\, |T - T_c|^{\nu} \right) \\ &\equiv \frac{1}{|\vec{\rho}|^{d-2+\eta}} f\left(|\vec{\rho}| / \xi \right) \end{aligned} \tag{9.15}$$

Table 9.1. *Set of thermodynamic exponents.*

Quantity	Behavior	Mean field value	3D value $O(n=0)$
Specific heat $c_h = \left(\frac{\partial Q}{\partial T}\right)_h$ $T > T_c, h = 0$	$\propto \lvert T - T_c\rvert^{-\alpha}$	$\alpha = 0$	$\alpha = 0.12$
Specific heat $c_h = \left(\frac{\partial Q}{\partial T}\right)_h$ $T < T_c, h = 0$	$\propto \lvert T - T_c\rvert^{-\alpha'}$	$\alpha' = 0$	$\alpha' = 0.12$
Magnetization $M(T \leq T_c, h = 0)$	$\propto \lvert T_c - T\rvert^{\beta}$	$\beta = 1/2$	$\beta = 0.308$
Susceptibility $\chi_T = \left(\frac{\partial M}{\partial h}\right)_T$ $h = 0, T > T_c$	$\propto \lvert T - T_c\rvert^{-\gamma}$	$\gamma = 1$	$\gamma = 1.19$
Susceptibility $\chi_T = \left(\frac{\partial M}{\partial h}\right)_T$ $h = 0, T < T_c$	$\propto \lvert T - T_c\rvert^{-\gamma'}$	$\gamma' = 1$	undefined ($\chi_T = \infty$)
Magnetization $M(T = T_c, h)$	$\propto \lvert h\rvert^{1/\delta}$	$\delta = 3$	$\delta = 4.865$
Correlation length $\xi^2 = \frac{\int C(T,\vec{h};\vec{\rho})\rho^2 \, \mathrm{d}^d\rho}{\int C(T,\vec{h};\vec{\rho}) \, \mathrm{d}^d\rho}$ $T > T_c, h = 0$	$\propto \lvert T - T_c\rvert^{-\nu}$	$\nu = 1/2$	$\nu = 0.602$
Correlation length $\xi = \frac{\int C(T,\vec{h};\vec{\rho})\rho^2 \, \mathrm{d}^d\rho}{\int C(T,\vec{h};\vec{\rho}) \, \mathrm{d}^d\rho}$ $T < T_c, h = 0$	$\propto \lvert T - T_c\rvert^{-\nu'}$	$\nu' = 1/2$	$\nu' = 0.602$
Anomalous dimension $C(T = T_c, h = 0, \vec{\rho})$	$\propto \rho^{-d+2-\eta}$	$\eta = 0$	$\eta = 0.023$

Equation (9.15) contains the definition of the *correlation length*, ξ. This important distance scale plays a fundamental role in the physics of the critical point. In fact, it is the divergence of this quantity as $T \to T_c$ that gives rise to much of the interesting behavior of the thermodynamic functions at and close to the critical point.

The full set of thermodynamic exponents is displayed in Table 9.1, along with two key exponents for the behavior of the correlation function.

As noted in Table 9.1, the "3D" exponents are those of a three-dimensional $O(n)$ system in the limit $n = 0$.

By and large, when two exponents are utilized to describe the behavior of a quantity above and below the critical temperature, those two exponents turn out to be equal. However, there are instances, exemplified by the case of the exponents γ and γ', in which the physics of the system rules out equality of the "mirror image" exponents. The general principle that there is a form of reflection symmetry about the critical point follows directly from the physics underlying critical phenomena. Equality of γ and γ' is ruled out in the case of interest to us by the fact that the susceptibility of an $O(n)$ system is, in general, infinite when $T < T_c$ and $h = 0$. This follows from the existence of gapless spin waves and is a consequence of a Coulomb-law-like "correlation hole" in that regime. Both of these phenomena will be discussed at greater length in Chapter 10.

9.2.3 Scaling laws

As it turns out, the thermodynamic exponents satisfy a number of relationships. These relationships have acquired the generic name of *scaling laws*. An example connects the three exponents α, β and γ:

$$\alpha + 2\beta + \gamma = 2 \tag{9.16}$$

This relation, known as the Rushbrooke scaling law, can be shown to hold for the mean field exponents displayed in Table 9.1, since

$$\alpha + 2\beta + \gamma = 0 + 2 \times \frac{1}{2} + 1 = 2$$

Another example of a scaling law is a "reflection symmetry" law, such as $\alpha = \alpha'$, or $\nu = \nu'$, or (when appropriate) $\gamma = \gamma'$.

There are also scaling laws that connect thermodynamic exponents with exponents for the correlation function. A very important example is

$$\gamma = \nu(2 - \eta) \tag{9.17}$$

Finally, a set of mathematical connections transcends the standard relations embodied in the scaling laws. These relationships are known as *hyperscaling laws*. Two key examples of this class are

$$\alpha = 2 - d\nu \tag{9.18}$$

$$\delta = \frac{d + 2 - \eta}{d - 2 + \eta} \tag{9.19}$$

A crucial characteristic of a hyperscaling relation is that it connects two, rather than three exponents, and it contains the spatial dimensionality, d, of the system.

Source of the thermodynamic scaling laws

The scaling law (9.16) and other laws relating thermodynamic exponents follow from the fact that the magnetic Gibbs free energy of the critical system takes on an asymptotic form, called a *generalized homogeneous* form. When $T \approx T_c$ and $\vec{h} \approx 0$, the Gibbs free energy acquires the following dependence on the thermodynamic parameters.

$$F(T, \vec{h}) \rightarrow |r|^{2-\alpha} f\left(|r|\, h^{-2+\alpha+\beta}, \frac{r}{|r|}\right) \tag{9.20}$$

In the above, the quantity r, called the *reduced temperature*, is given in terms of the absolute temperature by

$$r = \frac{T - T_c}{T_c}. \tag{9.21}$$

Note that the last argument in the expression on the right hand side of (9.20) is just the sign of the reduced temperature. This means that the function f is actually *two* functions: one relevant to $r > 0$ $(T > T_c)$ and the other appropriate to $r < 0$ $(T < T_c)$.

The specific heat at constant ordering field is given, in terms of the above thermodynamic functions, by

$$\begin{aligned} c_h &= T\left(\frac{\partial S}{\partial T}\right)_{\vec{h}} \\ &= -T\left(\frac{\partial^2 F}{\partial T^2}\right)_{\vec{h}} \\ &\rightarrow -\frac{\partial^2}{\partial r^2}|r|^{2-\alpha} f\left(|r|\, h^{-2+\alpha+\beta}, \frac{r}{|r|}\right) \\ &\propto |r|^{-\alpha} f\left(0, \frac{r}{|r|}\right) \end{aligned} \tag{9.22}$$

In the last line of (9.22), the limit $\vec{h} = 0$ has been taken. Notice that the above result implies a power-law for the specific heat. It also predicts equality for the exponents α and α'. Utilizing relationships (9.11) and (9.12), along with (9.20), we arrive at the exponent β for the spontaneous magnetization and the exponent $\gamma = 2 - \alpha - 2\beta$ for the isothermal susceptibility. Notice that the result for γ leads directly to the scaling law (9.16).

The asymptotic form (9.20), first introduced as a key to the behavior of a system near its critical point by Widom (1965), gives rise to an extensive set of relations between exponents. A comprehensive discussion of these relations is beyond the scope of this book.

Exercise 9.1
Given the following scaling form for the Gibbs free energy

$$G(T,h) = |r|^{2-\alpha} f\left(\frac{h}{|r|^{\Delta}}, \frac{r}{|r|}\right) \tag{9.23}$$

show that

$$\beta = 2 - \alpha - \Delta$$
$$\gamma = \Delta - \beta$$

Exercise 9.2
Using (9.23) for the Gibbs free energy show that at $T = T_c$ ($t = 0$) the magnetization behaves as

$$M \propto h^{1/\delta}$$

where $\beta(\delta - 1) = \gamma$.

Note that the last line of (9.15) implies a similar homogeneous form for the correlation function in the vicinity of the critical point.

Relationships coupling thermodynamic and correlation function exponents

The second set of scaling laws expresses the fact that many equilibrium properties reflect correlations in the system of interest. To see how this connection plays out in the case of the $O(n)$ model, we utilize (9.11), (9.9), (9.8), and (9.7) to derive a result for the mean magnetization. Performing the derivative with respect to $\vec{h}$,

$$\begin{aligned}
\langle \vec{M} \rangle &= -\frac{\partial}{\partial \vec{h}} F(T,\vec{h}) \\
&= k_B T \frac{\partial}{\partial \vec{h}} \ln\left(\sum_{\text{Conf.}} \exp\left(\frac{J}{k_B T}\sum_{i,j} \vec{S}_i \cdot \vec{S}_j + \frac{\vec{h}}{k_B T}\cdot \sum_i \vec{S}_i\right)\right) \\
&= \frac{\sum_{\text{Conf.}} \sum_i \vec{S}_i e^{-H\left[\{\vec{S}\},J,\vec{h}\right]/k_B T}}{Z\left(T,\vec{h}\right)} \\
&= \sum_i \langle \vec{S}_i \rangle
\end{aligned} \tag{9.24}$$

Note that the last line of (9.24) duplicates (9.5), with the thermal average as set forth in (9.10).

One further differentiation with respect to the ordering field $\vec{h}$ yields for the isothermal suseptibility

$$\begin{aligned}
\chi_T &= \frac{\partial}{\partial \vec{h}} \frac{\sum_{\text{Conf.}} \sum_i \vec{S}_i e^{-H\left[\left\{\vec{S}\right\},J,\vec{h}\right]/k_{\text{B}}T}}{Z\left(T,\vec{h}\right)} \\
&= \frac{\sum_{\text{Conf.}} \sum_{i,j} \vec{S}_i \vec{S}_j e^{-H\left[\left\{\vec{S}\right\},J,\vec{h}\right]/k_{\text{B}}T}}{Z\left(T,\vec{h}\right)} \\
&\quad - \frac{\sum_{\text{Conf.}} \sum_i \vec{S}_i e^{-H\left[\left\{\vec{S}\right\},J,\vec{h}\right]/k_{\text{B}}T}}{Z\left(T,\vec{h}\right)} \\
&\quad \times \frac{\sum_{\text{Conf.}} \sum_j \vec{S}_j e^{-H\left[\left\{\vec{S}\right\},J,\vec{h}\right]/k_{\text{B}}T}}{Z\left(T,\vec{h}\right)} \\
&= \frac{1}{k_{\text{B}}T} \sum_{i,j} \left(\langle \vec{S}_i \vec{S}_j \rangle - \langle \vec{S}_i \rangle \langle \vec{S}_j \rangle \right)
\end{aligned} \tag{9.25}$$

The last line of (9.25) is a sum over a generalized version of the correlation function defined in (9.14). This more general form defines a function with a vanishing amplitude at large separation of the locations i and j under all conditions, while the version displayed in (9.14) goes to a finite limit as $\vec{r} \to \infty$ when the average value of the magnetization is non-zero.

Invoking the scaling form for the correlation function indicated in Table 9.1, Equation (9.25) implies the following general result for the susceptibility, when the system under consideration possesses translational symmetry

$$k_{\text{B}}T\chi_T = \sum_{i,j} C(T,\vec{h};\vec{R}_i - \vec{R}_j) \tag{9.26a}$$

$$= V \sum_i C(T,\vec{h};\vec{R}_i) \tag{9.26b}$$

$$\to \int d^d r C(T,\vec{h};\vec{R}) \tag{9.26c}$$

$$\propto \int d^d R \frac{1}{\left|\vec{R}\right|^{d-2+\eta}} f\left(\left|\vec{R}\right| |T - T_{\text{c}}|^{\nu}\right) \tag{9.26d}$$

The quantity V in (9.26b) is the volume of the system; translational invariance has been assumed. In (9.26c) and (9.26d), the volume has been divided out. The resulting quantities remain finite in the thermodynamic ($V \to \infty$) limit. (9.26d) follows from the scaling form of the correlation function set forth in (9.15).

As a final step, we scale the quantity $|T - T_c|$ out of the integral. Writing $\vec{R} = |T - T_c|^{-\nu}\vec{x}$, one obtains the result

$$\chi_T \propto |T - T_c|^{-\nu(2-\eta)} \int d^d x f\left(|\vec{x}|\right)$$
$$= \mathcal{K}\,|T - T_c|^{-\nu(2-\eta)} \tag{9.27}$$

The scaling law (9.17) is thus derived. The constant $\mathcal{K}$ in the last line of (9.27) is just the integral in the first line.

Hyperscaling laws

Unfortunately, we are not yet in a position to pay the proper amount of attention to these very important relations. We will return to them once a more complete foundation has been laid.

9.3 Ginzburg–Landau–Wilson effective Hamiltonian

There is another, extremely useful way to frame the evaluation of the partition function of an $O(n)$ system. The new version of the equilibrium statistical mechanics of the system starts from a version of the Hamiltonian originally postulated by Ginzburg and Landau in the context of a model for superconductivity. In this approach the Hamiltonian in (9.7) is replaced by the following expression

$$\mathcal{H}_{\mathrm{GL}}\left(\left\{\vec{S}\right\}, T, \vec{h}\right) = \sum_{i=1}^{N}\left\{\frac{r}{2}\left|\vec{S}_i\right|^2 + \frac{1}{2}\left|\vec{\nabla}\vec{S}_i\right|^2 + \frac{u}{4}\left|\vec{S}_i\right|^4 - \vec{h}\cdot\vec{S}_i\right\}$$
$$\equiv \mathcal{H}_{\mathrm{GL}}^{(0)} - \vec{h}\cdot\vec{S}_i \tag{9.28}$$

The "gradient squared" term on the right hand side of (9.28) is shorthand for the discrete version of $\sum_{k,l}(\nabla_k S_{i,l})^2$. This term expresses the consequences of the ferromagnetic coupling between neighboring moments. The quantity r is the reduced temperature. It is equivalent to the quantity r defined in (9.21).

A very important difference between the system of moments for which the effective Hamiltonian in (9.28) is appropriate and the set to which the Hamiltonian in (9.7) applies is that in the latter case, the $\vec{S}$'s have a fixed magnitude, while the moments in (9.28) take on all possible amplitudes.[2] The partition function is as given in (9.8), except that the sum over configurations is now a multiple integral

[2] Another important characteristic of the Ginzburg–Landau spin variable is that it is classical, as opposed to quantum mechanical. This allows us to sum over configurations without regard to the underlying quantum structure of the spin degree of freedom that carries the magnetic moment.

over the individual $\vec{S}_i$'s. That is

$$Z = \prod_i \int \mathrm{d}^n S_i \mathrm{e}^{-\mathcal{H}\left(\left\{\vec{s}\right\},T,\vec{h}\right)} \tag{9.29}$$

Notice that the exponent no longer contains the factor $k_B T$. This is because the new Hamiltonian contains some of the effects of temperature. For this reason, it is more properly referred to as an *effective Hamiltonian*, and that is how we will denote it from now on.

9.3.1 Mean field approximation to the effective Hamiltonian

The mean field approximation to the critical point of the $O(n)$ model follows straightforwardly from an approximation to the partition function that replaces the multiple integral over the $\vec{S}_i$'s by an n-fold integration over the components of a dominant mode. To generate this approximation, and as a prelude to an analysis of various properties of the $O(n)$ model, we recast the effective Hamiltonian so that it is expressed in terms of the spatial Fourier transform of the $\vec{S}_i$'s. That is, we replace $\vec{S}_i$ as a variable by

$$\vec{S}(\vec{q}) = \frac{1}{N} \sum_i \vec{S}_i \mathrm{e}^{\mathrm{i}\vec{q}\cdot\vec{r}_i} \tag{9.30}$$

The inverse of the above transformation is

$$\vec{S}_i = \frac{V}{(2\pi)^d} \int_{\mathrm{BZ}} \vec{S}(\vec{q}) \mathrm{e}^{-\mathrm{i}\vec{q}\cdot\vec{r}_i} \mathrm{d}^d q. \tag{9.31}$$

As indicated in (9.31), the $\vec{q}$ integral is over the first Brillouin zone. Details about the Fourier transform and its inverse are contained in the second supplement to Chapter 2.

In terms of the new representation, the effective Hamiltonian takes on the form

$$\begin{aligned} &\mathcal{H}\left(\left\{\vec{S}(\vec{q})\right\}, T, \vec{h}\right) \\ &= \frac{V}{2} \int_{\vec{q}\in\mathrm{BZ}} \left(r + |\vec{q}|^2\right) \vec{S}(\vec{q}) \cdot \vec{S}(-\vec{q})\, \mathrm{d}^d q \\ &\quad + V \frac{u}{4} \int_{\vec{q}_1,\vec{q}_2,\vec{q}_3,\vec{q}_4\in\mathrm{BZ}} \left(\vec{S}(\vec{q}_1)\cdot\vec{S}(\vec{q}_2)\right)\left(\vec{S}(\vec{q}_3)\cdot\vec{S}(\vec{q}_4)\right) \\ &\quad \times \delta(\vec{q}_1 + \vec{q}_2 + \vec{q}_3 + \vec{q}_4)\, \mathrm{d}^d q_1\, \mathrm{d}^d q_2\, \mathrm{d}^d q_3\, \mathrm{d}^d q_4 - V\vec{h}\cdot\vec{S}(0) \end{aligned} \tag{9.32}$$

The quantity V in (9.32) is the volume of the system, as previously. To generate the mean field approximation one removes from the effective Hamiltonian all Fourier modes except $\vec{S}(\vec{q} = 0)$. The effective Hamiltonian reduces to

$$\mathcal{H}_{\mathrm{MF}} = V\left[\frac{r}{2}\left|\vec{S}(0)\right|^2 + \frac{u}{4}\left|\vec{S}(0)\right|^4 - \vec{h}\cdot\vec{S}(0)\right]$$
$$\equiv V\left[\frac{r}{2}|\vec{m}|^2 + \frac{u}{4}|\vec{m}|^4 - \vec{h}\cdot\vec{m}\right] \tag{9.33}$$

The last line of (9.33) can be taken as a definition of the magnetization per unit volume, $\vec{m}$. The mean field partition function is then the integral

$$Z_{\mathrm{MF}} = \int \mathrm{d}^n m e^{-\mathcal{H}_{\mathrm{MF}}(\vec{m})} \tag{9.34}$$

A full discussion of the mean field approximation and its extensions is deferred to the next chapter.

Exercise 9.3

Derive the expression (9.32) for the effective Hamiltonian of the $O(n)$ model from (9.28) for that quantity.

Exercise 9.4

Assume that the free energy as a function of reduced temperature, r, magnetization, m, and magnetic field, h, is given by the following expression

$$F = \frac{r}{2}m^2 + \frac{m^4}{12} - hm \tag{9.35}$$

This system is in thermal equilibrium when its free energy, as a function of m, is minimized. Thus, the equation of state of this system, which leads to a dependence on r and h for the equilibrium value of m is

$$\frac{\partial F}{\partial m} = 0$$

(a) Show that the solution to this equation of state is

$$m = |r|^{\beta}\mathcal{M}\left(\frac{h}{|r|^{\Delta}}, \frac{r}{|r|}\right)$$

What are the values of β and Δ?

(b) Use the above result to show that the minimized free energy has the following form:

$$F = r^{2-\alpha}\mathcal{K}\left(\frac{h}{|r|^{\Delta}}, \frac{r}{|r|}\right)$$

What is the value of the quantity α?

9.4 Scaling and the mean end-to-end distance; $\langle R^2 \rangle$

A detailed treatment of the $O(n)$ model and its predictions for the self-avoiding walks is the subject of the next chapter, but before taking up that study, we illustrate the utility of the spin analogy by deducing the power-law behavior of the mean end-to-end distance, asserted at the start of this chapter. To accomplish this result, we only need the invoke the scaling features of the magnetic system as summarized in Table 9.1.

Our starting point is, not unexpectedly, the formal definition of the correlation length

$$\xi^2 = \frac{\int C(T, \vec{R}) R^2 \, \mathrm{d}^d R}{\int C(T, \vec{R}) \, \mathrm{d}^d R} \tag{9.36}$$

Using the connection between the correlation function of the spin system and the generating function of self-avoiding walks[3] allows us to identify the numerator of the right hand side of (9.36) as the generator of the mean end-to-end distance, whilst the denominator is the susceptibility per unit volume

$$\begin{aligned} \xi^2 &= \frac{V}{\chi_T} \sum_{N=0}^{\infty} \int C(N; \vec{R}) R^2 \, \mathrm{d}^d R \left(\frac{z}{z_0}\right)^N \\ &= \frac{V}{\chi_T} \sum_{N=0}^{\infty} \langle R^2 \rangle \Gamma_N \left(\frac{z}{z_0}\right)^N \end{aligned} \tag{9.37}$$

z_0 being the mean field critical fugacity $(2dJ)^{-1}$ and Γ_N the total number of non-intersecting walks with fixed starting point. Near the critical point, the correlation length and susceptibility scale according to power-laws, $K_1/(z_c - z)^\nu$ and $K_2/(z_c - z)^\gamma$, respectively. The critical fugacity z_c is simply related to z_0 by the proportionality relation $z_c = z_0/\mu$. The constant μ is called the connectivity. Thus, in the critical region, we can write

$$\frac{K_1 K_2}{(z_c - z)^{2\nu+\gamma}} = V \sum_N \langle R^2 \rangle \frac{\Gamma_N}{z_0^N} z^N \tag{9.38}$$

In the supplemental notes of Chapter 1, the coefficient of z^N in an expansion of the function $(z_c - z)^{-s}$ was found to be $z_c^{-(N+s)} N^{\delta-1}/\Gamma(s)$, which permits the identification

$$V \langle R^2 \rangle \frac{\Gamma_N}{z_0^N} = \frac{K_1 K_2}{z_c^{N+\gamma+2\nu} \Gamma(\gamma + 2\nu)} N^{\gamma+2\nu-1} \tag{9.39}$$

[3] We establish this connection in Section 9.5.

Nothing can be concluded from (9.39) regarding the dependence of $\langle R^2 \rangle$ on N until we know how the total number of walks, Γ_N, depends on N. Fortunately, this quantity can be extracted directly from the susceptibility:

$$\begin{aligned}\chi_T &= V \int \mathrm{d}^d R C(T, \vec{R}) \\ &= V \sum_N \int \mathrm{d}^d R C(N; \vec{R}) \left(\frac{z}{z_0}\right)^N \\ &= V \sum_N \Gamma_N \left(\frac{z}{z_0}\right)^N \end{aligned} \tag{9.40}$$

Again, knowing that χ_T scales as $(z_c - z)^{-\gamma}$ near z_c, the coefficient of z^N in the expansion of the susceptibility in powers of z is, by the above argument, $K_2 N^{\gamma-1}/z_c^{N+\gamma}\Gamma(\gamma)$, which must be equal to $V\Gamma_N/z_0$ because of the uniqueness of power series. From this result, we have

$$\langle R^2 \rangle = \frac{K_1}{z_c^\gamma} \frac{\Gamma(\gamma)}{\Gamma(\gamma + 2\nu)} N^{2\nu} \tag{9.41}$$

and since the prefactor is independent of the number of steps we conclude that $\langle R^2 \rangle \propto N^{2\nu}$, precisely the observation of de Gennes.

From this simple calculation we only get a glimpse of the powerful analytical tools made available to us in transforming the statistics of self-avoiding walks to a statistical analysis of the critical phenomena of interacting magnetic moments. A modern analysis of the $O(n)$ model not only yields accurate numerical results for the correlation length exponent, ν, itself, but also provides insights and quantitative understanding of a host of other features of the self-avoiding walk problem as well, a task to which we now turn.

9.5 Connection between the $O(n)$ model and the self-avoiding walk

So far we have asserted the linkage between the $O(n)$ model in the limit $n = 0$ and the self-avoiding random walk. We have talked about the utility of that connection, in that one can extract the behavior of this kind of random walk from the thermodynamic properties of the associated magnetic system. In this final section, we will take the reader through the arguments that establish the mathematical equivalency of the two systems. The original derivation of this equivalency is due to de Gennes. This derivation is based on a comparison of the perturbation theoretical expansion for the partition function of the $O(n)$ model with the corresponding expansion for the generating function of the self-avoiding walk. The expansion parameter that regulates the convergence of the series is the combination $J/k_B T$, where J is the exchange energy. This argument requires that there be no significant effects that are non-perturbative in nature. Furthermore, perturbation theory in the case at hand is known to be an asymptotic expansion, having zero radius of convergence. Finally, the perturbation theoretical expansion is valid, even in an asymptotic sense, only at small values of $J/k_\mathrm{B}T$. The accessibility of the ordered phase of the magnetic system is not at all assured. The derivation of the connection presented here follows an argument due to Emery (1975), who demonstrated the equivalency between the spin system and the self-avoiding walk without recourse to perturbation theory.[4]

We recall the Ginzburg–Landau–Wilson effective Hamiltonian, (9.28). For purely formal reasons, which will become clear shortly, we imagine an ordering field $\vec{h}_i$ that varies from site to site and replace the Zeeman term in (9.28) by $-\sum_i \vec{h}_i \cdot \vec{S}_i$.

A generalized version of the partition function, Z, is then given by the multiple integral

$$Z = \prod_{i=1}^{N} \int d\vec{S}_i \mathrm{e}^{-\mathcal{H}_{\mathrm{GL}}^{(0)}\left[\vec{s}\right] - \sum_i \vec{h}_i \cdot \vec{S}_i} \tag{9.42}$$

This partition function is the source of all information, not only regarding the equilibrium thermodynamics of the system, but also with respect to correlations of the order parameter. For instance, a quantity of particular relevance is the two-point correlation function, as given in (9.14). This function is readily obtained from the partition function in (9.42) by taking derivatives with respect to the ordering field at different sites. To see how this works, take a derivative of the logarithm of the right hand side of the equation with respect to the αth component of $\vec{h}_i$.

$$\begin{aligned} \frac{\partial}{\partial h_{i,\alpha}} \ln(Z) &= \frac{\prod_{k=1}^{N} \int \mathrm{d}\vec{S}_k S_{i,\alpha} \mathrm{e}^{-\mathcal{H}_{\mathrm{GL}}^{(0)}\left[\vec{s}\right]}}{Z} \\ &= \langle S_{i,\alpha} \rangle \end{aligned} \tag{9.43}$$

Taking one further derivative,

$$\frac{\partial^2 Z}{\partial h_{i,\alpha} \partial h_{j,\alpha}} = \langle S_{i,\alpha} S_{j,\alpha} \rangle - \langle S_{i,\alpha} \rangle \langle S_{j,\alpha} \rangle \tag{9.44}$$

The right hand side is just the generalized version of the correlation function appearing in the last line of (9.25). This correlation function plays a key role in the connection between the spin model and the self-avoiding walk. As first pointed out

[4] For a somewhat modified version of the development, see Schäfer (1999).

by de Gennes, if

$$\lim_{n\to 0}\frac{1}{n}\sum_{\alpha=1}^{n}\langle S_{i,\alpha}S_{j,\alpha}\rangle = G(z;\vec{R}_i,\vec{R}_j)$$
$$\equiv \sum_{N=0}^{\infty} C(N;\vec{R}_i,\vec{R}_j)z^N \tag{9.45}$$

then $C(N;\vec{R}_i,\vec{R}_j)$ is equal to the number of N-step self-avoiding walks originating at $\vec{R}_i$ and ending at $\vec{R}_j$. It is this connection that we will now establish.

The principal difficulty in the evaluation of the partition function of the $O(n)$ model arises as a result of the quartic term $\frac{u}{4}|\vec{S}_i|^4$ in the exponent. If that term were eliminated, then the integrals in the right hand side of (9.42) would be generalized versions of a Gaussian integral, and such integrals can be evaluated fairly straightforwardly.[5] While the quartic term cannot be eliminated (in fact it is essential to the preservation of a convergent form for the partition function below the critical temperature), the introduction of an additional set of degrees of freedom allows us to reduce the integrals over the moments $\vec{S}_i$ to a Gaussian form. The key to the transformation of the integrals is the identity

$$\mathrm{e}^{-\frac{u}{4}|\vec{S}_i|^4} = \frac{1}{\sqrt{\pi u}}\int_{-\infty}^{\infty}\mathrm{d}\phi_i \exp\left(-\frac{\phi_i^2}{u} - i\phi_i|\vec{S}_i|^2\right) \tag{9.46}$$

Application of this identity at each point i replaces all quartic terms in the exponent by terms that are quadratic in the moments $\vec{S}_i$ – at a cost, however. The evaluation of the partition function now requires an integration over two sets of variable, $\vec{S}_i$ and ϕ_i. The essential difficulties attending the evaluation of the partition function remain, but, for our present purposes, there are distinct advantages in the new form of the partition function.

An immediate benefit of the transformation effected by the application of the relationship in (9.46) is that the limit $n \to 0$ is readily achieved. To see how this occurs, we utilize (9.46) in (9.42). Making use of the form (9.28) for the effective Hamiltonian, we are left with

$$Z = \prod_i \int \left(\frac{\mathrm{d}\phi_i}{\sqrt{\pi u}}\right)\mathrm{e}^{-\sum_i \frac{\phi_i^2}{u}} \prod_j \int \mathrm{d}^n S_j$$
$$\times \exp\left[-\sum_j\left(\left(\frac{r}{2}+\mathrm{i}\phi_j\right)|\vec{S}_j|^2 - \frac{1}{2}|\vec{\nabla}\vec{S}_j|^2\right)\right] \tag{9.47}$$

In the above, the ordering fields have been set equal to zero. Two features of the right hand side of (9.47) worth noting are as follows.

[5] For a discussion of the evaluation of general Gaussian integrals, see the supplement at the end of this chapter.

(1) The integral over one spin component at a given site is the same as the integral over any other spin component at that site.
(2) The N integrals over $S_{i,\alpha}$ can be done explicitly, since they are of a Gaussian form.

We have

$$\int \mathrm{d}^n S_i \exp\left[-\left(\frac{r}{2}+\mathrm{i}\phi_i\right)|\vec{S}|^2 - \frac{1}{2}|\vec{\nabla}\vec{S}_i|^2\right]$$
$$= \prod_{\alpha=1}^{n} \int \mathrm{d}S_{i,\alpha} \exp\left[-\left(\frac{r}{2}+\mathrm{i}\phi_i\right) S_{i,\alpha}^2 - \frac{1}{2}|\vec{\nabla} S_{i,\alpha}|^2\right] \tag{9.48}$$

which, in light of (1) above, reduces to

$$\left[\int \mathrm{d}S_i \exp\left[-\left(\frac{r}{2}+\mathrm{i}\phi_i\right) S_i^2 - \frac{1}{2}|\vec{\nabla} S_i|^2\right]\right]^n \tag{9.49}$$

In arriving at (9.49) the dummy variable of integration, $S_{i\alpha}$ has been replaced by S_i. The partition function factors nicely:

$$Z = \prod_{i=1}^{N} \int \frac{\mathrm{d}\phi_i}{\sqrt{\pi u}} \exp\left(-\sum_i \phi_i^2/u Z_n\right) \tag{9.50}$$

where

$$Z_n = \left[\prod_{i=1}^{N} \int \mathrm{d}S_i \exp\left(-\sum_i \left\{\left(\frac{r}{2}+\mathrm{i}\phi_i\right) S_i^2 + \frac{1}{2}|\vec{\nabla} S_i|^2\right\}\right)\right]^n \tag{9.51}$$

The integrals appearing in (9.51) are most readily performed by observing the following.

(1) The $\frac{1}{2}|\vec{\nabla} S_i|^2$ term can be replaced by $-\frac{1}{2} S_i \nabla^2 S_i$. This follows from a simple integration by parts. The integrated part vanishes in the thermodynamic limit.
(2) The exponent in (9.51) can be written

$$\sum_{i,j} S_i V_{ij} S_j \tag{9.52}$$

where

$$V_{ij} = \left(\frac{r}{2} - \frac{1}{2}\nabla^2\right)\delta_{ij} + i\phi_i\delta_{ij} \tag{9.53}$$

which can be written as the matrix element of the operator

$$\overset{\leftrightarrow}{V} = \left(\frac{r}{2} - \frac{1}{2}\nabla^2\right)\mathbf{I} + i\,\overset{\leftrightarrow}{\phi} \tag{9.54}$$

whose matrix elements in the coordinate basis, $|\vec{R}_i\rangle$ are given by

$$\langle \vec{R}_j|\,\overset{\leftrightarrow}{V}\,|\vec{R}_i\rangle \tag{9.55}$$

The integral over the S_i's now looks like

$$\left[\prod_{i=1}^{N}\int \mathrm{d}S_i \exp\left(-\sum_{i,j} S_i V_{ij} S_j\right)\right]^n \tag{9.56}$$

which readily integrates to

$$Z_n = \left[\frac{(2\pi)^{N/2}}{\det \overleftrightarrow{V}}\right]^n \tag{9.57}$$

Of course, the quantity in which we are really interested is the two-point correlation function and not the partition function, Z_n. Fortunately, the latter quantity yields readily to the kind of analysis that we have been applying here. The key to the extraction of the correlation function is the reintroduction of the Zeeman term $\sum_i \vec{h}_i \cdot \vec{S}_i$ into the effective Hamiltonian. Taking each of the components of site-dependent ordering field, $\vec{h}_i$, to be the same, i.e.

$$\vec{h}_i = h_i(1, 1, \ldots, 1) \tag{9.58}$$

and proceeding in exactly the same fashion, (9.56) becomes

$$Z_n = \left[\prod_{i=1}^{N}\int \mathrm{d}S_i \exp\left(-\sum_{i,j} S_i V_{ij} S_j + \sum_i h_i S_i\right)\right]^n \tag{9.59}$$

This integral is of the form considered in the supplemental notes at the end of this chapter. Using the results derived there, one ends up with

$$Z_n = \left[\frac{(2\pi)^{N/2}}{\det \overleftrightarrow{V}} \exp\left(-\frac{1}{2}\sum_{i,j} h_i \left(\overleftrightarrow{V}\right)^{-1}_{ij} h_j\right)\right]^n \tag{9.60}$$

In (9.60) $\overleftrightarrow{V}^{-1}$ is the inverse of the operator $\overleftrightarrow{V}$.

The two-point correlation function is found as indicated in (9.44).

$$\begin{aligned} G(\vec{R}_j, \vec{R}_k) &= \lim_{n\to 0} \sum_{\alpha=1}^{n} \langle S_{j,\alpha} S_{k,\alpha} \rangle \\ &= \lim_{n\to 0} \frac{\partial^2}{\partial h_j \partial h_k} \ln Z \end{aligned} \tag{9.61}$$

In taking the limit $n \to 0$, the trick is to expand $\ln Z$ to first order in n. All other terms vanish as $n \to 0$.

$$G(\vec{R}_j, \vec{R}'_k) = \frac{\partial^2}{\partial h_j \partial h_k} \prod_i \int \frac{\mathrm{d}\phi_i}{\sqrt{\pi u}} \mathrm{e}^{-\sum_i \frac{\phi_i^2}{u}} \frac{1}{2} \sum_{l,m} h_l V_{lm}^{-1} h_m$$
$$= \prod_{i'} \int \frac{\mathrm{d}\phi_{i'}}{\sqrt{\pi u}} V_{jk}^{-1} \mathrm{e}^{-\sum_i \frac{\phi_i^2}{u}} \tag{9.62}$$

For the components of the inverse matrix $\overleftrightarrow{V}^{-1}$,

$$T_{ij}^{-1} = \langle \vec{R}_i | \overleftrightarrow{V}^{-1} | \vec{R}_j \rangle$$
$$= \langle \vec{R}_i | \left[\left(\frac{r}{2} - \frac{1}{2} \nabla^2 \right) \overleftrightarrow{I} + \mathrm{i} \overleftrightarrow{\phi} \right]^{-1} | \vec{R}_j \rangle \tag{9.63}$$

Recall that the quantity $\overleftrightarrow{I}$ is the identity and the operator $\overleftrightarrow{\phi}$ is diagonal in the coordinate basis, $\langle \vec{R}_i | \overleftrightarrow{\phi} | \vec{R}_j \rangle = \phi_i \delta_{ij}$. Expressing the inverse operator $\overleftrightarrow{V}^{-1}$ as

$$\overleftrightarrow{V}^{-1} = \int_0^\infty \mathrm{e}^{-\frac{r}{2}t} \mathrm{e}^{\left(\frac{1}{2}\nabla^2 \overleftrightarrow{I} - \mathrm{i}\overleftrightarrow{\phi}\right)t} \, \mathrm{d}t \tag{9.64}$$

allows us the recast the correlation function in an extremely compact form

$$G(\vec{R}_i, \vec{R}_j) = \prod_i \int \frac{\mathrm{d}\phi_i}{\sqrt{\pi u}} \int_0^\infty \mathrm{d}t \mathrm{e}^{-\frac{r}{2}t} \langle \vec{R}_i | \mathrm{e}^{\left(\frac{1}{2}\nabla^2 \overleftrightarrow{I} - \mathrm{i}\overleftrightarrow{\phi}\right)t} | \vec{R}_j \rangle \tag{9.65}$$

Although the above expression is indeed compact, and its appearance is deceptively simple, it is not straightforward to evaluate. The problem stems from the fact that the operator $\overleftrightarrow{\phi}$ depends of the coordinates and therefore does not commute with ∇^2. The exponential operator appearing in (9.65) does not factor.

Exercise 9.5

Show that when $\overleftrightarrow{V}^{-1}$ is given by (9.64), $\overleftrightarrow{V}\overleftrightarrow{V}^{-1} = \overleftrightarrow{V}^{-1}\overleftrightarrow{V} = \overleftrightarrow{I}$.

9.5.1 Path integral reduction of the matrix element

The operator equation appearing in (9.65) closely resembles an imaginary time quantum-mechanical propagator, and, fortunately, Feynman taught us how to decompose operators of this sort into an appropriate path integral (Feynman and Hibbs, 1965; Kleinert, 1995). As we have mentioned, the two terms in the exponent in (9.65) do not commute, so it is not permissible to factorize the exponent. However, the following factorization is allowed when t is infinitesimal. Let $t = \epsilon$.

Then

$$e^{\left(\frac{1}{2}\nabla^2\overleftrightarrow{I}-i\overleftrightarrow{\phi}\right)\epsilon} = e^{-i\overleftrightarrow{\phi}\epsilon}e^{\frac{1}{2}\nabla^2\epsilon}e^{O(\epsilon^2)} \tag{9.66}$$

Exercise 9.6
Find the next order term, in terms of the quantity ϵ, of the expansion (9.66).

Therefore, if the range of t is decomposed into a large number of infinitesimals, i.e. $t = M\epsilon$, where $M \to \infty$ as $\epsilon \to 0$, in such a way as to keep the product $M\epsilon$ fixed, then the propagator can be factorized accordingly. Keeping the above limiting procedure in mind, we have

$$\langle \vec{R}_i| e^{\left(\frac{1}{2}\nabla^2-i\overleftrightarrow{\phi}\right)t} |\vec{R}_j\rangle = \langle \vec{R}_i| \left(e^{-i\overleftrightarrow{\phi}\epsilon}e^{\frac{1}{2}\nabla^2\epsilon}\right)\left(e^{-i\overleftrightarrow{\phi}\epsilon}e^{\frac{1}{2}\nabla^2\epsilon}\right)\cdots|\vec{R}_j\rangle \tag{9.67}$$

Introducing a complete set of intermediate states $\sum_i |\vec{R}_i\rangle\langle\vec{R}_i|$ between each term in the product of (9.67), we are able to represent the matrix element as a sum over all possible paths:

$$\sum_{\vec{R}_0,\ldots,\vec{R}_M} \langle\vec{R}_i|e^{-i\overleftrightarrow{\phi}\epsilon}|\vec{R}_M\rangle\langle\vec{R}_M|e^{\frac{1}{2}\nabla^2\epsilon}|\vec{R}_{M-1}\rangle\langle\vec{R}_{M-1}|e^{-i\overleftrightarrow{\phi}\epsilon}|\vec{R}_{M-2}\rangle\cdots|\vec{R}_j\rangle \tag{9.68}$$

The various matrix elements appearing in (9.68) are easily evaluated

$$\langle\vec{R}_M|e^{-i\overleftrightarrow{\phi}\epsilon}|\vec{R}_{M'}\rangle = \delta_{M,M'}e^{-i\epsilon\phi(\vec{R}_M)} \tag{9.69}$$

$$\langle\vec{R}_M|e^{\nabla^2\epsilon}|\vec{R}_{M'}\rangle = (2\pi)^{-d/2}\exp\left(-\frac{d}{2}\frac{\left|\vec{R}_M-\vec{R}_{M'}\right|^2}{\epsilon}\right) \tag{9.70}$$

In the above, d is the spatial dimensionality. Putting everything together, the propagator is seen to take on the form of a path integral.

$$\prod_l \int d^d R_l \exp\left[-\int_0^t \frac{d}{2}\left|\frac{d\vec{R}}{ds}\right|^2 ds - i\int_0^t \phi\left(\vec{R}(s)\right) ds\right] \tag{9.71}$$

with fixed end-points, $\vec{R}(0) = \vec{R}_i$ and $\vec{R}(t) = \vec{R}_j$.

The final step in establishing the result that the two-point correlation function is, indeed, the generating function for self-avoiding walks, is the substitution of (9.71) for the propagator in (9.65) and carrying out the integrations over the fields $\phi(\vec{R}_i)$.

The N integrations over the fields ϕ are all of the form

$$\int_0^\infty \frac{\mathrm{d}\phi(\vec{R}_l)}{\sqrt{\pi u}} \exp\left(-\frac{\phi(\vec{R}_l)^2}{u} - \mathrm{i}\phi(\vec{R}_l)\right) \tag{9.72}$$

which integrates to $\mathrm{e}^{-u/4}$, unless two of the $\vec{R}$'s are equal, in which case the integral becomes

$$\int \frac{\mathrm{d}\phi(\vec{R}_l)}{\sqrt{\pi u}} \exp\left(-\frac{\phi(\vec{R}_l)^2}{u} - 2\mathrm{i}\phi(\vec{R}_l)\right) \tag{9.73}$$

which yields the value e^{-u}. This eventually leads to an overall factor of $\mathrm{e}^{-u/2}$ for each intersection of the path. This factor can be written

$$\exp\left(-\frac{u}{4}\sum_{l,l'} \delta_{\vec{R},\vec{R}'}\right) \tag{9.74}$$

The factor of 2 is absorbed into the overcounting of the pairs. We see from this that the Ginzburg–Landau–Wilson version of the $O(n)$ model does not eliminate intersecting paths, but reduces their contribution to the generating function by the factor $\mathrm{e}^{-u/2}$. In the continuum limit, the Kronecker delta function goes over to the Dirac delta function. The final expression for the correlation function becomes

$$\begin{aligned} G(\vec{R}_i, \vec{R}_j) &= \int_0^\infty \mathrm{d}t \mathrm{e}^{-rt/2} \prod_l \int \mathrm{d}^d R_l \\ &\quad \times \exp\left(-\int_0^t \frac{\mathrm{d}}{2}\left|\frac{\mathrm{d}\vec{R}}{\mathrm{d}s}\right|^2 \mathrm{d}s \frac{u}{4}\int_0^t \int_0^t \delta(\vec{R}(s') - \vec{R}(s))\, \mathrm{d}s\, \mathrm{d}s'\right) \\ &= \int_0^\infty \mathrm{d}t \mathrm{e}^{-rt/2} C(t; \vec{R}_i, \vec{R}_j) \end{aligned} \tag{9.75}$$

The quantity $C(t; \vec{R}_i, \vec{R}_j)$ corresponds to the number of self-avoiding walks of length t with beginning and end-points $\vec{R}_i$ and $\vec{R}_j$, respectively. Clearly the correlation function and the generating function for self-avoiding walks are related. Indeed, (9.75) is precisely the form derived by de Gennes in his treatment, but to make the connection more transparent and to connect with the customary perturbative approach, consider the walk to be composed to discrete steps, N in number. Then, the integral over t is a sum over N

$$G(\vec{R}_i, \vec{R}_j) = \sum_{N=0}^{\infty} C(N; \vec{R}_i, \vec{R}_j) \mathrm{e}^{-rN/2} \tag{9.76}$$

Let's assume, in addition, that we are above the critical point. Recall $r/2 = (1 - T_c/T) \approx 0$, $e^{-r/2} \approx 1 - r/2 = T_c/T = J/T = z/z_c$. Then the result above takes on a familiar form. Perturbatively

$$G(\vec{R}_i, \vec{R}_j) = \sum_{N=0}^{\infty} C(N; \vec{R}_i, \vec{R}_j) \left(\frac{J}{T}\right)^N$$
$$= \sum_{N=0}^{\infty} C(N; \vec{R}_i, \vec{R}_j) \left(\frac{z}{z_c}\right)^N \tag{9.77}$$

Given that the left hand side of (9.77) is the correlation function of the $O(n)$ spin system and the right hand side is the generating function for self-avoiding walks, we have, at last, the desired connection.

9.6 Supplement: evaluation of Gaussian integrals

In this supplement, we outline a procedure by which integrals of the kind encountered in this chapter are evaluated. These integrals are of a Gaussian type with a canonical form

$$I = \prod_i \int_{-\infty}^{\infty} dx_i \exp\left(-\frac{1}{2} \sum_{i,j} x_i T_{ij} x_j + \sum_i h_i x_i\right) \tag{9:S-1}$$

The matrix $\hat{T}$ and the vector $\vec{h}$ are independent of the integration variables, and $\vec{h}$ is assumed to have real components, whereas $\hat{T}$ will not necessarily be real. The integration variables x_i may, likewise, be complex. Here, we assume that they are real. The integral can be cast into a compact form with the use of Dirac notation. We define a column vector $|\vec{x}\rangle$ as follows:

$$|\vec{x}\rangle = \begin{pmatrix} x_1 \\ \cdot \\ \cdot \\ \cdot \\ x_n \end{pmatrix} \tag{9:S-2}$$

and the adjoint vector $\langle \vec{x}|$

$$\langle \vec{x}| = (x_1, \ldots, x_n) \tag{9:S-3}$$

Equation (9:S-1) is then, in this notation

$$I = \prod_i \int_{-\infty}^{\infty} dx_i \exp\left(-\frac{1}{2} \langle \vec{x}| \hat{T} |\vec{x}\rangle + \left\langle \vec{h} | \vec{x} \right\rangle\right) \tag{9:S-4}$$

We further assume that $\hat{T}$ can be diagonalized by a unitary transformation, $\hat{U}$

$$\hat{U}^{\dagger}\hat{T}\hat{U} = \begin{pmatrix} \lambda_1 & 0 & \cdot & \cdot & 0 \\ 0 & \lambda_2 & 0 & \cdot & 0 \\ 0 & 0 & \cdot & & \\ & & & \cdot & 0 \\ 0 & 0 & \cdot & 0 & \lambda_N \end{pmatrix} \tag{9:S-5}$$

The λ's in (9:S-5) are the eigenvalues of the operator $\hat{T}$. In order for (9:S-5) to hold, the general matrix form of $\hat{T}$ must be normal, that is, it commutes with its adjoint. The next step is to transform to a representation that diagonalizes $\hat{T}$. Define a new set of variables according to the unitary transformation

$$|\vec{x}\rangle = \hat{U}\,|\vec{y}\rangle \tag{9:S-6}$$

and a new set of fields $|\vec{h}'\rangle$ by

$$\left|\vec{h}\right\rangle = \hat{U}\left|\vec{h}'\right\rangle. \tag{9:S-7}$$

The determinant of the unitary operator $\hat{U}$ is equal to one. Thus, the Jacobian accompanying this change of variables in (9:S-1) is also equal to one, and the integral transforms to

$$\begin{aligned} I &= \int \left(\prod_i \mathrm{d}y_i\right) \exp\left(-\frac{1}{2}\langle\vec{y}|\,\hat{U}^{\dagger}\hat{T}\hat{U}\,|\vec{y}\rangle + \left\langle\vec{h}'\right|\hat{U}^{\dagger}\hat{U}\,|\vec{y}\rangle\right) \\ &= \int \left(\prod_i \mathrm{d}y_i\right) \exp\left[\sum_i \left(-\lambda_i \frac{y_i^2}{2} + h_i' y_i\right)\right] \end{aligned} \tag{9:S-8}$$

The integrals are now uncoupled Gaussians, and therefore are easily evaluated. Of course, we are assuming that $\lambda_i > 0$. There are two cases to discuss. First, the vectors $|\vec{y}\rangle$ may be real. In that case one simply completes the squares and integrates. The result is

$$\begin{aligned} I &= \frac{(2\pi)^{N/2}}{\sqrt{\lambda_1\lambda_2\ldots\lambda_N}} \exp\left(\sum_i \frac{|h_i'|^2}{2\lambda_i}\right) \\ &= \frac{(2\pi)^{N/2}}{\sqrt{\det\hat{T}}} \exp\left(\frac{1}{2}\left\langle\vec{h}\right|\hat{U}\hat{\Lambda}^{-1}\hat{U}^{\dagger}\left|\vec{h}\right\rangle\right) \end{aligned} \tag{9:S-9}$$

where

$$\left(\Lambda^{-1}\right)_{ij} = \frac{1}{\lambda_i}\cdot\vec{\delta_{ij}} \tag{9:S-10}$$

Since $\hat{\Lambda}^{-1}$ is the inverse of the diagonalized version of $\hat{T}$, the operator $\hat{U}\hat{\Lambda}^{-1}\hat{U}^{\dagger}$ is the inverse of $\hat{T}$ in the original representation. We thus arrive at our final answer

$$I = \frac{(2\pi)^{N/2}}{\sqrt{\det \hat{T}}} \exp\left[\frac{1}{2}\sum_{i,j} h_i \left(T^{-1}\right)_{ij} h_j\right] \tag{9:S-11}$$

In the second case of interest, the vectors $|\vec{y}\rangle$ are complex. If this is so, the integral is carried over both the real and the imaginary parts of the components, which means $\mathrm{d}y$'s become $\mathrm{d}y_{\mathrm{R}}\,\mathrm{d}y_{\mathrm{I}}$'s. These integrals are of the same form as before and are evaluated accordingly.

9.6.1 Gaussian averages and Wick's theorem

Assuming that the probability density for random variables $x_1, x_2, \ldots, x_N$ is governed by the Gaussian form

$$\rho\left(x_1, x_2, \ldots, x_N\right) = \frac{\sqrt{\det \hat{T}}}{(2\pi)^{N/2}} \exp\left[-\frac{1}{2}\sum_{i,j} x_i T_{ij} x_j\right] \tag{9:S-12}$$

then the average of the function $F(x_1, \ldots, x_N)$ is given by

$$\langle F\rangle = \int \prod_{i=1}^{N} \mathrm{d}x_i F(x_1, x_2, \ldots)\rho(x_1, x_2, \ldots) \tag{9:S-13}$$

An average of particular interest (see, for instance, (6.30) in Chapter 6) is

$$\langle x_r x_s\rangle = \int x_r x_s \rho(x_1, x_2, \ldots) \prod_{i=1}^{N} \mathrm{d}x_i \tag{9:S-14}$$

We now introduce a fictitious "Zeeman-type" term into the exponential in the probability density. This allows us to take averages of the sort shown in (9:S-14) without the need to explicitly carry out integrations. Here is how this trick works. Consider the probability density

$$\rho_h(x_1, \ldots, x_N) = \rho(x_1, \ldots, x_N) \exp\left[\sum_k h_k x_k\right] \tag{9:S-15}$$

Then,

$$\langle x_r x_k\rangle = \lim_{\substack{h_i \to 0 \\ i=1,N}} \left\{\frac{\partial^2}{\partial h_r \partial h_s} \int \rho(x_1, \ldots, x_N) \prod_{i=1}^{N} \mathrm{d}x_i\right\} \tag{9:S-16}$$

From our previous results for the integral of the probability density ρ_h, (9:S-16) reduces to

$$\langle x_r x_s \rangle = \lim_{\substack{h_i \to 0 \\ i=1,N}} \left\{ \frac{\partial^2}{\partial h_r \partial h_s} \exp\left[\frac{1}{2} \sum_{i,j} h_i h_j \left(\hat{T}\right)_{ij}^{-1} \right] \right\} \qquad \text{(9:S-17)}$$

After taking the two derivatives indicated and setting each of the h_i's equal to zero, the average $\langle x_r x_s \rangle$ becomes

$$\begin{aligned} \langle x_r x_s \rangle &= \frac{1}{2} \left(\left(\hat{T}\right)_{rs}^{-1} + \left(\hat{T}\right)_{sr}^{-1} \right) \\ &= \left(\hat{T}\right)_{rs}^{-1} \end{aligned} \qquad \text{(9:S-18)}$$

The last equality holds for symmetric matrices $\hat{T}$. We will assume that this is the case for the matrices entering into the probability density through (9:S-12).

The general expression (9:S-17) can be generalized:

$$\begin{aligned} \langle x_r x_s x_t x_u \rangle &= \lim_{h_i \to 0} \left\{ \frac{\partial^4}{\partial h_r \partial h_s \partial h_t \partial h_u} \exp\left[\frac{1}{2} \sum_{i,j} h_i h_j \langle x_i x_j \rangle \right] \right\} \\ &= \lim_{h_i \to 0} \left\{ \frac{\partial^3}{\partial x_r \partial x_s \partial x_t} \left(\sum_j h_j \langle x_u x_j \rangle \exp\left[\frac{1}{2} \sum_{i,k} h_i h_k \langle x_i x_k \rangle \right] \right) \right\} \\ &= \lim_{h_i \to 0} \left\{ \frac{\partial^2}{\partial h_r \partial h_s} \left(\langle x_r x_s \rangle + h_j h_k \sum_{h_j h_k} \langle x_t x_j \rangle \langle x_u x_k \rangle \right) \right. \\ &\quad \left. \times \exp\left[\frac{1}{2} \sum_{l,m} h_l h_m \langle x_l x_m \rangle \right] \right\} \end{aligned} \qquad \text{(9:S-19)}$$

Taking the remaining two derivatives and setting all h_i's equal to zero, we are left with the final expression for $\langle x_r x_s x_t x_u \rangle$

$$\langle x_r x_s x_t x_u \rangle = \langle x_r x_s \rangle \langle x_t x_u \rangle + \langle x_r x_t \rangle \langle x_s x_u \rangle + \langle x_r x_u \rangle \langle x_t x_s \rangle \qquad \text{(9:S-20)}$$

This is a special case of *Wick's theorem*.

In general, Wick's theorem states that the Gaussian average of $\langle x_{i_1} x_{i_2}, \ldots, x_{i_N} \rangle$ is given by

$$\langle x_{i_1} x_{i_2}, \ldots, x_{i_N} \rangle = \sum_{\substack{\text{all} \\ \text{distinct} \\ \text{permutations}}} \langle x_{i_{p1}} x_{i_{p2}} \rangle \langle x_{i_{p3}} x_{i_{p4}} \rangle \cdots \langle x_{i_{p(N-1)}} x_{i_{pN}} \rangle \qquad \text{(9:S-21)}$$

Here, $p1$ is the first element in a given permutation of the integers $1, 2, \ldots, N$, and similarly for $p2$ and so on. The above equation holds only if there are an even number of x_{i_n}'s, or, in other words, if N is an even number. If $N = 2m$, then there are

$$\frac{(2m)!}{2^m m!} \tag{9:S-22}$$

permutations leading to distinct contributions to the right hand side of (9:S-21). The rule, as indicated on the right hand side of that equation, is that one only counts distinct contributions.

10

Walks and the $O(n)$ model: mean field theory and spin waves

10.1 Mean field theory and spin wave contributions

Now that we have established the link between self-avoiding random walks and the $O(n)$ model of magnetism, it is appropriate to look again at the magnetic system in the limit $n \to 0$. Of special interest to us is its behavior in the immediate vicinity of the critical point. We will make extensive use of the insights provided by the study of critical phenomena that have emerged over the past three decades. Our initial approach to the problem of the statistical mechanics of the $O(n)$ model will be to utilize the mean field ideas developed by Landau and others. Mean field theory will then be enhanced by the introduction of fluctuations which will be analyzed in a low order spin wave theory. The insights gained by this approach will lead us to a clearer picture of the phase transition as it pertains to the random walk problem. Finally, a full renormalization group calculation will be used to elucidate the scaling properties of this model. This final step in the analysis will be accomplished in Chapter 12.

10.1.1 Outline of the chapter

In this chapter we make use of the connection between the $O(n)$ model and the self-avoiding walk to discuss aspects of the statistics of such a walk. We begin by reviewing the relationship between the spin model described by the $O(n)$ energy function and the self-avoiding walk. Our initial focus will be on self-avoiding walks that are confined to a finite volume of space. The appropriate version of the $O(n)$ system consists of a finite collection of immobile spins occupying a fixed volume. When the number of spins is large – and the region containing them is extended – boundary conditions do not, in general, play a key role in the statistical mechanics of the system. On the other hand, boundary conditions exert a profound influence on the statistics of both self-avoiding and unrestricted walks, as we have already seen. In our discussion, we will assume periodic boundary conditions. These boundary conditions were introduced in Chapter 7, in the discussion of the mean field approximation for self-avoiding walk statistics. Here, we will detail the exact nature of these conditions. As it turns out, "periodic" boundaries have no influence on the total number of paths available to an unrestricted walker, in marked contrast to the case of absorbing boundaries.[1] While periodic boundaries do not, as a rule, correspond to any recognizable physical situation, they have the same general effect on key aspects of random walk statistics as reflecting boundaries, of which one can imagine physical realizations. Periodic boundary conditions possess the additional desirable attribute that calculations are, in general, considerably more tractable in periodic systems than in systems with real boundaries.

The issue of limits is especially important in the case at hand. The correspondence between the $O(n)$ model and the self-avoiding walk is firm only if the limit $n \to 0$ is taken before all other limiting procedures are implemented. In particular, it is important that n be taken to zero before the size of the system is allowed to become infinite. It is known that interchange of those limits introduces spurious results in other situations (Rudnick and Gaspari, 1986b).

An additional advantage flows from the tactic of keeping the size of the system finite. First, boundary conditions are automatically incorporated into the calculation. Second, one is led naturally to a useful interpretation of the "condensed" phase of the self-avoiding walk. This interpretation has been advanced by the authors of this book and others.

As a first step in our investigation of walks and the $O(n)$ model we work in the mean field approximation. In this approximation, key aspects of the statistics of the self-avoiding walk confined to a finite region of space depend on how many steps the walker has taken. If it has not taken too many, then the total number of walks available to it cannot be distinguished from the case in which the walker finds itself is an environment that extends to infinity in all directions. However, when there are enough steps in the walk, the self-interaction of the walk leads to a substantial reduction in the number of paths. This reduction is codified in terms of a kind of Boltzmann factor, $e^{-\varepsilon}$. The energy, ε, is given by the expression utilized by Flory in his argument for the size of a self-avoiding walk. The results of this analysis agree with those obtained in Chapter 7, where the mean field approximation was derived from a diagrammatic summation.

As the next step, fluctuations are added to the formulation; in the present context, the fluctuations take the form of spin waves. A calculation of the statistical mechanics of the model is carried out. In this calculation, the system is assumed to be finite. That is, we do not invoke the thermodynamic limit at the outset. This

[1] See Chapter 4 for a discussion of absorbing boundaries and their effect on unrestricted random walks.

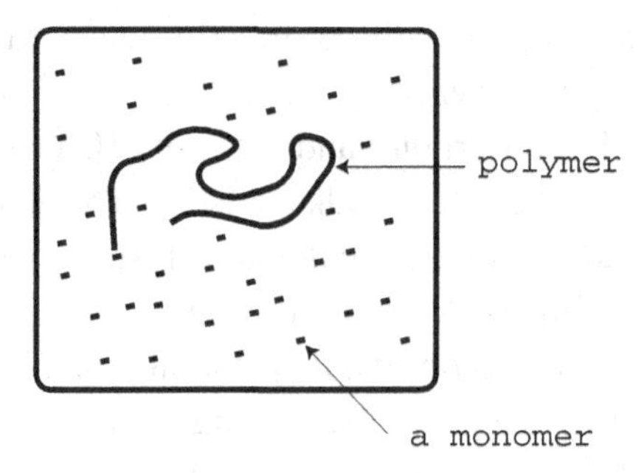

Fig. 10.1. A polymer in thermal equilibrium with a bath of monomers.

allows us to take limits in the proper order. We discover a "correlation hole" in the low-temperature phase, which we interpret in terms of an effective repulsion between the starting and ending points of the walk, or the two ends of the polymer. This repulsion plays an important role in the statistics of the self-avoiding polymer when it is tightly packed into a finite volume. We close the chapter with a more detailed discussion of the correlation hole, and, in particular, of the mechanisms giving rise to it. Before embarking on that project, we review a reinterpretation of the phase transition proposed by Redner and Reynolds (1981), which provides an intuitively appealing, if as yet less-than-rigorous, scaling field description of the process.

In this chapter, our primary focus will be on the generating function. That is, we will not concern ourselves with the N-step walk statistics that can be extracted from this quantity. To provide the discussions to follow with some intuitive underpinnings, we describe a physical system for which the generating function represents an immediate theoretical model. We have encountered this system, briefly and in broad outline, in Chapter 2. It consists of a polymer in thermodynamic equilibrium with a "reservoir" of monomers. We imagine that these monomers can attach to the ends of the polymer, thus adding to its length, and that, conversely, monomers can "evaporate" off the ends of the polymer, returning to the reservoir populated by their cohorts. A picture of such a collection of coexisting molecules is displayed in Figure 10.1. The statistics of the polymer are controlled by a grand partition function

$$G(z;\vec{x},\vec{y}) = \sum_N C(N;\vec{x},\vec{y})z^n \tag{10.1}$$

in which the quantity z plays the role of the fugacity. Here, one of the end-monomers of the collection of polymers in question is constrained to sit at the location $\vec{x}$ and the other end-monomer sits at $\vec{y}$. This restricted grand partition function is directly related to the statistical mechanical properties of the $O(n)$ magnet in the limit $n \to 0$. When we interpret the predictions that follow from calculations on the $O(n)$ model, our explications will be directly relevant to the statistics of the polymer in solution introduced here.

The functions on the left and right hand sides of (10.1) are, of course, also familiar from our discussion of random walk statistics. The quantity $G(z;\vec{x},\vec{y})$ is the generating function of walks subjected to the same restrictions that apply to the randomly coiling polymer, while $C(N;\vec{x},\vec{y})$ counts the N-step walks (corresponding to polymers consisting of N monomers) that start at $\vec{x}$ and end up at $\vec{y}$.

The $O(n)$ model describes a system that undergoes a phase transition. In the case of a magnetic system, this phase transition is between a disordered, paramagnetic phase at high temperatures and a phase that supports magnetism in the absence of an external field at low temperatures. The low-temperature ferromagnetic, state is known as the *ordered* phase of the system. The analogous system in the case of self-avoiding polymers is a collection of such polymers of various lengths in equilibrium with a "bath" of momomers. The phase transition in this system separates a phase in which the polymers are relatively well-separated (the analogy of the paramagnetic phase) and a phase in which the polymers fill space and intertwine like a plate of spaghetti (the analogy in this sytem of the ferromagnetic phase)(des Cloiseaux and Jannink, 1990). The particular case that we discuss in this chapter requires a slight reinterpretation of the phase transition. We will address this issue when it arises.

10.2 The mean field theory of the $O(n)$ model

The effective Hamiltonian describing an n-component magnetic moment system with rotationally invariant interactions in an external field is given by

$$\mathcal{H}[\vec{S}] = \sum_i \left[\frac{r}{2}|\vec{S}_i|^2 + \frac{1}{2}|\vec{\nabla}\vec{S}_i|^2 + \frac{u}{4}|\vec{S}_i|^4 - \vec{h}\cdot\vec{S}_i \right] \tag{10.2}$$

The spin $\vec{S}_i$ resides on the lattice site with label i. The partition function is given by the following multiple integral

$$Z = \prod_i \int \mathrm{d}\vec{S}_i \mathrm{e}^{-\mathcal{H}[\vec{S}]} \tag{10.3}$$

The partition function, Z, is the source of all information pertaining to the equilibrium thermodynamics of this $O(n)$ model.

The quantity of most direct relevance to the random walk is not Z but rather the spin–spin correlation function, $\langle S_{i\alpha} S_{j\beta} \rangle$. More precisely, the generating function of the self-avoiding walk is expressible as the following limit (see Chapter 9)

$$G(r; \vec{R}_i - \vec{R}_j) = \lim_{n \to 0} \frac{1}{n} \sum_{\alpha=1}^{n} \langle S_{i\alpha} S_{j\alpha} \rangle$$
$$= \sum_{N=0}^{\infty} C(N; \vec{R}_i - \vec{R}_j) \mathrm{e}^{-rN} \tag{10.4}$$

with $C(N; \vec{R}_i - \vec{R}_j)$ and $r = (T - T_c)/T_c$ having the usual meaning.

It proves convenient to work with the generator of walks with a fixed starting point that terminate at all possible end-points. We denote this quantity by Γ_N. The generating function of this quantity is related to $G(r; \vec{R}_i - \vec{R}_j)$ as follows

$$\Gamma(r) = \sum_j G(r; \vec{R}_i - \vec{R}_j)$$
$$= \sum_N \Gamma_N \mathrm{e}^{-rN} \tag{10.5}$$

From the above relationship we see that $\Gamma(r)$ is the counterpart for walks of the susceptibility per spin of the magnetic system:

$$\Gamma(r) = \frac{k_B T}{N_0} \chi_T \tag{10.6}$$

where N_0 is the total number of sites in the spin system. Accordingly,

$$\Gamma(r) = -\frac{(k_B T)^2}{N_0} \frac{\partial^2 \ln Z}{\partial h^2}\bigg|_{h=0, n=0} \tag{10.7}$$

The above relationships are true in general. The mean field results are obtained by assuming that the average magnetization is uniform and that fluctuations above the average can be neglected. That is, one writes

$$\vec{S}_i = \vec{M} \tag{10.8}$$

for all sites i. With this replacement, the partition function becomes

$$Z_{\mathrm{MF}} = \int \mathrm{d}^n M \mathrm{e}^{-N_0 \left\{ \frac{r}{2} |\vec{M}|^2 + \frac{u}{4} |\vec{M}|^4 - \vec{h} \cdot \vec{M} \right\}} \tag{10.9}$$

and our formula for $\Gamma(r)$ reduces to

$$\Gamma_{\mathrm{MF}}(r) = N_0 \lim_{n \to 0} \left\{ \frac{\int \mathrm{d}^n M M_1^2 \mathrm{e}^{-N_0 \left\{ \frac{r}{2} |\vec{M}|^2 + \frac{u}{4} |\vec{M}|^4 \right\}}}{\int \mathrm{d}^n M \mathrm{e}^{-N_0 \left\{ \frac{r}{2} |\vec{M}|^2 + \frac{u}{4} |\vec{M}|^4 \right\}}} \right\} \tag{10.10}$$

The ordering field, $\vec{h}$, has been taken to point in the "1" direction.

The complications arising from the $|\vec{M}|^4$ term can be dealt with in a variety of ways. One can, for example, note that

$$\Gamma_{\mathrm{MF}} = N_0 \lim_{n\to 0}\left[\frac{\int \mathrm{d}^n M M_1^2 f(|\vec{M}|^2)}{\int \mathrm{d}^n M f(|\vec{M}|^2)}\right]$$
$$= N_0 \lim_{n\to 0}\left[\frac{\int \mathrm{d}\lambda f(\lambda) \int \mathrm{d}^n M M_1^2 \delta(\lambda - |\vec{M}|^2)}{\int \mathrm{d}\lambda f(\lambda) \int \mathrm{d}^n M \delta(\lambda - |\vec{M}|^2)}\right] \tag{10.11}$$

The integrals over the components for the magnetization are easily performed with the use of the Fourier representation of the delta function $\delta(\lambda - |\vec{M}|^2)$. That is, replacing the delta function by

$$\frac{1}{2\pi}\int_{-\infty}^{\infty} \mathrm{d}S \mathrm{e}^{\mathrm{i}S\left(\lambda - |\vec{M}|^2\right)} \tag{10.12}$$

in both the numerator and denominator of (10.11), Γ_{MF} becomes

$$N_0 \lim_{n\to 0} \frac{\int \mathrm{d}\lambda\, \mathrm{d}S f(\lambda) \mathrm{e}^{\mathrm{i}S\lambda} \int \mathrm{d}M_1 M_1^2 \mathrm{e}^{-\mathrm{i}SM_1^2} \left\{\int \mathrm{d}M \mathrm{e}^{-\mathrm{i}SM^2}\right\}^{n-1}}{\int \mathrm{d}\lambda\, \mathrm{d}S f(\lambda) \mathrm{e}^{\mathrm{i}S\lambda} \left\{\int \mathrm{d}M \mathrm{e}^{-\mathrm{i}SM^2}\right\}^{n}} \tag{10.13}$$

In the limit $n \to 0$, the case of interest here, the denominator tends to $2\pi f(0) = 2\pi$, and we are left with

$$\frac{N_0}{2\pi}\int\int \mathrm{d}\lambda\, \mathrm{d}S \mathrm{e}^{\mathrm{i}S\lambda} f(\lambda) \frac{\int \mathrm{d}M_1 M_1^2 \mathrm{e}^{-\mathrm{i}SM_1^2}}{\int \mathrm{d}M_1 \mathrm{e}^{-\mathrm{i}SM_1^2}} = \frac{N_0}{2\pi}\int\int \mathrm{d}\lambda\, \mathrm{d}S f(\lambda)\frac{\mathrm{e}^{\mathrm{i}S\lambda}}{2\mathrm{i}S} \tag{10.14}$$

The integral over S is simply $\mathrm{i}\pi$, whence our final result for the mean field partition function:[2]

$$\Gamma_{\mathrm{MF}} = \frac{N_0}{4}\int_0^{\infty} \mathrm{d}\lambda f(\lambda)$$
$$= \frac{N_0}{4}\int_0^{\infty} \mathrm{e}^{-N_0\left(\frac{r}{2}\lambda + \frac{u}{4}\lambda^2\right)} \mathrm{d}\lambda \tag{10.15}$$

And we're done. The result in (10.15) is essentially identical to the result displayed in (7.57) in Chapter 7.[3] The equivalence between (10.15) and an identical result obtained directly from the path integral formulation of the random walk

[2] An important detail is that the integral over S yields this result only if λ is positive. A proper evaluation of the integral over M_1 in (10.14) requires that we replace $-\mathrm{i}sM_1^2$ in the exponent on the left hand side of the equation by $(-\mathrm{i}s - \epsilon)M_1^2$, where ϵ is a real, positive infinitesimal. This induces a pole in the integral on the right hand side of the equation just above the real axis, and the integral over S is now non-zero only when $\lambda > 0$.

[3] Full correspondence is achieved if we replace N_0 in (10.15) by N, λ by 4λ, $4u$ by f and $2r$ by $(1 - z/z_c)$. We forge the final link in the chain of connection by noting that $G_{\mathrm{MF}}(z; \vec{x}, \vec{y})$ does not depend on the starting and ending points, x and y, which means that the sum over end-points results in a multiplication by the total number of sites in the finite system.

provides additional evidence for the validity of the assertion that the magnetic analogy is appropriate for the discussion of the statistics of the self-avoiding walk and buttresses the general result established at the end of Chapter 7.

Recapitulating Chapter 7, it was found that in the mean field theory limit and above the critical temperature, the generating function is that of the unrestricted random walk. In the condensed phase, self-avoidance severely reduces the number of walks. This reduction is manifested in the factor $e^{Nr^2/4u}$. It was also found that, upon inverting the generating function to find the total number of walks in the condensed phase, the reduction factor in the mean field result is identical to the effect produced by the repulsive interaction term in Flory's argument for the mean radius of a self-avoiding walker. This corresponds to the standard interpretation of the phase transition in the context of self-avoiding polymer statistics.

10.3 Fluctuations: low order spin wave theory

To improve on mean field theory, one incorporates fluctuations into the calculation, and we do that here. Recall (10.2). In the continuum limit, the effective Hamiltonian becomes

$$\mathcal{H}[S(\vec{x})] = \int \mathrm{d}^d x \left\{ \frac{r}{2}|\vec{S}(\vec{x})|^2 + \frac{1}{2}|\vec{\nabla}\vec{S}(\vec{x})|^2 + \frac{u}{4}|\vec{S}(\vec{x})|^4 - \vec{h}\cdot\vec{S}(\vec{x}) \right\} \tag{10.16}$$

Low order spin wave theory amounts to expressing the magnetization $\vec{S}(\vec{x})$ in terms of a mean field contribution, $\vec{M}(\vec{x})$ and a fluctuating, "spin wave," field $\vec{\sigma}(\vec{x})$. We then retain terms in the effective Hamiltonian to second order in the field σ.

The final ingredient in the calculation is to impose appropriate boundary conditions on the spin field $\vec{S}(\vec{x})$. In the present context we have two choices for these boundary conditions: periodic or free. We choose the former to maintain consistency with the mean field result. Free boundary conditions possess the desirable attribute of direct relevance to the statistics of random walks. The nature of the boundary conditions. Chapter 4 contains a discussion of the effects of boundary conditions on unrestricted walks; here, we will touch upon the implications of boundary conditions for self-avoiding walks.

The first consequence of free boundary conditions arises at the mean field level. The mean field solution for the magnetization, $\vec{M}(\vec{x})$, is no longer unifom, as free boundary conditions require that the magnetization vanishes at the surface. The mean field equation of state takes the form

$$-\nabla^2\vec{M}(\vec{x}) + r\vec{M}(\vec{x}) + u\vec{M}(\vec{x})|\vec{M}(\vec{x})|^2 - \vec{h} = 0 \tag{10.17}$$

The fact that $\vec{M}$ depends on $\vec{x}$ complicates matters. However, deep in the ordered phase (i.e. at low enough temperatures, or when $r \ll 0$) $\vec{M}(\vec{x})$ ought to be uniform throughout the interior of the system, except in the immediate vicinity of the bounding surface. The effects of fluctuations can then be assessed with relative ease. One might well expect that the same situation holds well above the critical point ($r \gg 0$) as well. This turns out to be, indeed, the case. It is in the immediate vicinity of the critical point ($r \approx 0$) that things become complicated. Analytical solutions to the problem of the $O(n)$ model in this regime require a more advanced technology than presented in this chapter.

Exercise 10.1

Show that (10.17) follows from the minimization of the effective Hamiltonian in (10.16) with respect to the field $S(\vec{x})$.

We now return to the case of periodic boundary conditions and the calculation at hand. We start, as advertised, by expanding the spin field into a mean field and a spin wave contribution

$$\vec{S}(\vec{x}) = \vec{M}(\vec{x}) + \vec{\sigma}(\vec{x}) \tag{10.18}$$

Our notation implies a position dependence of $\vec{M}$, which is absent, given the periodic boundary conditions that apply here. We maintain the notation in the interest of generality. The effective Hamiltonian in (10.16) splits into contributions that are zeroth order, first order, second order, and so on, in the fluctuating field $\vec{\sigma}(\vec{x})$. All terms higher than second order are ignored. The contributions to the effective Hamiltonian are as follows.

(1) Zeroth order in $\vec{\sigma}$

$$\mathcal{H}_{\mathrm{MF}} = \int \mathrm{d}^d x \left\{ \frac{r}{2} |\vec{M}(\vec{x})|^2 + \frac{1}{2} |\vec{\nabla} \vec{M}(\vec{x})|^2 + \frac{u}{4} |\vec{M}(\vec{x})|^4 - \vec{h} \cdot \vec{M}(\vec{x}) \right\} \tag{10.19}$$

(2) First order in $\vec{\sigma}$

$$\mathcal{H}_1 = \int \mathrm{d}^d x \left(r\vec{M}(\vec{x}) - \nabla^2 \vec{M}(\vec{x}) + u\vec{M}(\vec{x}) |\vec{M}(\vec{x})|^2 - \vec{h} \right) \cdot \vec{\sigma}(\vec{x}) \tag{10.20}$$

(3) Second order in $\vec{\sigma}$

$$\mathcal{H}_2 = \int \mathrm{d}^d x \left\{ \frac{1}{2} |\vec{\nabla} \vec{\sigma}(x)|^2 + \frac{r}{2} |\vec{\sigma}(\vec{x})|^2 + \frac{u}{2} |\vec{M}(\vec{x})|^2 |\vec{\sigma}(\vec{x})|^2 + u \left(\vec{M}(\vec{x}) \cdot \vec{\sigma}(\vec{x}) \right)^2 \right\} \tag{10.21}$$

The first order term vanishes because $M(\vec{x})$ satisfies the mean field equation of state (10.17).

Exercise 10.2

Find the third order term in the expansion of the effective Hamiltonian with respect to $\sigma(\vec{x})$.

We now set $\vec{M}(\vec{x})$ equal to a constant. The second order Hamiltonian, $\mathcal{H}_2$, then decouples simply with the introduction of Fourier modes. One writes

$$\vec{\sigma}(\vec{x}) = \sum_{\vec{q}} \vec{\sigma}_{\vec{q}} \psi_{\vec{q}}(\vec{x}) \tag{10.22}$$

where

$$\psi_{\vec{q}}(\vec{x}) = \sqrt{\frac{1}{L_1}} e^{iq_1x_1} \sqrt{\frac{1}{L_2}} e^{iq_2x_2} \cdots \sqrt{\frac{1}{L_d}} e^{iq_dx_d} \tag{10.23}$$

The normalized modes in (10.23) are appropriate to a d-dimensional rectangular solid. The boundary conditions are satisfied if the q_i's take on the following discrete values

$$q_i = \frac{2n_i\pi}{L_i} \tag{10.24}$$

the n_i's being positive and negative integers.

When it is expressed in terms of the Fourier amplitudes, $\vec{\sigma}_{\vec{q}}$, of the spin wave, the second order effective Hamiltonian separates into a sum of contributions, each due to an individual mode. The decoupling is as follows

$$\mathcal{H}_2 = \sum_{\vec{q}} \left\{ \left(\frac{r}{2} + \frac{1}{2}q^2 + \frac{1}{2}u|\vec{M}|^2 \right) \vec{\sigma}_{\vec{q}} \cdot \vec{\sigma}_{\vec{q}} + u \left(\vec{M} \cdot \vec{\sigma}_{\vec{q}} \right)^2 - \vec{h} \cdot \vec{\sigma}_{\vec{q}} \langle \phi | \psi_{\vec{q}} \rangle \right\} \tag{10.25}$$

In (10.25) the external field is written in the form $\vec{h}(\vec{x}) = \vec{h}\phi(\vec{x})$, and $\langle \phi | \psi_{\vec{q}} \rangle$ represents the usual integration of a product of functions. The partition function then factors nicely

$$Z = \int d^n M e^{-\mathcal{H}_{\mathrm{MF}}} \int \prod_{\vec{q}} d^n \sigma_{\vec{q}} \exp\left[- \left(\vec{\sigma}_{\vec{q}} \cdot \overleftrightarrow{X}_{\vec{q}} \cdot \vec{\sigma}_{\vec{q}} + \vec{h} \cdot \vec{\sigma}_{\vec{q}} \langle \phi | \psi_{\vec{q}} \rangle \right) \right] \tag{10.26}$$

where the tensor operator $\overleftrightarrow{X}_{\vec{q}}$ is given by

$$\overleftrightarrow{X}_{\vec{q}} = \left(\frac{r}{2} + \frac{1}{2}q^2 + \frac{1}{2}u|\vec{M}|^2 \right) \overleftrightarrow{I} + u|\vec{M}\rangle\langle\vec{M}| \tag{10.27}$$

The quantity $\overleftrightarrow{I}$ in (10.27) is the identity operator.

According to the prescription outlined in the supplement at the end of Chapter 9, the generalized Gaussian integrals are easily performed once the eigenvalues and the inverse of the matrix $\overleftrightarrow{X}_{\vec{q}}$ are known. Both are simply found. The matrix $\overleftrightarrow{X}_{\vec{q}}$ has two distinct eigenvalues:

$$\lambda_1 = \frac{r}{2} + \frac{1}{2}q^2 + \frac{1}{2}u|\vec{M}|^2 \tag{10.28}$$

$$\lambda_2 = \frac{r}{2} + \frac{1}{2}q^2 + \frac{3}{2}u|\vec{M}|^2 \tag{10.29}$$

with degeneracies $n-1$ and 1, respectively.

Exercise 10.3
Verify that the eigenvalues of the operator $\overleftrightarrow{X}_{\vec{q}}$ are as given by (10.28) and (10.29). Derive the degeneracies of those eigenvalues.

The inverse of this operator is given by

$$\begin{aligned}\overleftrightarrow{X}_{\vec{q}}^{-1} &= \frac{\overleftrightarrow{I}}{\lambda_1} - \frac{u|\vec{M}\rangle\langle\vec{M}|}{\lambda_1\lambda_2} \\ &= \frac{\overleftrightarrow{I}}{r/2 + q^2/2 + u|\vec{M}|^2/2} \\ &\quad - \frac{u|\vec{M}\rangle\langle\vec{M}|}{\left(r/2 + q^2/2 + u|\vec{M}|^2/2\right)\left(r/2 + q^2/2 + 3u|\vec{M}|^2/2\right)}\end{aligned} \tag{10.30}$$

Exercise 10.4
Verify that

$$\overleftrightarrow{X}_{\vec{q}}^{-1}\overleftrightarrow{X}_{\vec{q}} = \overleftrightarrow{X}_{\vec{q}}\overleftrightarrow{X}_{\vec{q}}^{-1} = \overleftrightarrow{I} \tag{10.31}$$

where $\overleftrightarrow{X}_{\vec{q}}^{-1}$ is as defined in (10.30).

Armed with these results and the general formulas for Gaussian integrals, we are able to perform the indicated integrals. The result of these integrations is

$$\prod_{\vec{q}} \frac{\pi^{n/2}}{\sqrt{\lambda_2}\sqrt{\lambda_1^{n-1}}} \exp\left[-\frac{1}{4}\vec{h}\langle\phi|\psi_{\vec{q}}\rangle \cdot \overleftrightarrow{X}_{\vec{q}}^{-1} \cdot \vec{h}\langle\phi|\psi_{\vec{q}}\rangle\right] \tag{10.32}$$

We are interested in the value of the above expression in the limit $n \to 0$. The factor in (10.32) that depends on the λ's is then replaced by

$$\sqrt{\frac{\lambda_2}{\lambda_1}} = \sqrt{\frac{r + q^2 + u|\vec{M}|^2}{r + q^2 + 3u|\vec{M}|^2}} \tag{10.33}$$

The final result for the partition function is, then

$$\prod_{\vec{q}} \sqrt{\frac{r + q^2 + u|\vec{M}|^2}{r + q^2 + 3u|\vec{M}|^2}} \int \mathrm{d}^n M \exp\left[-N_0\left(\frac{r}{2}|\vec{M}|^2 + \frac{u}{2}|\vec{M}|^4\right) + \int \vec{M} \cdot \vec{h}(\vec{x})\, \mathrm{d}^d x\right]$$

$$\times \exp\left[\int\int \mathrm{d}^d x\, \mathrm{d}^d x' \frac{\vec{h}(\vec{x}) \cdot \vec{h}(\vec{x}')}{4} G_0(\vec{x}, \vec{x}') - \frac{1}{4}\int\int \mathrm{d}^d x\, \mathrm{d}^d x' \left(\vec{h}(\vec{x}) \cdot \vec{M}\right)\left(\vec{h}(\vec{x}') \cdot \vec{M}\right) D(\vec{x}, \vec{x}')\right] \tag{10.34}$$

where

$$G_0(\vec{x}, \vec{x}') = \sum_{\vec{q}} \frac{\psi_{\vec{q}}(\vec{x})\psi_{\vec{q}}(\vec{x}')}{r + q^2 + 3u|\vec{M}|^2} \tag{10.35}$$

and

$$D(\vec{x}, \vec{x}') = \sum_{\vec{q}} \frac{\psi_{\vec{q}}(\vec{x})\psi_{\vec{q}}(\vec{x}')}{\left(r + q^2 + u|\vec{M}|^2\right)\left(r + q^2 + 3u|\vec{M}|^2\right)} \tag{10.36}$$

We will deal with each of the terms appearing in (10.34) after we have obtained the final result for the generating function. To arrive at that result, we take two steps. First, the coefficient of the quadratic term in the expansion of the partition function in powers of $\vec{h}$ must be found. Then, it is necessary to evaluate the $n \to 0$ limit of the expression obtained as the result of the first step. Both steps are readily taken, given the equation

$$G(z; \vec{x}, \vec{x}') = \lim_{n=0} \frac{1}{n} \frac{\partial^2 \ln Z}{\partial h(\vec{x}) \partial h(\vec{x}')} \tag{10.37}$$

relating the partition function Z to the generating function for self-avoiding random walks. All n components of the ordering field, $\vec{h}$, are equal. For details of the connection see Chapter 9.

The first step involves a straightforward expansion of the exponential. Three terms are generated, leading to the following result:

$$\int \mathrm{d}^n M \exp\left[-N_0\left(\frac{r}{2}|\vec{M}|^2 + \frac{u}{4}|\vec{M}|^4\right)\right]$$

$$\times \left\{\frac{1}{2}\int \vec{h}(\vec{x}) \cdot \vec{M}\, \mathrm{d}^d x + \frac{1}{4}\int\int \vec{h}(\vec{x}) \cdot \vec{h}(\vec{x}') G_0(r; \vec{x}, \vec{x}')\, \mathrm{d}^d x\, \mathrm{d}^d x' - \frac{1}{4}\int\int \left(\vec{h}(\vec{x}) \cdot \vec{M}\right)\left(\vec{h}(\vec{x}') \cdot \vec{M}\right) D(\vec{x}, \vec{x}')\, \mathrm{d}^d x\, \mathrm{d}^d x'\right\} \tag{10.38}$$

The final step is also straightforward. We recall that the multidimensional integral over the components of $\vec{M}$ can be reduced to an integral over a single scalar variable,

λ, in the limit $n \to 0$. The two types of integral occurring in (10.38) are of the form

$$\int f(|\vec{M}|^2)\, \mathrm{d}^n M \qquad (10.39)$$

and

$$\int \left(\vec{h}(\vec{x})\cdot\vec{M}\right)\left(\vec{h}(\vec{x}')\cdot\vec{M}\right) f(|\vec{M}|^2)\, \mathrm{d}^n M \qquad (10.40)$$

With the above substitution for $(|\vec{M}|^2)$, and with the use of the integral representation of the delta function $\delta(\lambda - |\vec{M}|^2)$, the $n \to 0$ limit of the above two integrals reduce to

$$\int_{-\infty}^{\infty} f(\lambda)\delta(\lambda)\, \mathrm{d}\lambda = 1 \qquad (10.41)$$

and

$$\frac{\vec{h}(\vec{x})\cdot\vec{h}(\vec{x}')}{2}\int_{-\infty}^{\infty} f(\lambda)\, \mathrm{d}\lambda \qquad (10.42)$$

respectively, where $f(\lambda)$ is as given in (10.15). Reassembling the terms in (10.38) after the integrations over $\vec{M}$ have been taken, we are left with

$$\frac{\vec{h}(x)\cdot\vec{h}(x')}{4}\left\{G_0(\vec{x},\vec{x}') - \left(\int_0^{\infty} f(\lambda)\, \mathrm{d}\lambda\right) D(\vec{x},\vec{x}')\right\} \qquad (10.43)$$

We have now reduced the calculation of the correlation function of the $O(n)$ model, equivalent to the generating function of the self-avoiding walk, to the evaluation of second derivative with respect to the amplitude of an n-dimensional magnetic field that has all coefficients equal.[4]

To summarize, there are now two cases to consider.

Case 1: $r \gg 0$

Here we are high above the critical point. The dominant contribution comes from the first integral, (10.41), and the number of walks starting out at $\vec{x}$ and ending at $\vec{x}'$ is essentially given by

$$G_0(\vec{x},\vec{x}') = \sum_{\vec{q}} \frac{\psi_{\vec{q}}(\vec{x})\psi_{\vec{q}}(\vec{x}')}{r+q^2}$$
$$\to \frac{1}{(2\pi)^d}\int \mathrm{d}^d q\, \frac{\mathrm{e}^{\mathrm{i}\vec{q}\cdot(\vec{x}-\vec{x}')}}{r+q^2} \qquad (10.44)$$

[4] Again, see Chapter 9.

As we will see, the condition $r \gg 0$ implies that the linear extent of the walk is much smaller than the size of the container in which it finds itself. The confining volume places no constraints on the walk. The fact that the $\psi_{\vec{q}}(\vec{x})$'s are exponential functions instead of plane waves reflects the periodic boundary conditions that apply.

Case 2: $r \ll 0$. **The condensed phase and the correlation hole**

Now the walks have a linear extent that greatly exceeds the dimensions of the confining volume. Self-avoidance dominates. The major contribution to the integral over λ occurs in the region $\lambda \approx |r|/u$, where the integral is sharply peaked. Thus, the integral in (10.41) does not contribute, while the integral in (10.42) is accurately represented by a steepest-descent calculation, similar to those performed earlier. The starting point of the integration contour is $\lambda = |r|/u$. Combining this result with the mean field term appearing in (10.34) gives for the number of walks beginning at $\vec{x}$ and ending at $\vec{x}'$, in the condensed phase the result

$$\int_0^\infty \mathrm{e}^{-N_0(r\lambda+u\lambda^2)}\,\mathrm{d}\lambda\,\left(1 - D(\vec{x},\vec{x}')\right) \tag{10.45}$$

We recognize the first term in (10.45) as the mean field contribution, so the effects of fluctuations, at least within low order spin wave theory, give rise to the second term. This term – which we call a correlation hole – suggests the following picture of the condensed phase. The volume occupied by the walk is comparable to the volume of the box in which it is confined. We therefore expect that the walker will try to avoid regions in which there is a high concentration of crossings, thereby stretching out the walk. Recall Flory's excluded volume argument for polymers. This "stretching" effect manifests itself through the two-point correlation function $D(\vec{x},\vec{x}')$, which *increases* with increasing separation. One can think of this as giving rise to an effective repulsion between endpoints of the walk. Asymptotically, the repulsive potential has a Coulomb spatial dependence. That is, $D(\vec{x},\vec{x}') \propto |\vec{x}-\vec{x}'|^{2-d}$. This can be seen simply as a change in variables in the expansion for $D(\vec{x},\vec{x}')$:

$$D(\vec{x},\vec{x}') = \sum_{\vec{q}} \frac{\psi_{\vec{q}}(\vec{x})\psi_{\vec{q}}(\vec{x}')}{q^2\left(2|r|+q^2\right)} \tag{10.46}$$

Replacing the discrete set of $\vec{q}$'s by a continuum, we find that $D(\vec{x},\vec{x}')$ is proportional to the integral

$$\int \mathrm{d}^d q \frac{\mathrm{e}^{\mathrm{i}\vec{q}\cdot(\vec{x}-\vec{x}')}}{q^2\left(2|r|+q^2\right)} \tag{10.47}$$

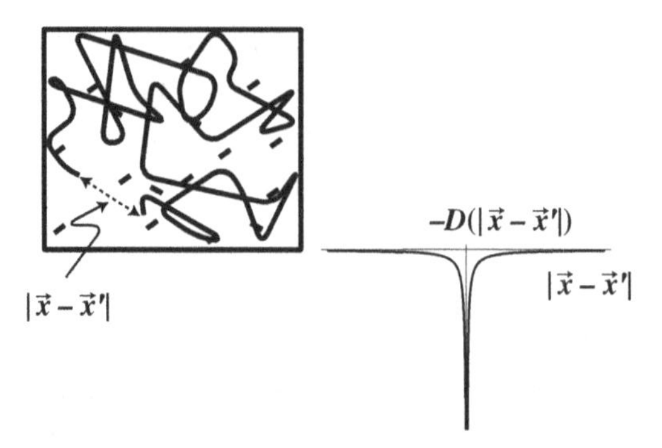

Fig. 10.2. Showing how the correlation hole affects configurations in which the end of the self-avoiding walk – or the excluded-volume polymer – approaches the point at which it starts. Also shown in this figure are the monomeric units that coexist with the longer polymer.

When $|\vec{x} - \vec{x}'| \gg 1/|r|$, the integral is dominated by small $|\vec{q}|$ and the spatial dependence scales out of the integral, yielding the asymptotic form

$$D(\vec{x}, \vec{x}') \propto \frac{1}{|\vec{x} - \vec{x}'|^{d-2}} \tag{10.48}$$

as stated. This picture of the condensed state emerges from more general arguments than the specific calculation presented in this section. Figure 10.2 illustrates how the correlation hole emerges. The walk, or the polymer, is long compared to the linear dimensions of the container to whose interior it is restricted. If the ends of the walk are not too close, the generating function is independent of the start and end-points. However, if the origin and point of termination of the walk approach each other, the repulsive interaction, real or effective, that enforces self-avoidance acts to reduce the weight of this configuration. A correlation hole corresponding to a three-dimensional walk is shown alongside the figure.

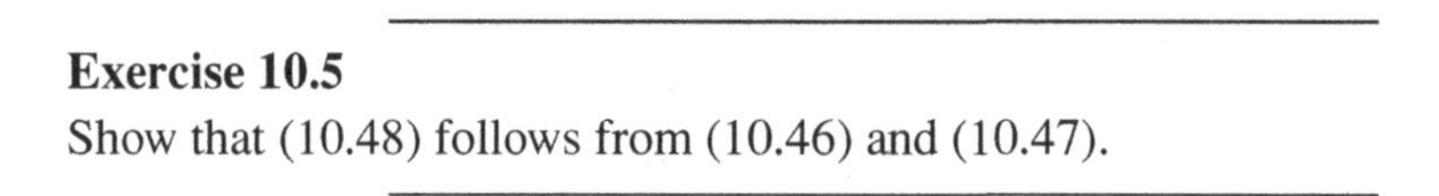

Exercise 10.5

Show that (10.48) follows from (10.46) and (10.47).

10.3.1 The condensed phase – compact walks

From our discussion of the correlation effects displayed in the results of a spin wave calculation, the following picture of the condensed phase emerges. Above the critical point, where $r \gg 0$, the extent of a walk is small in comparison to the dimensions of the box, and mean field theory adequately describes the gross features. Below the critical point, where $r \ll 0$, the walks are very compact, occupying a substantial fraction of the available volume. Self-avoidance now plays a significant role in the statistics of the walk. One important manifestation of this effect is a strong statistical repulsion between the end-points of the walk, which takes the form of a "correlation hole." The third regime of interest, the immediate vicinity of the critical point, in which $r \approx 0$, separates the two above. Here, the walks have a "natural" extent that is the order of the linear dimensions of the box. Modifications of the mean field approximation, due to fluctuations become very important, and a low order spin wave calculation is inadequate. More advanced calculational approaches are necessary to tackle this problem.

This picture of the condensed phase is consistent with a description of the phase transition put forth by Redner and Reynolds (1981). Though their model lacks a complete theoretical justification, it has intuitive appeal and we present it here.

Recall that the generating function for the total number of walks with fixed starting point has the form

$$\Gamma(r) = \sum_N \Gamma_N e^{-rN} \tag{10.49}$$

Now, Γ, as suggested by the mathematical similarity of the expression in (10.49) to the grand canonical distribution of statistical mechanics can be viewed as the partition function of an ensemble of a *single random walk* with number of steps, N, the statistics of which are governed by a grand canonical probability distribution. Here, r plays the role of the chemical potential. As in statistical mechanics, we define a grand potential

$$\Omega = \frac{1}{V} \ln \Gamma \tag{10.50}$$

from which the "average" properties of the walk can be determined. Of particular interest will be the average number of steps taken by the walker, $\langle N \rangle$. Obviously,

$$\langle N \rangle = -\frac{\partial}{\partial r} \ln \Gamma = -V \frac{\partial \Omega}{\partial r} \tag{10.51}$$

We further define the average density of steps by

$$\frac{\langle N \rangle}{V} = -\frac{\partial \Omega}{\partial r} \tag{10.52}$$

It is instructive to evaluate $\langle N \rangle$ with the use of the mean field treatment of self-avoidance. In Chapter 7, our mean field result for $\Gamma(r)$ was found to be

$$\begin{aligned} \Gamma_{\mathrm{MF}} &= \frac{V}{a^d} \frac{1}{2r} \qquad r > 0 \\ &= \frac{1}{2} \left(\frac{\pi}{u} \right)^{1/2} \left(\frac{V}{a^d} \right)^{3/2} e^{-\frac{V}{a^d} \frac{r^2}{4u}} \qquad r < 0 \end{aligned} \tag{10.53}$$

Differentiating $\ln \Gamma_{\mathrm{MF}}$, one ends up with a grand potential of the form

$$\Omega_{\mathrm{MF}} = \frac{1}{V} \{\ln V - \ln r + \text{const.}\} \quad r > 0 \tag{10.54}$$

$$= \frac{1}{V} \left\{ \ln V - \frac{V}{a^d} \frac{r^2}{4u} + \text{const.} \right\} \quad r < 0 \tag{10.55}$$

Note the interesting behavior of the density above and below the transition:

$$\frac{\langle N \rangle}{V} = \frac{1}{V} \left(\frac{1}{r} \right) \quad r > 0 \tag{10.56}$$

$$= \frac{|r|}{2a^d u} \quad r < 0 \tag{10.57}$$

In the thermodynamic limit (N and V going to infinity such that N/V approaches a constant) the density behaves as one might expect an appropriate order parameter characterizing the phase transition might behave:

$$\rho_{\mathrm{MF}} = 0 \quad r > 0$$
$$\neq 0 \quad r < 0 \tag{10.58}$$

suggesting that in the condensed phase the volume occupied by the walk is of the order of the volume of the confining space. Above the transition, the walk occupies a vanishingly small portion of the available volume. This behavior of the quantity ρ indicates that it might serve the purpose of an order parameter for the "transition" between the limiting case of a self-avoiding walk whose natural extent is small compared to the spatial region to which it is confined and the alternate limiting case of self-avoiding walk that when unconstrained would occupy a significantly larger volume than is contained in the boundaries surrounding the region in which it propagates. Given this definition, we will demonstrate that the way in which the new order parameter behaves can be quantified in terms of a set of critical exponents, and that the scaling relations that connect the new critical exponents are understandable in terms of the scaling form that is taken by the correlation function from which the density is derived. In the remainder of this section we will establish the critical exponents of the new order parameter. Then we will explore the source of these exponents by looking at the behavior of the "thermodynamic" system near the critical point. In this way we will establish the relationship between the two sets of critical exponents and, at least in the mean field approximation, determine the value of the new set.

Let's take up the latter task first. From (10.58) we see that $\rho_{\mathrm{MF}} \to |r|$, when $r < 0$. From this we immediately discover that the critical exponent β' is equal to 1. Note that this value differs from the mean field value of the magnetic exponent β, which has the value 1/2 in the mean field approximation. We will establish the relationship between the two sets of exponents later on in this section. For the moment we distinguish between the two by using primes to indicate exponents of the self-avoiding walk and unprimed exponents for the magnetic system.[5] The other critical exponents of the walk are evaluated in the usual manner: observe that a second derivative of the grand potential, Ω, leads to the fluctuation formula

$$\frac{\partial^2 \Omega}{\partial r^2} = \frac{1}{V} \left\{ \langle N^2 \rangle - \langle N \rangle^2 \right\} \tag{10.59}$$

which, with the use of the standard fluctuation response relation,[6] allows us to define the susceptibility of the walk system. We thus find

$$\chi' = \frac{\partial^2 \Omega}{\partial r^2} \to |r|^{-\gamma'} \tag{10.60}$$

which in mean field approximation leads to $\gamma'_{\rm MF} = 0$. However, the expression $\partial^2 \Omega / \partial r^2$ can equally well be interpreted, when the variable r is recognized as the temperature, as the specific heat, with its own critical exponent α'. Thus,

$$\frac{\partial^2 \Omega}{\partial r^2} \to |r|^{-\alpha'} \tag{10.61}$$

and we see that $\alpha' = \gamma'$, quite generally. The equality between the specific heat and the susceptibility exponents follows directly from the fact that there exists a single scaling parameter for our system; temperature and magnetic field play no role... at least as long as we restrict the system to represent a single self-avoiding walk. When there are several walks, all of them having an undetermined number of steps, both the temperature and magnetic field are relevant to a description of the statistical properties of the ensemble.

The final critical exponent of our fictitious thermodynamic system describes the behavior of the correlation length. Here things are simple. The exponent ν' is the same for both the thermodynamics of the walk and the magnetic system. In the mean field approximation, $\nu'_{\rm MF} = 1/2$.

In summary, the new mean field exponents are

$$\beta'_{\rm MF} = 1 \tag{10.62}$$

$$\gamma'_{\rm MF} = \alpha'_{\rm MF} = 0 \tag{10.63}$$

$$\nu'_{\rm MF} = 1/2 \tag{10.64}$$

[5] We do this in the full realization that that there is the possibility of confusion, in that primes are also used for thermodynamic exponents in the low-temperature phase.

[6] See, e.g. (9.25) in Chapter 9.

As we are about to show, it is no coincidence that these new exponents satisfy the usual scaling laws

$$\alpha' + 2\beta' + \gamma' = 2 \tag{10.65}$$

$$2 - \alpha' = d_c \nu' \tag{10.66}$$

which tell us that the upper critical dimensionality, d_c is 4, as expected. The scaling relations (10.65) and (10.66) are quite general and follow in a straightforward way from the scaling properties of the correlation function. As a by-product, we are able with a minimum of effort to arrive at the relation between the two sets of critical exponents – those of the self-avoiding walk and those of the magnetic system.

We have already established that the generating function $G(r; \vec{R}_i - \vec{R}_j)$ was simply the two-point correlation function for the spins. We know how the correlation function scales near the critical point, and, therefore, how G scales:

$$G \sim \frac{1}{R^{d-2+\eta}} F\left(r L^{1/\nu}, \frac{R}{L}\right) \tag{10.67}$$

where $R = |\vec{R}_i - \vec{R}_j|$, and L is the characteristic linear extent of the system.

From our definition of Γ, we see that

$$\Gamma(r) = \sum_{\vec{R}_j} G\left(r; \vec{R}_i - \vec{R}_j\right) \tag{10.68}$$

Equation (10.68) implies that Γ scales as follows

$$\Gamma(r) = L^{2-\eta} f(r L^{1/\nu}) \tag{10.69}$$

Since $\log \Gamma$ scales as the function f in (10.69), we can write

$$\Omega(r) \sim L^{-d} H(r L^{1/\nu}) \tag{10.70}$$

from which we arrive at the following scaling form for the order parameter ρ

$$\rho = \frac{\partial \Omega}{\partial r} = L^{1/\nu - d} H'(r L^{1/\nu}) \tag{10.71}$$

Now, we argue that the order parameter must remain finite in the "ordered" phase, as L becomes arbitrarily large. This requires that $H'(r L^{1/\nu}) \to r^{\nu d - 1} L^{(\nu d - 1)/\nu}$. This leads to a cancellation of the L-dependence of ρ in (10.71) and yields

$$\rho \to r^{\nu d - 1} \tag{10.72}$$

From this we are able to extract β' as given by

$$\beta' = d\nu - 1 \tag{10.73}$$

Table 10.1. *The new critical exponents and their relationship to previously defined quantities.*

$\beta' = 1 - \alpha$
$\gamma' = \alpha$
$\alpha' = \alpha$
$\nu' = \nu$
$d_c = 4$

The remaining exponents follow by noting the following two points:

(1) νd is also equal to $2 - \alpha$
(2) $\alpha = \alpha'$, since both are derived from the specific heat. This also leads to $\gamma' = \alpha$.

These results are summarized in Table 10.1.

From Table 10.1, we see immediately that the new exponents for the walk satisfy the usual scaling laws

$$\alpha' + 2\beta' + \gamma' = 2 \tag{10.74}$$

$$2 - \alpha' = d_c \nu' \tag{10.75}$$

which hold in general. This establishes the consistency of the grand canonical ensemble interpretation of the generating function. By defining a thermodynamic system corresponding to the ensemble of walks in a natural way we are able to deduce a description of the condensed phase as one in which walks are extremely compact and fill the allowed volume. Our thermodynamic analysis of the phase transition, based in part on the scaling properties of the generating function, and the picture that emerges is in complete accord with what was found with the use of an independent approach.

10.4 The correlation hole

In order to understand a striking contribution to the random-walk-generating function engendered by the fluctuation contribution – in particular the "correlation hole" – it is useful to consider the way in which the self-avoiding walk "sees" itself as it propagates in the highly condensed phase. We start by noting that the effects of self-avoidance are concentrated in energetic effects, arising from a Boltzmann-like factor associated with the energy cost of self-intersection. On the average, one expects that the probability that a walker will have stepped on a particular location,

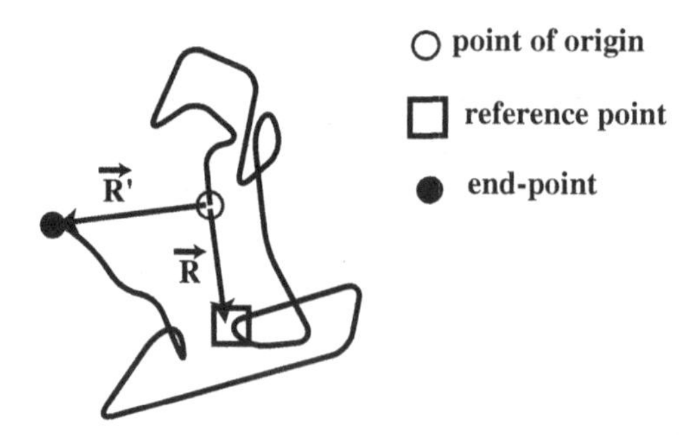

Fig. 10.3. The number of steps a distance R away from the starting point of the random walk.

$\vec{r}$, will be weighted by the factor $e^{-u\rho(\vec{r})}$, where $\rho(\vec{r})$ is the number of steps in the immediate vicinity of the point in question. If the strength of the interaction is not great, there will be a modification of the walk statistics that goes like $\sim -u\rho(\vec{r})$. In order to see just how this modification goes, it is necessary to calculate the number of steps there will be in a random walk at a certain distance from the point of origin of that walk.

We consider an unrestricted walk that starts out at some point in space. For convenience, that point will be given the position vector $\vec{r} = 0$. We then ask how many steps there will be at a distance R away from the starting point. The number of walks that end a distance R away from where the walker starts can be obtained directly from the generating function

$$G(z; \vec{R}) = \int \frac{e^{i\vec{k}\cdot\vec{R}}}{k^2 + z_c - z} \, d^d k \tag{10.76}$$

We are interested in a more general quantity here: the number of walks for which one or more of the steps takes the walker a distance R away from the point of origin. The generating function that yields this result is

$$\int d^d R' \int d^d k_1 \int d^d k_2 \frac{e^{i\vec{k}_1\cdot\vec{R}}}{k_1^2 + z_c - z} \times \frac{e^{i\vec{k}_2\cdot\left(\vec{R}'-\vec{R}\right)}}{k_2^2 + z_c - z} \tag{10.77}$$

(see Figure 10.3). Integrating over $\vec{R}'$ we obtain a delta function that restricts the vector $\vec{k}_2$ to be equal to zero. We are left with

$$\left(\int d^d k_1 \frac{e^{i\vec{k}_1\cdot\vec{R}}}{k_1^2 + z_c - z}\right) \times \frac{1}{z_c - z} \tag{10.78}$$

The coefficient of z^N in the above expression is the coefficient of z^N in

$$\int \mathrm{d}_1^k \sum_{N_1=0}^{\infty} \mathrm{e}^{\mathrm{i}\vec{k}_1\cdot\vec{R}} \left(\frac{z^{N_1}}{\left(k_1^2 + z_c\right)^{N_1+1}} \right) \times \sum_{N_2=0}^{\infty} \left(\frac{z^{N_2}}{z_c^{N_2+1}} \right) \tag{10.79}$$

In order to extract the desired coefficient, we must evaluate the following sum

$$\frac{1}{z_c} \frac{1}{z_c + k_1^2} \sum_{n=0}^{N} \left(\frac{z_c}{z_c + k_1^2} \right)^n = \frac{1}{z_c} \frac{1}{k_1^2} \left[1 - \left(1 + k_1^2/z_c\right)^{-(N+1)} \right] \tag{10.80}$$

In the limit of large N, this expression becomes

$$\frac{1}{z_c k_1^2} \left[1 - \mathrm{e}^{-(N+1)k_1^2/z_c} \right] \tag{10.81}$$

The number of steps a distance R away from the starting point is, then, given by

$$\int \mathrm{d}^d k_1 \frac{\mathrm{e}^{\mathrm{i}\vec{k}_1\cdot\vec{R}}}{z_c k_1^2} \left[1 - \mathrm{e}^{-(N+1)k_1^2/z_c} \right] \tag{10.82}$$

Carrying out the integration over $\vec{k}_1$, we obtain for the number of steps a distance R away from the walker's point of origin

$$\frac{1}{4\pi z_c R} \operatorname{erfc} \left(\frac{z_c R^2}{2(N+1)} \right) \tag{10.83}$$

where $\operatorname{erfc}(x)$ is the complementary error function, defined by

$$\operatorname{erfc}(x) = \frac{2}{\sqrt{\pi}} \int_x^{\infty} \mathrm{e}^{-t^2} \mathrm{d}t \tag{10.84}$$

The function in (10.83) falls off as $1/R$ when $R \leq \sqrt{2N/z_c}$, when $R > \sqrt{2N/z_c}$, the fall off is as a Gaussian.

Now, when the walk is densely packed, the region to which it is confined is small compared to the walk's natural extent, so we may assume that the inequality $R \ll \sqrt{N/z_c}$ is always satisfied. This means that $\rho(R) \sim R^{-1}$ in three dimensions ($\rho(R) \sim R^{-(d-2)}$ in general). There will thus be a correction to the number of walks going as $-u/R$, which is precisely the result that we find for the effect of fluctuations on the statistics of the random walk.

11

Scaling, fractals, and renormalization

We are almost ready to fully exploit the connection, established in earlier chapters, between the statistics of a self-avoiding random walk and the statistical mechanics of a magnet near the phase transition from its paramagnetic and ferromagnetic states. Because of the mathematical similarity between the two systems, we will be able to make use of an array of calculational strategies that, collectively, represent realizations of the *renormalization group*. This generic method for the study of systems with long-range correlations has fundamentally altered the way in which physicists view the world around them. The method is so powerful and so widespread in its application, that it seems worthwhile to do a little more than simply explain how to use it in the present context. This chapter consists of a discussion of the philosophy underlying the renormalization group and of a general description of the way in which it is applied. We will finish off by taking the reader through a simple calculation that is relevant to random walks and the associated magnetic system. Then, we will generalize the method to encompass a wide class of systems, the $O(n)$ model being one of them. In the next chaper, the reader will be subjected to a full-blown introduction to the method, as it applies to the self-avoiding walk. Those already familiar with the renormalization group may wish to skip directly to Chapter 12.

11.1 Scale invariance in mathematics and nature

The notion of scale invariance is not exactly new. A famous poem by Jonathan Swift goes as follows:

So, naturalists observe, a flea
Has smaller fleas that on him prey,
And these have smaller still to bite 'em
And so proceed *ad infinitum*.[1]

[1] This notion was enlarged upon by the mathematician Augustus De Morgan, who wrote:

Great fleas have little fleas upon their backs to bit 'em
And little fleas have lesser fleas, and so *ad infinitum*
And the great fleas themselves, in turn, have greater fleas to go on,
While these again have greater still, and greater still, and so on.

That the world is replicated on smaller and smaller scales – *ad infinitum*, as the poem asserts – is a seductive one. Of course, we now know that life on the scale of whatever bites a flea is different from life on the scale of a flea, and that life on the flea's scale, in fact, differs in important respects from life on the scale of the animals that are bitten by them. Nevertheless, there are cases in which phenomena are replicated on smaller and smaller, or larger and larger, length scales. in these instances, there is said to be *scale invariance*, and it is on such cases that we will begin our discussion.

11.1.1 Mathematical scale invariance: fractals

Let's start with the simplest possible case in which scale invariance manifests itself. Consider a point. This mathematical fiction is a zero-dimensional object that has no spatial extent in any direction. You cannot acually see a point, as it is vanishingly small, and is thus beyond the range of the unaided eye or any apparatus built for magnifying small objects. Nevertheless, we'll talk about what you "see" when you look at a point. The important thing about a point is that it is a point no matter the magnification with which it is viewed. In this sense, a point exhibits scale invariance. It is looks the same under all magnifications. This is trivial scale invariance.

Another example is the case of a straight line. Again, because the ideal line is infinitesimally thin, you cannot see it, but if you could, the straight line – which extends out to infinity in one dimension, but which has no width – would look like a straight line under any magnification. Here again, there is essentially trivial scale invariance.

If we curve the line, by making it, for instance, the circumference of a circle, then what we end up with is something that does not look the same under all magnifications. The closer you look at a curved line, the more it appears to straighten. This reflects the mathematical fact that a curve possesses a length scale. In the case at hand, the length scale is established by the radius of the circle of which the curve is the circumference. The more you magnify the curve, the greater the radius appears to be. In the limit of infinite magnification, under which the circle appears to be infinitely large, the curve has straightened out and becomes a straight line. On the other hand, as you pull further and further back, so that the circle becomes smaller and smaller, the curve eventually starts to resemble a point.

In general, a curved line is an example of a mathematical construction that has an intrinsic scale. Are there cases of curves that possess scale invariance, and for which the scale invariance is a little more complicated than in the example of the

Fig. 11.1 The first step in the construction of a Koch curve.

Fig. 11.2. The next step in the construction of a Koch curve.

point or the line? The answer is yes. These curves are known as *fractals*. Just what is a fractal? Mathematically speaking, a fractal is an entity that has a fractional dimensionality. We'll see what that means after we've constructed a fractal. We'll start simply, with the Koch curve.[2]

The Koch curve

Recall what happened to the straight line under magnification. Its apparent curvature vanished. This means that any "features" that a curve exhibit are magnified out of visibility as our view of the curve is taken to smaller and smaller length scales. We can see to it that the curve itself does not straighten out into a line by placing smaller features on top of features, and even smaller features on top of them, much in the same way as Jonathan Swift imagined littler and littler fleas living on each other. The feature in the case of the Koch curve is triangular, and it is placed on the middle third of a straight line segment. Figure 11.1 illustrates the process. Notice that one of the consequences of the addition of the triangular peak to the center of the line segment is that the length of the curve has increased by a factor of $4/3$. This is because the new curve consists of four segments, each of which has a length equal to a third of the original line. The next step in the construction of the Koch curve is to modify the four line segments that we now have in exactly the same way as we did the original line segment (see Figure 11.2). Instead of four segments there are now $4 \times 4 = 16$ of them, and the length has increased by another factor of 4/3. At the next step, we decorate each of these 16 segments in the same way as we have decorated the segments previously, and we continue in the fashion... *ad infinitum*. The result of this procedure is a curve-like object that displays interesting structure at any degree of magnification. The object has another interesting property. Its length is infinite. This is because at every stage in its construction you have increased the length of the curve by the factor 4/3, and

[2] For a discussion of the Koch curve, and fractals in general, see (Mandelbrot, 1982).

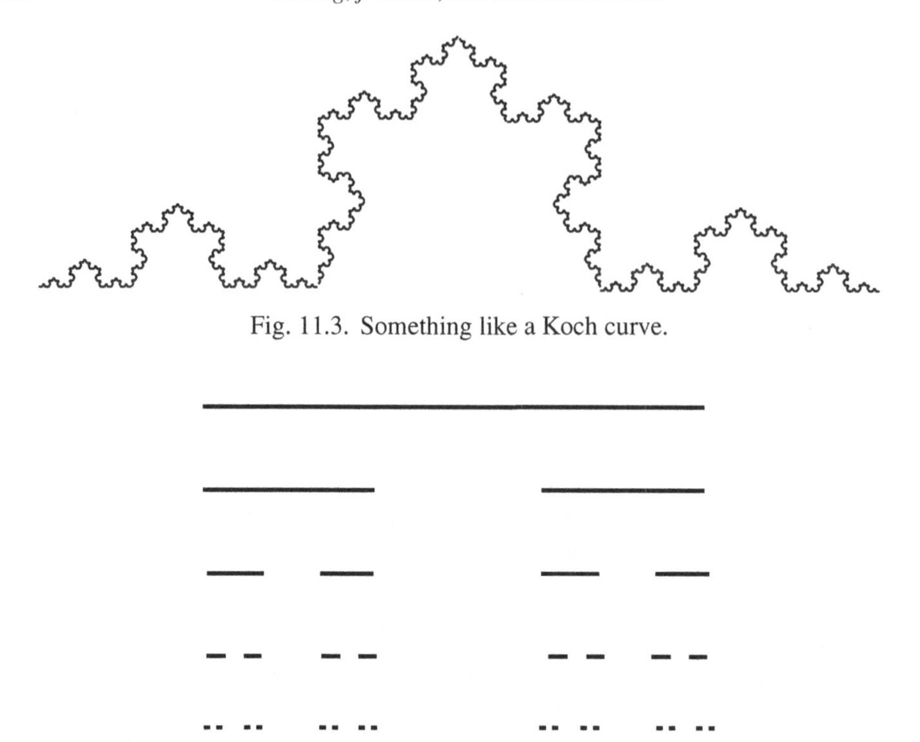

Fig. 11.3. Something like a Koch curve.

Fig. 11.4. The steps in the construction of the Cantor set.

you have done this an infinite number of times. Figure 11.3 is an attempt to depict what a Koch curve looks like. It is impossible to draw, just as it would be impossible to see, all the structure in the curve. There is the strong intimation of details upon details that would show up under sufficient magnification.

The Koch curve is one of the simplest examples of a fractal curve. This particular fractal is one of a class that goes by the name deterministic fractal. As another example of this class, consider the Cantor set, constructed in the following manner. First, remove the middle third sction of a straight line segment. There are now two segments, each one third the length of the original line. Then, eliminate the middle third of each of these two segments to generate a total of four segments, each a ninth of the length of the original line. Continue in this process, again, *ad infinitum*. What you are left with constitutes a fractal, similar in properties to the Koch curve, except that now its total length is zero. The successive steps in the construction of the Cantor set are illustrated in Figure 11.4.

Fractals need not be geometrically precise – they can also be generated by random processes. Examples of such processes include the shoreline of a section of coast, of the trails left by a collection of random walkers. Fractals belonging to this class are known as random fractals. Other examples of both deterministic and random

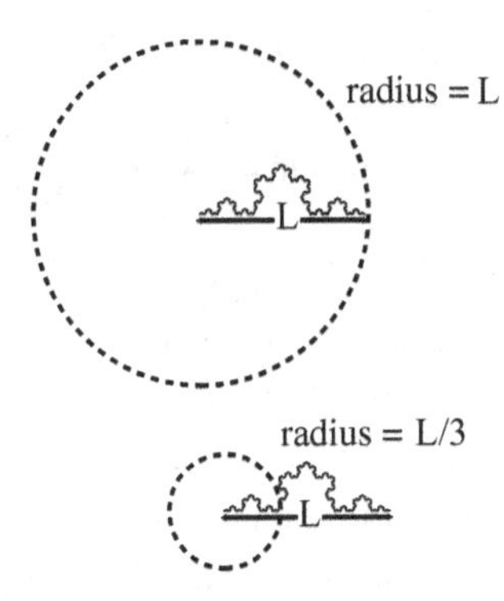

Fig. 11.5. Finding the fractal dimensionality of the Koch curve.

fractals can be found in books by Mandelbrot, which provide a lucid introduction to the general topic (e.g. Mandelbrot, 1982).

Now we turn to the etymology of the term fractal, which is short for fractional dimensionality. As it turns out, there are quite a few ways to define the dimensionality of a mathematical set. We will use one here, corresponding to what is now called the *Hausdorff dimensionality*. It corresponds to the "amount" of a curve or other structure that is contained in a compact volume of some linear extent. In the case of the Koch curve, imagine drawing a circle of radius r centered at one end of the curve. We ask how much of the Koch curve is inside that circle. Let the end-to-end distance of the Koch curve be L. Then if $R \geq L$, all the Koch curve is contained in the circle. Start with $R = L$. Now shrink the radius of the circle to $R/3$. One quarter of the Koch curve is now contained in the circle (see Figure 11.5). If the radius of the circle is shrunk by another factor of three, another three quarters of the Koch curve is eliminated from the interior of the circle. Now, we define the amount of the Koch curve that is contained within a circle of radius R by $n(R)$. The exercise that we have just performed indicates that $n(R)$ has the property

$$n(R/3) = n(R)/4 \tag{11.1}$$

Suppose we make the assumption that $n(R)$ is in the form of a power law, that is

$$n(R) = KR^p \tag{11.2}$$

Equation (11.1) then takes the form

$$K\left(\frac{R}{3}\right)^p = K\frac{R^p}{4} \tag{11.3}$$

Eliminating the multiplicative constant K and the factor R^p from both sides of (11.3), we find

$$3^{-p} = 1/4 \tag{11.4}$$

Taking the logarithm of both sides of (11.4), we find

$$-p \ln 3 = -\ln 4 \rightarrow p = \frac{\ln 4}{\ln 3} \tag{11.5}$$

The power connecting R, the size of the region, and $n(R)$, the amount of the Koch curve inside the region is $\ln 4/\ln 3$. This exponent p in the power-law dependence of n on R is the Hausdorff dimensionality of the Koch curve.

The reason that the exponent is called a dimension has to do with what one finds if one performs the corresponding calculation for simple curves and surfaces. Suppose we were interested in the amount of a line segment that would be contained in a circular region when the circle is centered at the end of the line segment and has a radius equal to R. If R is less than the length of the line, the amout of line segment inside the circle is just equal to the length of the portion of the line that lies inside the circle, and that length is R. Thus $n(R) = R$, and the power law that one inserts in an equation such as (11.2) is one, which is the dimensionality of a line.

Exercise 11.1

Show that if the object of interest is a two-dimensional surface, then the amount of the object inside a circle of radius R scales as R^2.

Seen from the above perspective, the fractal dimensionality is a generalization of the notion of spatial dimensionality, with the added property that fractal dimensionalities need not be integral. The fractal dimensionality of a Koch curve lies between one and two, so in a sense, the Koch curve represents a kind of interpolation between a line and a plane.

Exercise 11.2

Repeating the argument above in the case of the Cantor set, determine the fractal dimensionality of this object.

Another quick way to calculate the fractal dimensionality of certain geometric fractals is to count the replicas, $\mathcal{R}$, of the similar objects, and the magnification, $\mathcal{M}$, of one of them that is needed to reproduce the object from which the replicas were derived. The fractal dimensionality is then given by the formula

$$p = \frac{\ln \mathcal{R}}{\ln \mathcal{M}} \tag{11.6}$$

Applying this procedure to the Koch curve and the Cantor set immediately yields the results just presented for their fractal dimensionalities. This scaling feature of fractals is made prominent by denoting their Hausdorff dimension as the scaling dimension.

Exercise 11.3

Show that the method just described yields the same fractal dimensionalities for the Koch curve and the Cantor set as the method described earlier in this section.

To review the determination of the fractal dimensionality of the Koch curve, we established a way of determining how much of the curve there is within a given length of a point. We also saw how that quantity varies as the length is changed. We found that when the length is changed by one factor, the amount changes by another. On the basis of this finding, we are able to infer a power-law relationship between the length and the amount of the fractal. This all works for an unrestricted (non-excluded volume) random walk. Here, the amount of the walk inside a region of linear extent R ought to go as R^2. This is because the distance that a walker travels in N steps goes like $N^{1/2}$, and we can interpret the "amount" of a random walker in a region as the number of steps within that region. The fractal dimensionality of an unrestricted random walk is, then equal to two. Our goal is the fractal dimensionality of the self-avoiding walk.

The renormalization group is based on a generalization of the method described above. One looks at the relationships between the properties of a system as viewed on one length scale and the same properties as observed on another one. The relationships are generally expressible in terms of ratios. On the basis of the specific ratios discovered, it is possible to construct power-laws for the dependences of quantities of interest on underlying variables.

11.1.2 Examples of the statistical mechanical renormalization group. The mean field magnet and the Gaussian model

Mean field theory, as discussed in Chapter 9, asserts the following power-law dependences of the magnetic susceptibility

$$\chi_T(T) \propto (T - T_c)^{-1} \quad (h = 0, T > T_c) \tag{11.7}$$

$$\chi_T(T) \propto (T_c - T)^{-1} \quad (h = 0, T < T_c) \tag{11.8}$$

Here, h is the externally applied magnetic field, and the quantity $\chi_T(T) = (\partial M/\partial h)_T\big|_{h=0}$ is the isothermal magnetic susceptibility at zero applied magnetic field. Is there a way in which this power law can be inferred from the way in which the system looks at different length scales? Recall the Landau theory result for the effective Hamiltonian of the magnetic system:

$$\mathcal{H}(r, h, m) = V\left[rm^2 + um^4 - hm\right] \tag{11.9}$$

where r is the reduced temperature, $\propto (T - T_c)$, of this magnetic system, and m is the magnetization per unit volume, which is equal to ML^{-d}. The quantity V in (11.9) is the volume of the system, equal to L^d, where L is the linear extent of the system and d is the system's spatial dimensionality. Now, imagine that we change our frame of reference by altering the length scale on which we view the system. If we magnify the system by the factor b, then $L \to bL$. Each term in (11.9) is then multiplied by b^d, because of the change in the common factor V. We can absorb this change in the effective Hamiltonian by making $m \to m' = mb^{d/4}, r \to r' = rb^{d/2}$, and $h \to h' = hb^{3d/4}$. The equilibrium magnetization is the solution to the magnetic equation of state: $\partial\mathcal{H}/\partial m = 0$, or, equivalently, $\partial\mathcal{H}/\partial m' = 0$. This means that the magnetization per unit volume of the system has the property

$$m(r, h)b^{d/4} = F\left(rb^{d/2}, hb^{3d/4}\right) \tag{11.10}$$

or

$$m(r, h) = b^{-d/4}F\left(rb^{d/2}, hb^{3d/4}\right) \tag{11.11}$$

Because parameter b is extraneous to the actual physical behavior of the system, the right hand side of (11.11) ought to be independent of b. Utilizing the fact that we are able to construct the functional dependence of m on the variables r and h, (11.11) reduces to

$$m(r, 0) = b^{-d/4}F\left(rb^{d/2}, 0\right) \tag{11.12}$$

As $r \to 0$, if $m(r, 0)$ is to be independent of the scaling parameter b, then this function must obey the power law

$$m(r, 0) = \kappa r^{1/2} \tag{11.13}$$

as it should. In the other limiting case, when we are at the critical temperature $r = 0$, m is consistent with the functional form

$$m(0, h) = b^{-d/4}F\left(0, hb^{3d/4}\right) \tag{11.14}$$

Again, the b-dependence drops out if $m(0, h)$ has the following power-law behavior

$$m(0, h) = \kappa' h^{1/3} \tag{11.15}$$

Quite generally, if the magnetization is to be independent of the scale factor b, its functional dependence on the variables r and h must be of the form

$$m(r, h) = r^{1/2}f(h/r^{3/2}) \tag{11.16}$$

with $f(h/r^{3/2}) \to h^{1/3}r$ as $r \to 0$. It is straightforward to establish that the dependence on r and h of m displayed in (11.16) is consistent with the relationship (11.11). If we take the first derivative of m with respect to h, and then set $h = 0$, we find

$$\left(\frac{\partial m}{\partial h}\right)_r\bigg|_{h=0} = r^{-1} f'(0) \tag{11.17}$$

This is consistent with the results displayed in (11.7) and (11.8). We are thus able to infer the critical exponent for the susceptibility from the way in which variables transform under changes of scale.

Let's look at another example of the statistical mechanics of a magnetic system. In the so-called Gaussian model the statistical mechanics of the system is controlled by the effective Hamiltonian

$$\mathcal{H} = \int \mathrm{d}^d x \left[\frac{r}{2} S(\vec{x})^2 + \left|\vec{\nabla} S(\vec{x})\right|^2 - hS(\vec{x})\right] \tag{11.18}$$

Here the quantity $S(\vec{x})$ is the spin at the location $\vec{x}$. If the length scale is adjusted by the factor b, then the volume element $\mathrm{d}^d x$ becomes $b^d \mathrm{d}^d x$. On the other hand, the derivative $\vec{\nabla}$ goes to $b^{-1}\vec{\nabla}$. This is because taking a spatial derivative is, algebraically, the same as dividing by a distance. The changes in the effective Hamiltonian can be subsumed into the following changes in the spin field $S(\vec{x})$, the reduced temperature r and the magnetic field h as follows:

$$S(b\vec{x}) = b^{(2-d)/2} S'(\vec{x}) \tag{11.19}$$

$$r = b^{-2} r' \tag{11.20}$$

$$h = b^{-(d+2)/2} h' \tag{11.21}$$

This means that the magnetization per unit volume, which is equal to the average $\langle S \rangle$, has the following dependence on r, h and b

$$m = b^{(2-d)/2} f\left(rb^2, hb^{(d+2)/2}\right) \tag{11.22}$$

The form displayed in (11.22) is easily arrived at once the scaled variables, (11.19)–(11.21), are substituted into the expression

$$m(r, h) = \frac{\int S(\vec{x}) \mathrm{e}^{-\beta\mathcal{H}} \mathcal{D}\left[S(\vec{x})\right]}{\int \mathrm{e}^{-\beta\mathcal{H}} \mathcal{D}\left[S(\vec{x})\right]} \tag{11.23}$$

Again, the variable b is not relevant to the physics of the system, so the magnetization must have a form that eliminates that dependence. Repeating the steps leading to

(11.16) for the mean field case, we arrive at the form

$$r^{(d-2)/4} F\left(\frac{h}{r^{(d+2)/4}}\right) \tag{11.24}$$

for the Gaussian case. Taking the derivative of m with respect to h, we find for the isothermal susceptibility

$$\chi_T = r^{-1} F'\left(\frac{h}{r^{(d+2)/4}}\right) \tag{11.25}$$

In the limit $h \to 0$, we regain the critical exponent displayed in (11.7) and (11.8).

Although we obtained the same susceptibility exponent in the case of the Gaussian model as we did for the mean field approximation, it is important to note that the combinations of r and h in the "scaling" functions in (11.16) and (11.24) are different. Note, however, that the combinations coincide when the spatial dimensionality, d, of the system is equal to four. This coincidence is not unexpected, given the special role that $d = 4$ was found to play in the self-avoiding walk discussed in Chapter 8.

Exercise 11.4

For the Gaussian model, calculate the critical exponents β, γ, and δ.

11.2 More on the renormalization group: the real space method

When we investigated the Koch curve to determine its Hausdorff dimensionality, we utilized a "passive" renormalization group method. We performed no operation on the curve other than to make measurements on it. This is appropriate to the nature of a curve. It is an object that, once constructed, does nothing but sit there. In the case of statistical mechanical systems, we are dealing with variables that fluctuate as the result of the thermal motion that all physical systems execute. The scrutiny to which we subject them must take this motion into account, and because of that, the way in which one determines their properties as viewed on different length scales is somewhat more complicated.

As a gentle introduction to the renormalization group method for statistical mechanical systems, we'll look at the Gaussian model in one dimension. There, the partition function is the multiple integral

$$Z = \int \cdots \int \mathrm{e}^{-\mathcal{H}[S_i]} \Pi_i \, \mathrm{d}S_i \tag{11.26}$$

where the exponent contains the effective Hamiltonian $\mathcal{H}[S_i]$, given by

$$\mathcal{H}[S_i] = \sum_i \left[\frac{r}{2}S_i^2 + \frac{a}{2}(S_i - S_{i+1})^2\right] \tag{11.27}$$

The system that we are looking at here is somewhat different from the Gaussian model discussed in the previous section, in that the spin variables now occupy a one-dimensional lattice, in contrast to the spin field $S(\vec{x})$, which is defined on a continuum of points. The statistical mechanics of the two models do not differ when it comes to the behavior close to and at the critical point.

The way in which we are going to see how this system behaves as we scrutinize it on different length scales is with the use of the so-called "decimation" method.[3] What we will do is hold fixed the spin variables on every other site, allowing the remaining spin variables to flucuate freely. In this way, we will generate a "restricted" partition function, depending on half of the degrees of freedom of the system. From this new partition function, we will extract a new effective Hamiltonian, in which the variables are the degrees of freedom that we have held fixed, which are fewer. For obvious reasons this process has been identified as "thinning out" degrees of freedom. The relationships of interest will be those that couple the parameters in the new free energy to the parameters in the the original effective Hamiltonian.

The degrees of freedom that will be held fixed are the spins on the even-numbered sites. Those that are allowed to fluctuate are the spin variables on the odd-numbered sites. We'll start by focusing on the spin on site number 3. The contribution to the partition function that depends on this spin variable is of the form

$$\int dS_3 \exp\left(-\frac{r}{2}S_3^2 - \frac{a}{2}(S_3 - S_2)^2 - \frac{a}{2}(S_3 - S_4)^2\right) \tag{11.28}$$

Rearranging terms in the exponential, we are left with the contribution

$$\int dS_3 \exp\left[-\left(\frac{r}{2} + a\right)S_3^2 + aS_3(S_2 + S_4) - \frac{a}{2}\left(S_2^2 + S_4^2\right)\right] \tag{11.29}$$

The integral is carried out in the usual manner, by completing squares, yielding

$$\sqrt{\frac{2\pi}{r + 2a}} \exp\left[-\frac{a^2}{2(r + 2a)}(S_2 - S_4)^2 - \frac{ar}{2(r + 2a)}\left(S_2^2 + S_4^2\right)\right] \tag{11.30}$$

Combining the results in (11.30) with the remaining terms in the partition function, we find for the new partition function

$$\left(\frac{2\pi}{r + 2a}\right)^{N/2} \exp\left[\sum_i \left(-\frac{r'}{2}S_{2i}^2 - \frac{a'}{2}\left(S_{2i} - S_{2(i+1)}\right)^2\right)\right] \tag{11.31}$$

[3] This method is not, strictly speaking, decimation, which refers to a reduction by 10%. The terminology was introduced into statistical mechanics with the method, and, for better or worse, it has stuck.

where

$$r' = \frac{r(r+4a)}{r+2a} \tag{11.32}$$

$$a' = \frac{a^2}{r+2a} \tag{11.33}$$

Exercise 11.5

Fill in the steps leading to (11.31).

The exponential in the restricted partition function looks a lot like the exponential in the original parition function. We will improve the resemblance a little by rescaling the spin variables so as to reset the coupling coefficient to its original value. Replacing S_{2i} by $S_{2i}\sqrt{(r+2a)/a}$, i.e.

$$S_{2i} \to S_{2i}\sqrt{\frac{r+2a}{a}} \tag{11.34}$$

we generate an exponent in the restricted partition function that has the form

$$\sum_i \left(-\frac{r''}{2}S_{2i}^2 - \frac{a''}{2}\left(S_{2i} - S_{2(i+1)}\right)^2\right) \tag{11.35}$$

where

$$r'' = \frac{r(r+4a)}{a} \tag{11.36}$$

$$a'' = a \tag{11.37}$$

We now relabel the site indices ($2i \to i$), so that the effective Hamiltonian of the new system is of the same appearance as the original effective Hamiltonian, except for the change in the reduced temperature, r, which has been altered as per (11.36).

What if we had included in the model the ordering field term $h\sum_i S_i$? The effect of this term is to alter the factor multiplying the exponential. The ordering field, after the decimation procedure, has the following relationship to the original ordering field:

$$h'' = h\frac{r+4a}{\sqrt{a(r+2a)}} \tag{11.38}$$

Exercise 11.6

Show that if an ordering field, h, is included in the Gaussian effective Hamiltonian in (11.27), through the additional term

$$-\sum_i hS_i$$

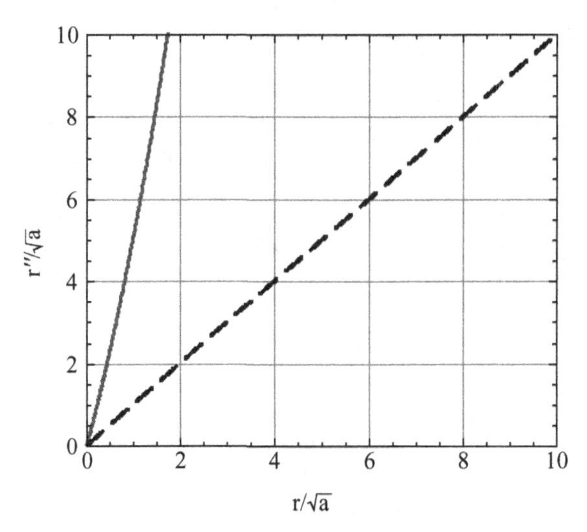

Fig. 11.6. The recursion relation between r and r''.

then the additional recursion relation (11.38) results. Additionally, show that this term has no effect on the recursion relation (11.36).

The relationship between the effective Hamiltonians of the original spin system and the spin system that is the result of the decimation procedure is $\mathcal{H}(r, h) \rightarrow \mathcal{H}(r'', h'')$. We infer the statistical mechanics of the system on the basis of this connection, and in particular on the basis of the relationships between the new thermodynamic parmeters and the original ones. This is because the decimation procedure can be applied to the new partition function in exactly the same way as it was to the original partition function, yielding yet another effective Hamiltonian of further altered thermodynamic parameters. The relations that connect the triple-primed parameters to the double-primed ones are identical to (11.36) and (11.37). That is, we have obtained a set of transformations that yield modifications of the reduced temperature and ordering field via the recursion relations

$$r^{(n)} = \mathcal{R}_1\left(r^{(n-1)}\right) \tag{11.39}$$

$$h^{(n)} = \mathcal{R}_2\left(h^{(n-1)}\right) \tag{11.40}$$

The connection between the reduced temperature, r, before the decimation procedure and the reduced temperature, t'', after decimation is illustrated in Figure 11.6. The dashed line in the figure is the 45° line $r'' = r$. To find the new reduced temperature associated with an original value of r, one locates r on the x axis, and then reads value on the y axis corresponding to the solid curve.

How further iterations of the relationship (11.36) between the original reduced temperature and the reduced temperature after decimation are graphically

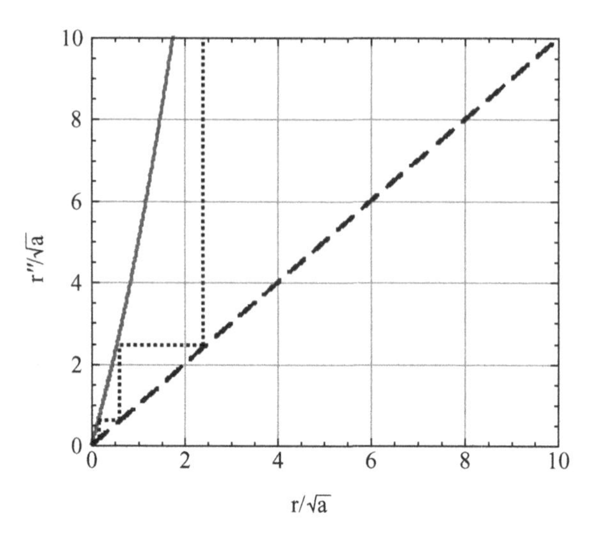

Fig. 11.7. Graphical iteration of the recursion relation between r and r''.

accomplished is illustrated in Figure 11.7. To find the reduced temperature at the first iteration of the relation, one projects vertically on the solid curve, as indicated by the leftmost vertical dotted line in the figure. To find the reduced temperature at the second iteration, one projects horizontally from the solid curve to the 45° line, along the lowest horizontal dotted line and then again projects vertically onto the curve along the next vertical dotted line. As is clear from the figure, an initial reduced temperature is increased more and more as the recursion relation is iterated over and over again.

When the reduced temperature is small enough, the recursion relation between r and r'' is well-approximated by

$$r'' = 4r \tag{11.41}$$

It r'' is also small, then the approximation in (11.41) can also be used at the next iteration. Continuing as long as the reduced temperature is small, one ends up with a reduced temperature, $r^{(n)}$, after n iterations, that is given by

$$r^{(n)} = 4^n r \tag{11.42}$$

Again, under the assumption of small r, the value of the new ordering field is given by

$$h'' = 2^{3/2} h \tag{11.43}$$

Then, the result of n iterations is to generate an $h^{(n)}$, that is given by

$$h^{(n)} = 2^{3n/2} h \tag{11.44}$$

Again, this result holds as long as $r^{(n)}$ is not too large. By the same reasoning, the rescaling of the spin field in (11.34) yields a total spin rescaling factor equal to

$$b^{(n)} = 2^{n/2} \tag{11.45}$$

We can now apply the same reasoning to the new effective Hamiltonian as we did in the discussion in Section 11.1.2 on the scaling behavior of the mean field and continuous spin Gaussian models. According to these arguments, the rescaled expectation value of the magnetization per site is the same function of the rescaled reduced temperature and ordering field as the original magnetization is of the original reduced temperature and magnetic field. Mathematically,

$$m2^{n/2} = f\left(r2^{2n}, h2^{3n/2}\right) \tag{11.46}$$

or

$$m = 2^{-n/2} f\left(r2^{2n}, h2^{3n/2}\right) \tag{11.47}$$

As the number of iterations of the recursion relations is arbitrary, the net magnetization cannot depend on n, which is consistent with

$$f = r^{-1/4} F\left(\frac{h}{r^{3/4}}\right) \tag{11.48}$$

Using this result, we find

$$\left(\frac{\partial m}{\partial h}\right)_r = r^{-1} F'\left(\frac{h}{r^{3/4}}\right) \tag{11.49}$$

and, once again, we recover for the critical exponent γ the mean field value of one. Note also that the general scaling form in (11.49) is consistent with the scaling function for the magnetic susceptibility in (11.25) with $d = 1$.

The form of the magnetization when the renormalized reduced temperature is not small is more complicated than the scaling form displayed in (11.48). This is because the solution of the general recursion relation does not have the simple form displayed in (11.41). We'll pursue the construction of the general partition function in more detail once we've investigated a variant of the above decimation procedure

---

Worked-out example: the one-dimensional Ising model

Here, we formulate the one-dimensional Ising model and set up the renormalization group for it. The Ising model Hamiltonian for an array of spins equally spaced along a line is

$$H = -J \sum_{i=1}^{N} S_i S_{i+1} \tag{11.50}$$

where the spin variables, S_i can take on the values ± 1. The coupling constant J is positive, and we will assume that periodic boundary conditions hold, in that $S_{N+1} = S_1$. The partition function for this Ising chain at temperature T is

$$Z = \sum_{S_1=\pm 1} \cdots \sum_{S_i=\pm 1} \cdots e^{K S_1 S_2 + K S_2 S_3 + \cdots} \tag{11.51}$$

where $K = J/k_B T$.

For the moment, let's sum over all configurations of the spin S_3. We isolate the only terms in the exponential in (11.51) that depend on this degree of freedom, and obtain

$$\sum_{S_3=-1}^{1} e^{K(S_2 S_3 + S_3 S_4)} = 2 \cosh K(S_2 + S_4) \tag{11.52}$$

Because S_2 and S_4 take on the values ± 1, the function $2 \cosh K(S_2 + S_4)$ assumes two values as well. What we have is

$$\cosh K(S_2 + S_4) = \begin{cases} 2 \cosh 2K & S_2 = S_4 \\ 2 & S_2 = -S_4 \end{cases} \tag{11.53}$$

We can replace this function by the new one $A e^{K' S_2 S_4}$, if the quantities A and K' are properly chosen.

Exercise 11.7

Show that the correct choices for A and K' are

$$A = 2\sqrt{\cosh 2K} \tag{11.54}$$

$$K' = \frac{1}{2} \ln \cosh 2K \tag{11.55}$$

Exercise 11.8

If an ordering field is introduced, so that the Ising model Hamiltonian contains the additional term

$$-h \sum_i S_i$$

repeat the above analysis and show that summing over the configurations of the spin S_3 results in a function that can be recast into the form

$$A e^{K_1' S_2 S_4 + K_2' (S_2 + S_4)} \tag{11.56}$$

Determine the values of A, K_1' and K_2' in terms of $K_1 = J/k_B T$ and $K_2 = h/k_B T$.

Returning to the system with no ordering field, we have

$$\sum_{S_3=-1}^{1} e^{K(S_2S_3+S_3S_4)} = A(K)e^{K'S_2S_4} \tag{11.57}$$

By summing over the configurations of all odd-indexed spins, S_1, S_3, $S_5, \ldots$, we eliminate half of the degrees of freedom in the one-dimensional Ising system. By this process we thin out the degrees of freedom. The partition function then takes the form

$$Z_N = \sum_{S_2=\pm1} \sum_{S_4=\pm1} \cdots A(K)^{N/2} e^{K'(S_2S_4+S_4S_6+\cdots)} \tag{11.58}$$

Relabeling the sites, $S_2 \to S_1$, $S_4 \to S_2, \ldots$ we have

$$Z_N(K) = A(K)^{N/2} Z_{N/2}(K') \tag{11.59}$$

The partition function $Z_{N/2}(K')$ has the same form as the original partition function, except that it is for a system with half as many degrees of freedom as the system with which we started out, and the coupling between adjacent spins is altered. This tells us that we can write

$$Z_{N/2} = A(K')^{N/4} Z_{N/4}(K'') \tag{11.60}$$

where the relationship between K'' and K' is the same as the relation connecting K' and K. The full partition function is thus given by

$$Z_N(K) = A(K)^{N/2} A(K')^{N/4} \sum_{S_1=\pm1} \cdots \sum_{S_i=\pm1} \cdots e^{K''(S_1S_2+S_2S_3\ldots)} \tag{11.61}$$

The process can clearly be continued until all the spins are eliminated.[4] What this all means is that the evaluation of the partition function is tantamount to solving the recursion relation

$$K_{n+1} = \frac{1}{2} \ln\cosh 2K_n \tag{11.62}$$

where K_1 is the original value of J/k_BT. The partition function is then given by

$$Z = A(K_1)^{N/2} A(K_2)^{N/4} \cdots A(K_j)^{N/2^{j+1}} \cdots \tag{11.63}$$

From this result, an expression for the free energy emerges naturally.[5]

[4] This is, strictly speaking, true only if the total number of spins is a power of two. Otherwise there will be spins "left over" in the decimation process. We ignore this complication, which turns out not to affect the results obtained with this approach.

[5] The solution of the one-dimensional Ising model via the renormalization group is discussed in (Maris and Kadanoff, 1978; Nauenberg, 1975; Nelson and Fisher, 1973).

Exercise 11.9
Express the renormalization group equation for the one-dimensional Ising model in terms of the variabel $x = e^{-2K} = e^{-2J/k_B T}$.

(a) Find the fixed points of this recursion relation. Discuss their physical significance. The fixed point of this recursion relation is a K such that that $\frac{1}{2} \ln \cosh 2K = K$.
(b) Linearize the renormalization group equation about the low-temperature (large K) fixed point and from this show that the correlation length, $\xi(T)$, goes as $e^{2J/k_B T}$ as $T \to 0$.
(c) Calculate the partition function, $Z(K)$, in the limit $T \to 0$, and from your result show that the free energy tends to the limit $E = -NJ$.

11.2.1 The momentum shell renormalization group method

The decimation method works reasonably well for the one-dimensional Gaussian and Ising models. There are, however, complications when it is applied to these models in higher dimensions. For one thing, interactions are introduced between sites that are further and further apart, which means that the coupling terms become more and more complex. In addition, while the real space is extremely well-suited for other systems, the self-avoiding walk is not, alas, one of them. We now discuss a renormalization procedure which is quite general and which we'll then apply to the $O(n)$ model.

There is a way to approach the elimination of degrees of freedom that generalizes quite nicely. One makes use of the fact that one can represent the spins with the use of spatial Fourier transforms, as discussed in Supplement 2 to Chapter 2. To recapitulate: in a lattice in which the distance between neighboring sites is c, the spin degrees of freedom, s_i, can be represented in terms of Fourier-transformed variables, $s(\vec{q})$, as follows:

$$S_i = \frac{1}{\sqrt{N}} \sum_{\vec{q}} S(\vec{q}) e^{i\vec{q}\cdot\vec{R}_i} \tag{11.64}$$

where $\vec{R}_i$ is the position vector of the ith site, and the wave-vector $\vec{q}$ is confined to a Brillouin zone, the linear extent in q space of which is inversely proportional to the distance, c between sites in real space. That is, if Q is the width of the Brillouin zone, then $Q \propto 1/c$. The quantity N in (11.64) is the total number of sites on the lattice, or the total number of independent wave-vectors in the Brillouin zone.

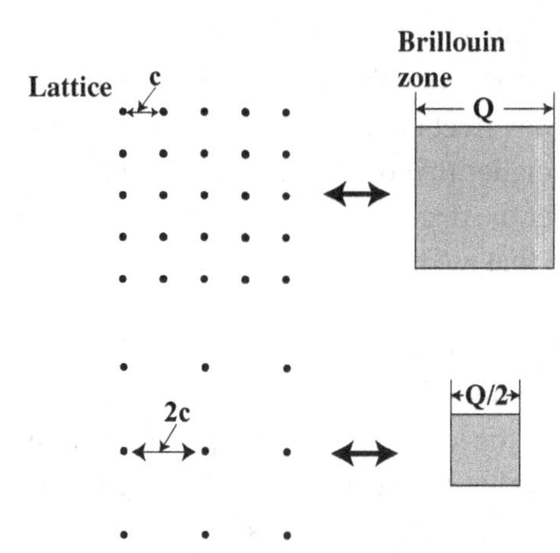

Fig. 11.8. The connection between the lattice and the associated Brillouin zone.

The elimination of spin degrees of freedom on sites is, in a sense, equivalent to the elimination of $S(\vec{q})$'s in the region between the larger Brillouin zone appropriate to the original lattice and the smaller Brillouin zone associated with the lattice containing the remaining spins. The equivalence is illustrated in Figure 11.8.

The Gaussian model is especially simple when expressed in terms of Fourier-transformed variables. All that is required is to substitute (11.64) into (11.27), and to make use of the orthogonality of the plane waves:

$$\sum_{i=1}^{N} \mathrm{e}^{\mathrm{i}(\vec{q}+\vec{q}')\cdot\vec{R}_i} = N\delta_{\vec{q},-\vec{q}'} \tag{11.65}$$

as deduced in Supplement 2 of Chapter 2. The effective Hamiltonian has the form

$$\mathcal{H}\left[S(\vec{q})\right] = \sum_{\vec{q}} \left[\left[\frac{r}{2} + \alpha(\vec{q})\right] S(\vec{q})S(-\vec{q})\right] - \sqrt{N}S(0)h \tag{11.66}$$

where $\alpha(q) = -2a(1 - \cos qc)$ in one dimension and for small q the coupling term takes on the approximate value

$$\alpha(\vec{q}) \sim ac^2q^2 = Jq^2 \tag{11.67}$$

a result that holds in d dimensions.

To simplify matters, we'll assume that $\alpha(\vec{q})$ is as given by the right hand side of (11.67). The elimination of $S(\vec{q})$'s with wave-vectors lying between the smaller Brillouin zone associated with the "decimated" lattice and the Brillouin zone appropriate to the full lattice is straightforward, as the coupling in (11.66) is between $S(\vec{q})$'s for which the magnitude of the wave-vector is the same. The mathematical elimination of these degrees of freedom is accomplished by performing a set of

Gaussian integrations. However, because of the lack of coupling between Fourier-transformed spin modes for which the wave-vectors have different magnitudes, this set of integrations has no effect on the dependence of the effective Hamiltonian on the spin modes in the smaller Brillouin zone. This means that we can effectively ignore the integration over the degrees of freedom that are averaged out and focus on the degrees of freedom that are held fixed in this realization of the renormalization group method.

At first sight, it appears as if there is no change at all in the effective Hamiltonian of the reduced set of degrees of freedom. However, it is also necessary to take into account the fact that the length scale appropriate to the system has changed. In particular, if a decimation has been carried out, the length scale has doubled. In light of the reciprocal relationship between scales in real and q space, the scale against which wave-vectors are measured has been halved. To take this into account, we replace all wave-vectors, $\vec{q}$ in the new effective Hamiltonian by $\vec{q}/2$. Furthermore, it is necessary to express the number of degrees of freedom, N, in terms of the number of degrees of freedom in the reduced set of thermodynamic variables. Again, if the procedure that has been carried out is decimation, then, we ought to replace N by $2^d N$, where d is the system's spatial dimensionality. Having carried out these rescalings, we are left with an effective Hamiltonian that is of the form

$$\sum_{\vec{q}}\left[\left(\frac{r}{2}+\frac{Jq^2}{4}\right)S(\vec{q})S(-\vec{q})\right]-\sqrt{N}2^{d/2}S(0)h \tag{11.68}$$

We now rescale the spin field so as to preserve the magnitude of the coupling coefficient, J. This is accomplished by the scale change

$$S(\vec{q}) \rightarrow 2S(\vec{q}) \tag{11.69}$$

The effective Hamiltonian of the reduced set of degrees of freedom is, then

$$\sum_{\vec{q}}\left[\left(\frac{4r}{2}+Jq^2\right)S(\vec{q})S(-\vec{q})\right]-\sqrt{N}2^{(d+2)/2}S(0)h \tag{11.70}$$

This is the same as the original effective Hamiltonian, except that the reduced temperature and the externally imposed magnetic field are altered as follows:

$$r \rightarrow 2^2 r \tag{11.71}$$

$$h \rightarrow 2^{(d+2)/2}h \tag{11.72}$$

Now, this procedure can be generalized, in that the change in the size of the Brillouin zone need not have been by a factor of two. If we replace the 2 by b in Eqs. (11.69), (11.71) and (11.72), we see, following previous arguments, that the relationship between the magnetization, the reduced temperature, and the applied magnetic field must obey the following scaling requirement:

$$m(r, h) = b^{(2-d)/2} m\left(rb^2, hb^{(d+2)/2}\right) \tag{11.73}$$

The factor $b^{(2-d)/2}$ comes about because m, which is essentially $\langle S_i \rangle$, scales as the spin per site, which, in turn, scales as $S(\vec{q})/\sqrt{N}$. As we have seen, this introduces the factor $b/b^{d/2}$, or $b^{(2-d)/2}$. The functional form expressed in (11.73) is precisely the scaling requirement that was inferred for the Gaussian model earlier with the use of a "passive" renormalization group argument.

11.2.2 The continuous renormalization group

One particular advantage of the use of the shrinking Brillouin zone to determine the scaling properties of a statistical mechanical system is that the alterations in the length scale can be performed continuously. This simplifies portions of the analysis, and it greatly enhances the power and versatility of the method. As a start, we'll reanalyze the Gaussian model by imagining that the Brillouin zone's width is equal to $e^{-\ell}$, where ℓ is a continuous variable. Suppose, now, that in the process of changing length scales, the parameter ℓ changes by an infinitesimal amount, $\delta\ell$. The new scale on which wave-vectors are measured is now $e^{-\ell-\delta\ell} = (1 - \delta\ell)e^{-\ell}$. Notice that this corresponds to a *fractional* change in the size of the Brillouin zone. The rescaling of the wave-vector results in the replacement

$$\begin{aligned} \vec{q} &\rightarrow \vec{q}\frac{(1-\delta\ell)e^{-\ell}}{e^{-\ell}} \\ &= \vec{q}(1-\delta\ell) \end{aligned} \tag{11.74}$$

Then, the coupling term in the effective Hamiltonian becomes

$$q^2(1-\delta\ell)^2 S(\vec{q})S(-\vec{q}) = q^2 S(\vec{q})S(-\vec{q})(1-2\delta\ell) \tag{11.75}$$

The coupling term is kept constant by a rescaling of the spin field:

$$S(\vec{q}) \rightarrow (1+\delta\ell)S(\vec{q}) \tag{11.76}$$

The term containing the reduced temperature, r, is then changed as follows:

$$rS(\vec{q})S(-\vec{q}) \rightarrow rS(\vec{q})S(-\vec{q})(1+2\delta\ell) \tag{11.77}$$

The number of degrees of freedom, N, is altered as well:

$$\begin{aligned} N &\rightarrow N\frac{e^{d(\ell+\delta\ell)}}{e^{d\ell}} \\ &= N(1+d\delta\ell) \end{aligned} \tag{11.78}$$

This means that the term involving the external magnetic field alters as follows:

$$\sqrt{N}S(0)h \rightarrow \left(1 + \frac{d+2}{2}\delta\ell\right)\sqrt{N}S(0)h \tag{11.79}$$

The results above indicate that the results of the change in length scales are the following alterations of the reduced temperature and applied magnetic field

$$r \rightarrow r(1 + 2\delta\ell) \tag{11.80}$$

$$h \rightarrow h\left(1 + \frac{d+2}{2}\delta\ell\right) \tag{11.81}$$

Calculating the infinitesimal changes in r and h, and then dividing by the infinitesimal fractional increment in the size of the Brillouin zone, $\delta\ell$, we end up with the following differential equations for the changing reduced temperature and applied field:

$$\frac{\mathrm{d}r}{\mathrm{d}\ell} = 2r \tag{11.82}$$

$$\frac{\mathrm{d}h}{\mathrm{d}\ell} = \frac{d+2}{2}h \tag{11.83}$$

We would like to know the relationship between the magnetization per site on the original lattice and the variables appropriate to the "thinned out" lattice. To obtain this relationship, we recall the connection between spin variables on a lattice and their spatial Fourier transform, given by (11.64). Let $N(\ell)$ be equal to the number of degrees of freedom in the reduced lattice, and define the rescaled spins on that lattice as $S_i^{(\ell)}$, with spatial Fourier transforms $S^{(\ell)}(\vec{q})$. Then, the contribution of the mode with zero wave-vector to the magnetization per lattice site on the original lattice is given by

$$\begin{aligned} S_i &= \frac{S(0)}{\sqrt{N}} \\ &= \frac{S(0)}{S^{(\ell)}(0)} \frac{\sqrt{N(\ell)}}{N} S_i^{(\ell)} \\ &= \mathrm{e}^{((2-d)/2)\ell} S_i^{(\ell)} \end{aligned} \tag{11.84}$$

The magnetization per site on the thinned out lattice will be some function of the renormalized parameters appearing in the new effective Hamiltonian, i.e.

$$m^{(\ell)} = f\left(r(\ell), h(\ell)\right) \tag{11.85}$$

This means that the magnetization per site on the original lattice is given by

$$m = \mathrm{e}^{((2-d)/2)\ell} f\left(r(\ell), h(\ell)\right) \tag{11.86}$$

By the same arguments that we have previously invoked, the left hand side of (11.86) is independent of ℓ. Taking the derivative with respect to ℓ of (11.86) we end up with the following differential equation for $m(r, h)$

$$0 = \left(-\frac{d-2}{2} + \frac{dr}{d\ell}\frac{\partial}{\partial r} + \frac{dh}{d\ell}\frac{\partial}{\partial h} \right) m(r, h)$$
$$= \left(-\frac{d-2}{2} + 2r\frac{\partial}{\partial r} + \frac{d+2}{2}h\frac{\partial}{\partial h} \right) m(r, h) \quad (11.87)$$

One demonstrates by substitution that the scaling form (11.24) for $m(r, h)$ satisfies (11.87). Alternatively, a solution to this partial differential equation can be constructed with the use of the method of characteristics. The result is the same.

This last development of the scaling form of the magnetization of the Gaussian model may seem unnecessarily complicated. However, it is a prototypical version of relationships that are central to the field-theoretical approach to scaling in the vicinity of the critical point.

11.3 Recursion relations: fixed points and critical exponents

In this section, we will describe the general philosophy underlying the renormalization group method, along with the assumptions that are central to its application. We will also introduce the reader to some of the terms that have been coined to describe operations and entities in the context of the renormalization group method. This having been done, the reader will be prepared to face the thickets of mathematics to be confronted in the next chapter.

It should be noted that the discussion to follow could have easily been placed at the beginning of this chapter, in that nothing that has been previously discussed is essential to the understanding of what follows. However, the fact that the reader has already gone through some simple examples of the application of the renormalization group method ought to allow for the rapid assimilation of the ideas that will be outlined in the next few pages. We follow closely the conceptual point of view laid out in the pioneering work of Kenneth Wilson, the inventor of renormalization group as applied to critical phenomena (Wilson, 1971a).

11.3.1 Block spins

The key property of a critical point is that there is a diverging length scale, which is a result of the fact that the correlation length is infinite at the critical point. The kind of system of interest to us actually possesses additional length scales, because of the system's inevitable finite size, and the fact that it is made up of molecules. However, the special properties of a critical point are controlled by the correlation length, and in particular by the consequences of that length's becoming larger and larger as the critical point is approached. At the critical point the correlation length has become infinite. Ignoring all other length scales, the system is now *scale-invariant*, in that there is no way to determine the scale on which the system is being viewed on the basis of the (statistical mechanical) properties of the system alone. On the other hand, when the system is not at its critical point, the connection between the systems as viewed under different length scales is non-trivial. They key to understanding how the system behaves both at and away, but not too far away, from the critical point, lies in the "mappings" that can be established between the properties of the system under changes of length scale.

Once the nature of a system's degrees of freedom have been established, the statistical mechanics of that system are entirely determined by the system's energy, or, in the present context, by the system's effective Hamiltonian. Now, imagine that we are interested in a system in which the degrees of freedom are magnetic moments, sitting on the sites of a lattice. The effective Hamiltonian of this system will be of the general form

$$H\left[\{S_i\}, \{\kappa_i\}\right] \tag{11.88}$$

where $\{\kappa_i\}$ represents the complete set of parameters needed to quantify the effective Hamiltonian, and $\{S_i\}$ stands for the complete set of degrees of freedom in the system. In the case at hand, the variables are spins on the lattice sites.

Imagine, now that one alters the system in some way so that the appropriate set of statistical mechanical variables are now "block spins" on a larger lattice. The arguments in the new effective Hamiltonian will be those variables and a new set of coupling parameters. It is assumed that one can make a one-to-one correspondence between the variables in the new system and those in the original one. This is clearly not possible if the system is formally finite, because in the process of thinning out one has necessarily reduced the number of degrees of freedom. That this is so follows from the fact that the number of sites in a lattice of finite extent is controlled by the ratio of the total volume occupied by the lattice to the "primitive" volume surrounding a site. The greater the spacing between lattice points, the larger this volume, and the smaller the number of sites that will fit in a given total volume. We will assume that the system is in the thermodynamic limit, which allows us to eliminate a finite fraction of the degrees of freedom and still create a one-to-one connection between the original set of degrees of freedom and the set we are left with after the elimination.

Because we have not, by design, changed the set of thermodynamic variables in the process that we are imagining, the only change in the effective Hamiltonian is in the values of the parameters. The "block spinning" process that is envisioned here thus gives rise to a set of mappings of the form

$$\kappa_i \to R_i(\kappa_1, \kappa_2, \kappa_3, \ldots) \tag{11.89}$$

and so on. The renormalization group procedure consists of the construction and analysis of the relations in (11.89).

Now, we have already discussed ways in which degrees of freedom can be thinned out in the context of a scaling analysis of the Gaussian model, and we have generated relationships that correspond to (11.89). In the next section we will further explore the construction of those relations, which are, in general, called *recursion relations*. We will look at what one ought to expect in the vicinity of a critical point, and how one utilizes the recursion relations to extract critical exponents.

The recursion relations need not be of the discrete form depicted in (11.89). We now know from our scrutiny of the Gaussian model that one can, at least in principle, construct differential recursion relations, of the form

$$\frac{d\kappa_i}{d\ell} = W_i(\kappa_1, \kappa_2, \kappa_3, \ldots) \tag{11.90}$$

which implies that the scaling properties of the system can be extracted from a set of first order differential equations.

Now, it is well-known that equations of the form (11.90) can give rise to exceedingly complex behavior. The motion of a turbulent fluid is contained in the Navier–Stokes equation, which can be expressed as an infinite number of first order differential equations. Even when the number of variables and equations is quite small, chaos may result. There are, in fact, examples in statistical mechanics in which the recursion relations give rise to chaotic behavior. They are rare, however, and, in some cases, contrived. We will ignore this possibility, which turns out to be irrelevant to the random walk. As good fortune has it, the behavior of interest to us is reasonably simple.

They key to the interpretation of the implications of the recursion relations is the realization that the properties of the system are controlled by the correlation length, and by the fact that the correlation length is reduced in magnitude with each iteration of the recursion relation. We will take the quantity ℓ in (11.90) to be the same as it was in the case of the Gaussian model. This means that the length scale, s, on which the system is being scrutinized, is given by

$$s = e^{\ell} \tag{11.91}$$

and that we can append to the differential recursion relations (11.90) the following differential equation for the evolution of the correlation length as viewed on the

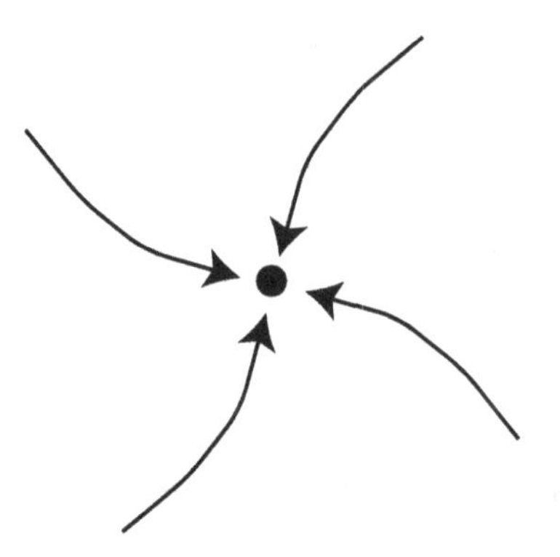

Fig. 11.9. Flow of parameters to their fixed point values on the critical hypersurface.

changing length scale, $\xi^{(\ell)}$,

$$\frac{\mathrm{d}\xi^{(\ell)}}{\mathrm{d}\ell} = -\xi^{(\ell)} \tag{11.92}$$

What should one expect when the system is at its critical point? First, we know that the correlation length is infinite. This means that (11.92) will have no effect on it, in that it remains infinite, no matter the length scale on which it is measured. If a single set of thermodynamic parameters describes a system that has this property, then the right hand sides of the equations in the set (11.90) will all be equal to zero. On the other hand, the more reasonable expectation is that a system is at its critical point when the parameters take on a restricted set of values. The particular values of the parameters are determined by requiring the correlation length, which is a function of the parameters, to be infinite ($\xi(\kappa_1, \kappa_2, \ldots) = \infty$). We'll call this restricted set the *critical hypersurface* in the space spanned by the full set of parameters. The recursion relations will necessarily take a point on the critical hypersurface to another point on the same subspace.

The next question is: where will a point on the critical hypersurface end up as it propagates under the influence of (11.90)? The general theory of first order differential equations allows for a number of possibilities. However, we will assume that there is only one outcome of the movement of the evolution of the parameter set under the influence of the recursion relations, and that is propagation to a fixed point. In other words, we assume that a point on the critical hypersurface will "flow" to a point from which it will not depart. Figure 11.9 illustrates this flow to the fixed point.

Let's look in a little more detail at this fixed point. Mathematically, the fixed point, $(\kappa_1^*, \kappa_2^*, \ldots)$, corresponds to the set of coupling parameters, κ_i at which the right hand side of the continuous recursion relation (11.90) is equal to zero, i.e. $W_i(\kappa_1^*, \kappa_2^*, \ldots) = 0$. We'll make the assumption that the right hand sides of the differential recursion relations do not have any non-analytic behavior in the vicinity of this fixed point. This means that we can expand the right hand sides of these equations as a power series in the difference between the parameters and the values that they take at the fixed point. Let κ_i^* be the value of the parameter κ_i at the fixed point of interest. Then, writing $\kappa_i(\ell) = \kappa_i^* + \delta\kappa_i(\ell)$, we have the following set of equations for the ℓ-dependence of the $\delta\kappa_i$'s

$$\frac{\mathrm{d}\delta\kappa_i(\ell)}{\mathrm{d}\ell} = \sum_j A_{ij}\delta\kappa_j(\ell) + \sum_{j,k} A_{ijk}\delta\kappa_j(\ell)\delta\kappa_k(\ell) + \cdots \tag{11.93}$$

The right hand side of (11.93) contains no terms independent of $\delta\kappa_i(\ell)$, to ensure consistency with the solution $\delta\kappa_i(l) = 0$.

When the deviations of the parameters from their fixed point values are small, the right hand side of (11.93) can be truncated at first order in the $\delta\kappa_i$'s. We are left with the linear differential equations

$$\frac{\mathrm{d}\delta\kappa_i(\ell)}{\mathrm{d}\ell} = \sum_j A_{ij}\delta\kappa_j(\ell) \tag{11.94}$$

This set of equations is solved with the use of standard methods, i.e. by linear transformation to a new set of variables in which the matrix A is diagonal. One finds

$$\delta\kappa_i(\ell) = \sum_\alpha C_{i,\alpha}\mathrm{e}^{\Lambda_\alpha \ell} \tag{11.95}$$

where the "rates" Λ_α are the eigenvalues of the matrix

$$\begin{pmatrix} A_{11} & A_{12} & A_{13} & \\ A_{21} & A_{22} & A_{23} & \cdots \\ A_{31} & A_{32} & A_{33} & \\ & \vdots & & \ddots \end{pmatrix} \tag{11.96}$$

If the fixed point is stable, in that any deviation from it decays under the action of the recursion relations, then all the Λ_α's are negative.[6] We expect that any deviation from the fixed point along the critical hypersurface will, indeed, decay. On the other hand, if one displaces the parameter set off of the critical hypersurface, then the correlation length is finite, which reduces under the renormalization group procedure, and this change must be reflected in a change in the set of thermodynamic parameters in terms of which the correlation length can be expressed. In other

[6] The deviations from the fixed point associated with negative eigenvalues are termed "irrelevant," and the deviations associated with eigenvalues equal to zero are known as "marginal." By contrast, a deviation associated with a positive eigenvalue is called "relevant."

words, it is expected that a few of the Λ_α's are positive. How many depends on how many dimensions are needed to fill out the space of parameters, given the critical hypersurface. In the case of a spin system, in which there are two ways of displacing the system from the critical point, corresponding to a change in temperature and the application of a magnetic field, the codimension of the critical hypersurface is two, which means that two of the eigenvalues of the matrix (11.96) are positive. Let's call the two positive eigenvalues Λ_r and Λ_h, and let's parameterize the distance off of the critical hypersurface in terms of the reduced temperature and the magnetic field. In addition, we'll associate the eigenvalue Λ_h with displacements from the fixed point caused by the imposition of a magnetic field.

By the symmetry of the system, we know that the absence of an applied magnetic field at one length scale implies the absence of a symmetry-breaking field at any length scale, or so one hopes. Suppose that the reduced temperature, r, is not equal to zero, but that otherwise all parameters are set at their fixed point values. If r is small, then we can write

$$\delta\kappa_i(0) = rX_i \tag{11.97}$$

which means that

$$\delta\kappa_i(\ell) = rX_i\left(C_{i,t}\mathrm{e}^{\Lambda_r\ell} + \sum_{\alpha'} C_{i,\alpha'}\mathrm{e}^{\Lambda_{\alpha'}\ell}\right) \tag{11.98}$$

The symbol α' in (11.98) refers to the set of negative eigenvalues of the matrix (11.96). The important behavior of the parameters is contained in the first contribution in parentheses on the right hand side of the equation. For the moment, let's discard the others, so that the evolution of the parameter set is controlled by the contribution from the exponentially growing terms. Their distance from the critical hypersurface is then proportional to $r\mathrm{e}^{y\ell}$.

At the same time as this "distance from criticality" is growing, the correlation length is shrinking. This quantity is given by

$$\xi^{(\ell)}(r) = \xi^{(0)}(r)\mathrm{e}^{-\ell} \tag{11.99}$$

as can be readily intuited from (11.92). We can determine the value of ℓ, which we'll call ℓ^*, at which the rescaled correlation length is equal to one by setting the right hand side of (11.99) equal to unity. In this way we find

$$\ell^* = \ln \xi^{(0)}(r) \tag{11.100}$$

Choosing ℓ in this way imposes no loss in generality, since the thermodynamic functions are independent of the scale factor. Thus, the value of ℓ is arbitrary. Now, the set of thermodynamic parameters for which the correlation length is

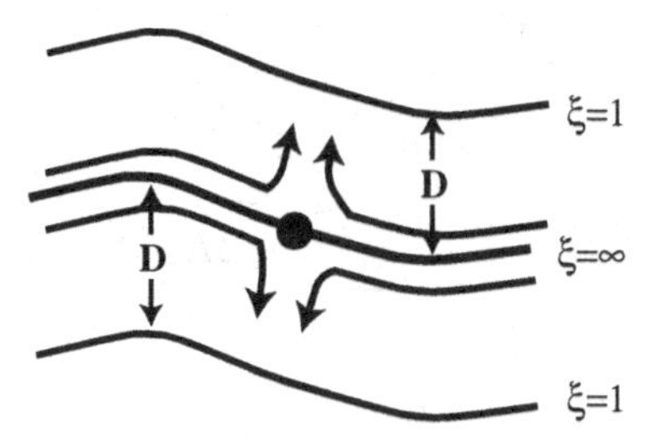

Fig. 11.10. The critical hypersurface, on which the correlation length, ξ, is infinite and hypersurfaces on which the correlation length is equal to some fixed value. Here, the value is one. Also shown on this figure is the flow in parameter space under the action of the continuous renormalization group. The heavy dot on the critical hypersurface is the fixed point there.

equal to one ought to constitute another hypersurface in parameter space, separated from the critical hypersurface by some distance, D. The situation is illustrated in Figure 11.10. This means that in order for the correlation length to be unity, we also require

$$Kre^{\Lambda_r \ell^*} = D \tag{11.101}$$

where $K = X_i C_{i,t}$, the quantities X_i and $C_{i,t}$ being those appearing in (11.98).

Equation (11.101) tells us that

$$e^{\ell^*} = \left(\frac{Kr}{D}\right)^{-1/\Lambda_r} \tag{11.102}$$

Inserting this result into the logarithm of (11.100), we find

$$\xi^{(0)}(r) = \left(\frac{Kr}{D}\right)^{-1/\Lambda_r} \tag{11.103}$$

Because the right hand side is the correlation length on the original length scale, we have the result that the correlation length is proportional to r^{-1/Λ_r}, or, in other words, that the correlation length exponent is given by $\nu = 1/\Lambda_r$. This is an important result. It tells us that the critical exponent ν can be *calculated* directly from the renormalization group equations – once they are known. We can go further. Again, for our magnetic system with only two parameters, r and h, the free energy per spin takes on the scaling form

$$f(r, h) = e^{-\ell d} f\left(re^{\Lambda_r \ell}, he^{\Lambda_h \ell}\right) \tag{11.104}$$

The factor $e^{-\ell d}$ arises because the degrees of freedom, i.e. number of spins, is reduced by this factor upon renormalization. All the critical exponents can be determined according to the prescription discussed earlier in this chapter. A straightforward calculation yields

$$\beta = \frac{d - \Lambda_h}{\Lambda_r} \tag{11.105}$$

$$-\gamma = \frac{d - 2\Lambda_h}{\Lambda_r} \tag{11.106}$$

$$2 - \alpha = \frac{d}{\Lambda_r} \tag{11.107}$$

$$\delta = \frac{\Lambda_h}{d - \Lambda_h} \tag{11.108}$$

and, of course, $\nu = 1/\Lambda_r$. As promised, all the critical exponents are expressed in terms of the two relevant eigenvalues, Λ_r and Λ_h.

As a final exercise before bringing this chapter to a close, we apply the general argument just presented to the Gaussian model. Recall the renormalization group equations for the reduced temperature, r and the external field, h, which are given in (11.82) and (11.83).

$$\frac{dr(\ell)}{d\ell} = 2r(\ell) \tag{11.109}$$

$$\frac{dh(\ell)}{d\ell} = \left(\frac{d+2}{2}\right) h(\ell) \tag{11.110}$$

We see from these equations that the matrix A is diagonal with eigenvalues $\Lambda_r = 2$ and $\Lambda_h = (d+2)/2$. Placing these values into the general expressions (11.105)–(11.108), we have for the critical exponents of the Gaussian model

$$\beta = \frac{d-2}{4} \tag{11.111}$$

$$\gamma = 1 \tag{11.112}$$

$$2 - \alpha = \frac{d}{2} \tag{11.113}$$

$$\delta = \frac{d+2}{d-2} \tag{11.114}$$

and $\nu = 1/2$, completing our analysis of this system.

Arriving at the renormalization group equations and analyzing their solutions can be a formidable task for more complicated effective Hamiltonians, such as the one describing the $O(n)$ model. The strategy driving the procedure, however, is the one we have outlined here.

12

More on the renormalization group

12.1 The momentum-shell method

The method discussed in Section 11.3.1 will now be pursued further, in that it will be applied to the full effective Hamiltonian of an $O(n)$ spin system, in which the effective Hamiltonian contains terms that are linear, quadratic, and of fourth order in the spin field. It is in the consideration of the higher order terms in the effective Hamiltonian (higher order than quadratic, that is) that the complications arise. The calculations that will be outlined here are not especially challenging in execution, but we will hint at extensions and generalizations that can become so.

In this chapter, the reader will be introduced to the field-theoretical version of the renormalization group, and to its first effective realization, the ϵ expansion for critical exponents. The approach will be that of the momentum-shell method developed in the previous chapters. The application of the method to the full $O(n)$ Hamiltonian will be more complicated due to the coupling terms, which were neglected before. A straightforward, though somewhat tedious, calculation will lead to a modified set of renormalization equations. These differential equations will then be solved to lowest order in the variable $\epsilon = 4 - d$, where d is the system's spatial dimensionality (three in the cases of interest to us). Using scaling arguments, the critical exponents will be obtained. Their relevance to the self-avoiding random walk will also be discussed.

The techniques to be discussed here are descendants of the original renormalization group method developed by Kenneth Wilson (Wilson, 1971a; Wilson, 1971b; Wilson and Kogut, 1974). They have fallen into disuse, but remain powerful because of their intuitive underpinnings. Indeed, in our opinion, they represent the fastest route to understanding what the renormalization group is all about.

12.2 The effective Hamiltonian when there is fourth order interaction between the spin degrees of freedom

The form of the $O(n)$ effective Hamiltonian that we have been looking at is

$$\mathcal{H}\left[\vec{S}(\vec{q})\right] = \sum_{\vec{q}} \left[\frac{1}{2}q^2|\vec{S}(\vec{q})|^2 + \frac{r}{2}|\vec{S}(\vec{q})|^2\right] + \frac{u}{N}\sum_{\vec{q}_1\ldots\vec{q}_4}\left(\vec{S}(\vec{q}_1)\cdot\vec{S}(\vec{q}_2)\right)\left(\vec{S}(\vec{q}_3)\cdot\vec{S}(\vec{q}_4)\right)\delta_{\vec{q}_1+\vec{q}_2+\vec{q}_3+\vec{q}_4} \quad (12.1)$$

This effective Hamiltonian follows from the application of spatial Fourier transforms to the real-space version of the Ginzburg–Landau–Wilson effective Hamiltonian

$$\mathcal{H}\left[\vec{\sigma}_i\right] = \sum_i \left[\frac{1}{2}|\vec{\nabla}\vec{\sigma}_i|^2 + \frac{r}{2}|\vec{\sigma}_i|^2\right] + u\sum_i \left(|\vec{\sigma}_i|^2\right)^2 \quad (12.2)$$

The derivation of (12.1) from (12.2) is detailed in Chapter 9.

In order to proceed in as direct a way as possible and at the same time retain the complications caused by the coupled spin interactions, we will carry through the momentum-shell calculation appropriate to a system in which the spins have one component only. This is the $O(1)$ model, also known as an Ising-like system. In the strict version of the Ising model, the spin degrees of freedom are scalars that can take on the values ± 1. As we will see, the calculations carry over, with only slight modifications, to the $O(n)$ model with arbitrary n, including $n = 0$.

12.2.1 Renormalized Ising Hamiltonian

For a one-component system, $\vec{S}(\vec{q})$ appearing in (12.1) reduces to a scalar, $S(\vec{q})$, and the Ginzburg–Landau–Wilson effective Hamiltonian becomes

$$\mathcal{H}\left[S(\vec{q})\right] = \sum_{\vec{q}} \frac{1}{2}\left(r + q^2\right) S(\vec{q})S(-\vec{q}) + \frac{u}{N}\sum_{\vec{q}_1,\ldots,\vec{q}_4} S(\vec{q}_1)\cdots S(\vec{q}_4)\delta_{\vec{q}_1+\cdots+\vec{q}_4} \quad (12.3)$$

Recall that a lattice, such as the square lattice shown in Figure 12.1 has a reciprocal lattice that is also square, shown in Figure 12.2. If the characteristic spacing in our real space lattice is a, then the characteristic spacing of the lattice in reciprocal space is $2\pi/a$. The distances between lattice points in the direct and reciprocal lattices scale inversely. From this it follows that the spacing between q's is inversely proportional to the characteristic linear dimension of the system, L. The length, L, is proportional to $N^{1/d}$, where N is the number of lattice sites, and d is the spatial dimension of the system. The constant of proportionality depends on the shape of the system, but we will ignore such factors.

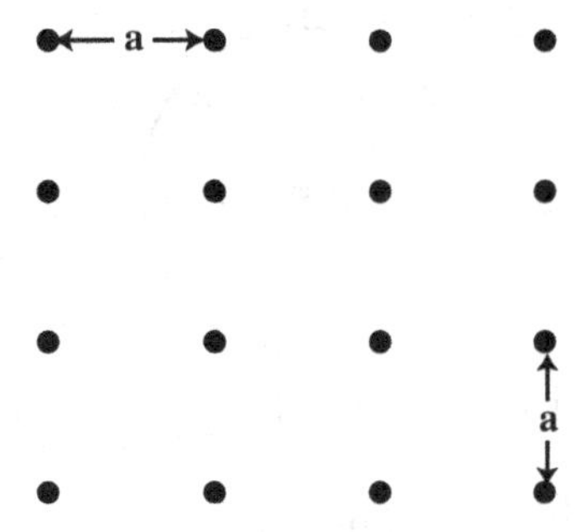

Fig. 12.1. The lattice in real space. The spacing between nearest neighbors is a.

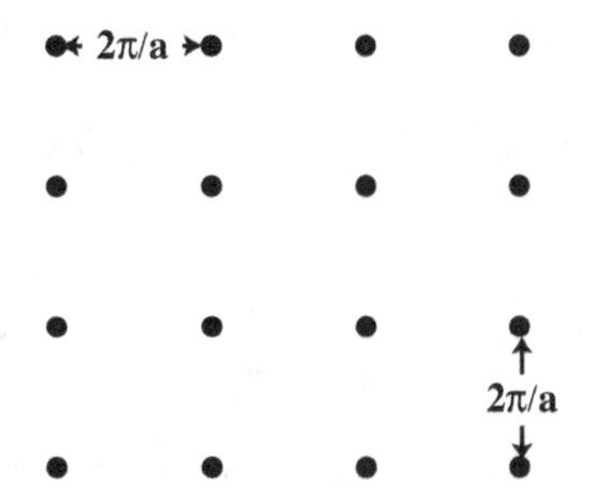

Fig. 12.2. The reciprocal lattice. Note that the lattice spacing is $2\pi/a$.

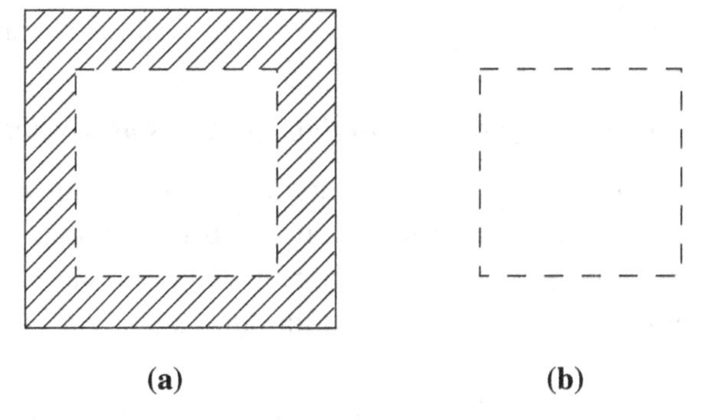

Fig. 12.3. Thinning out the degrees of freedom. The $s(\vec{q})$'s that are eliminated reside in the shaded region between the outer boundary of the reduced Brillouin zone, shown in the right hand side of the figure, and the outer boundary of the original Brillouin zone.

12.2.2 Momentum-shell thinning out of degrees of freedom

As we discussed in Chapter 11, thinning out the degrees of freedom is accomplished by eliminating $S(\vec{q})$'s in a shell just below the surface of the Brillouin zone (BZ). Eliminating $S(\vec{q})$'s in the shaded region in Figure 12.3(a) leaves one with a system in which the degrees of freedom are contained in the reduced zone shown in Figure 12.3(b). This Brillouin zone corresponds to a system in which the spins sit

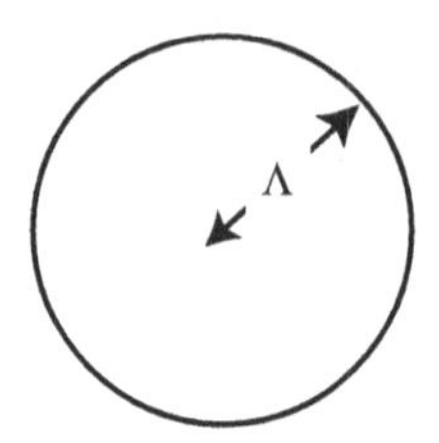

Fig. 12.4. The spherical Brillouin zone with radius Λ.

on the points of a lattice with a larger distance between neighboring points. In fact, if the linear dimension of the BZ is reduced by the factor $1/b$, with $b > 1$, then the spacing in the corresponding real space lattice is increased by the factor b.

12.2.3 Continuous renormalization

Our effective Hamiltonian is

$$\frac{1}{2}\sum_{\vec{q}}\left(r+q^2\right)S\left(\vec{q}\right)S\left(-\vec{q}\right)+\frac{u}{N}\sum_{\vec{q}_1,\ldots,\vec{q}_4}S\left(\vec{q}_1\right)\cdots S\left(\vec{q}_4\right)\delta_{\vec{q}_1+\cdots+\vec{q}_4} \tag{12.4}$$

The elimination procedure will be carried through for a Brillouin zone in the shape of a sphere to exploit the spherical symmetry of the quadratic energy coefficient $r+q^2$. We take the radius of the sphere to be $\Lambda = \mathrm{e}^{-\ell}$ (see Figure 12.4). We eliminate all $S(\vec{q})$'s within the shell $\mathrm{e}^{-\ell} > |\vec{q}| > \mathrm{e}^{-\ell-\delta\ell}$. We keep $\delta\ell$ finite but small.

12.2.4 How to properly eliminate degrees of freedom

What follows is an outline of the general arguments which allow us to eliminate $S(\vec{q})$'s as a procedure leading to a renormalization transformation. Suppose the energy of the system depends on two types of variables and a single coupling constant. Call them S_i, S_i', and K. The partition function is

$$Z(K) = \sum_{\{S_i\},\{S_i'\}} \mathrm{e}^{-H(K,S_i,S_i')} \tag{12.5}$$

where $1/k_{\mathrm{B}}T$ has been absorbed into the constant K. Suppose, further, that for a given configuration of S_i's we are able to sum over the set of variables $\{S_i'\}$. We could then define a partial partition function

$$\mathcal{Z}\left[S_i\right] = \sum_{\{S_i'\}} \mathrm{e}^{-H(K,S_i,S_i')} \tag{12.6}$$

The above equation can also be written in the following form:

$$A\left(K'\right)\mathrm{e}^{-H'(K',S_i)} = \sum_{\{S_i'\}} \mathrm{e}^{-H(K,S_i,S_i')} \tag{12.7}$$

where $H'(K', S_i)$ is an effective Hamiltonian with a reduced number of degrees of

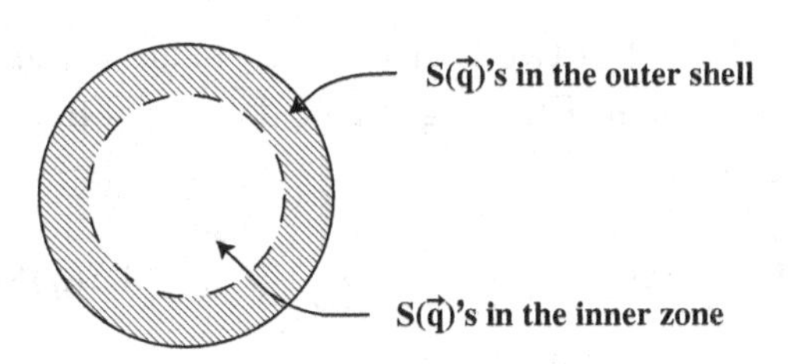

Fig. 12.5. The spherical Brillouin zone, the inner portion and the outer shell.

freedom and a changed coupling constant, K'. If, through a rescaling of amplitudes of the degrees of freedom and length scales, the Hamiltonian H' can be made to take on the same form as the original Hamiltonian, H, we can write

$$Z_N = \sum_{\{S_i\},\{S_i'\}} \mathrm{e}^{-H(K,S_i,S_i')} = \sum_{\{S_i\}} A(K')\mathrm{e}^{-H'(K',S_i')} \tag{12.8}$$

$$Z_N(K) = A(K')Z_{N'}(K') \tag{12.9}$$

Summing over the variables $\{S_i'\}$ and rescaling yields the renormalization group equation

$$K' = \mathcal{R}(K) \tag{12.10}$$

This is the procedure that we will follow in eliminating the $S(\vec{q})$'s in the outer shell of the Brillouin zone. Starting with (12.4), we split the contributions to H accordingly. (see Figure 12.5).

$$\begin{aligned} H = {} & \frac{1}{2} \sum_{|\vec{q}|<\mathrm{e}^{-\ell-\delta\ell}} (r+q^2)\, S(\vec{q})\, S(-\vec{q}) + \frac{u}{N} \sum_{\substack{|\vec{q}_1|<\mathrm{e}^{-\ell-\delta\ell} \\ \vdots \\ |\vec{q}_4|<\mathrm{e}^{-\ell-\delta\ell}}} S(\vec{q}_1)\cdots S(\vec{q}_4)\, \delta_{\vec{q}_1+\cdots+\vec{q}_4} \\ & + \frac{1}{2} \sum_{\mathrm{e}^{-\ell}<|\vec{q}|<\mathrm{e}^{-\ell-\delta\ell}} (r+q^2)\, S(\vec{q})\, S(-\vec{q}) \\ & + \frac{4u}{N} \sum_{\substack{\vec{q}_1 \text{ in shell} \\ \vec{q}_2,\ldots,q_4 \text{ in zone}}} S(\vec{q}_1)\cdots S(\vec{q}_4)\, \delta_{\vec{q}_1+\cdots+\vec{q}_4} \\ & + \frac{6u}{N} \sum_{\substack{\vec{q}_1,\vec{q}_2 \text{ in shell} \\ \vec{q}_3,\vec{q}_4 \text{ in zone}}} S(\vec{q}_1)\cdots S(\vec{q}_4)\, \delta_{\vec{q}_1+\cdots+\vec{q}_4} \\ & + \sum_{\substack{\vec{q}_1,\vec{q}_2,\vec{q}_3 \text{ in shell} \\ \vec{q}_4 \text{ in zone}}} S(\vec{q}_1)\cdots S(\vec{q}_4)\, \delta_{\vec{q}_1+\cdots+\vec{q}_4} \\ & + \frac{u}{N} \sum_{\vec{q}_1,\ldots,\vec{q}_4 \text{ in shell}} S(\vec{q}_1)\cdots S(\vec{q}_4)\, \delta_{\vec{q}_1+\cdots+\vec{q}_4} \end{aligned} \tag{12.11}$$

In the limit $\delta\ell \to 0$, the last two terms in (12.11) are negligibly small, and we will thus be able to drop them. Thinning out the degrees of freedom in this way, the partition function becomes

$$
\begin{aligned}
Z = \int \prod_{\vec{q}} \mathrm{d}S(\vec{q}) \exp\Bigg[& -\frac{1}{2} \sum_{\vec{q} \text{ in zone}} (r+q^2)\, S(\vec{q})S(-\vec{q}) \\
& - \frac{u}{N} \sum_{\text{all } \vec{q}\text{'s in zone}} S(\vec{q}_1) \cdots S(\vec{q}_4)\, \delta_{\vec{q}_1+\cdots+\vec{q}_4} \Bigg] \\
\times \exp\Bigg[& -\frac{1}{2} \sum_{\vec{q} \text{ in shell}} (r+q^2) S(\vec{q})S(-\vec{q}) \\
& - \frac{4u}{N} \sum_{\substack{\vec{q}_1 \text{ in shell} \\ \vec{q}_2,\ldots,\vec{q}_4 \text{ in zone}}} S(\vec{q}_1) \cdots S(\vec{q}_4)\, \delta_{\vec{q}_1+\cdots+\vec{q}_4} \Bigg] \\
\times \exp\Bigg[& -\frac{6u}{N} \sum_{\substack{\vec{q}_1,\vec{q}_2 \text{ in shell} \\ \vec{q}_3,\vec{q}_4 \text{ in zone}}} S(\vec{q}_1) \cdots S(\vec{q}_4)\, \delta_{\vec{q}_1+\cdots+\vec{q}_4} \Bigg]
\end{aligned}
\tag{12.12}
$$

If the outer shell is thin, and we take $\mathrm{e}^{-\ell} = 1$, which is consistent with choosing the coefficient of the q^2 term equal to unity, e.g. $(r + Jq^2) = (r + q^2)$, in the notation of the previous chapter, then $(r + q^2)s(\vec{q})s(-\vec{q})$ can be replaced by $(r + 1)s(\vec{q})s(-\vec{q})$ in the outer shell.

Let's focus on the spins in the shell, and suppose we expand the exponents with respect to the quartic terms involving $s(\vec{q})$'s within the shell. We have

$$
\begin{aligned}
& \exp\Bigg[-\frac{1}{2} \sum_{\vec{q} \text{ in shell}} (r+1) S(\vec{q})S(-\vec{q}) \Bigg] \\
& \times \Bigg[1 - \frac{4u}{N} \sum_{\substack{\vec{q}_1 \text{ in shell} \\ \vec{q}_2,\ldots,\vec{q}_4 \text{ in zone}}} S(\vec{q}_1) \cdots S(\vec{q}_4)\, \delta_{\vec{q}_1+\cdots+\vec{q}_4} \\
& \quad + \frac{1}{2}\left(\frac{4u}{N}\right)^2 \Bigg(\sum_{\substack{\vec{q}_1 \text{ in shell} \\ \vec{q}_2,\ldots,\vec{q}_4 \text{ in zone}}} S(\vec{q}_1) \cdots S(\vec{q}_4)\, \delta_{\vec{q}_1+\cdots+\vec{q}_4} \Bigg)^2 + \cdots \Bigg] \\
& \times \Bigg[1 - \frac{6u}{N} \sum_{\substack{\vec{q}_1,\vec{q}_2 \text{ in shell} \\ \vec{q}_3,\vec{q}_4 \text{ in zone}}} S(\vec{q}_1) \cdots S(\vec{q}_4)\, \delta_{\vec{q}_1+\cdots+\vec{q}_4} + \cdots \Bigg]
\end{aligned}
\tag{12.13}
$$

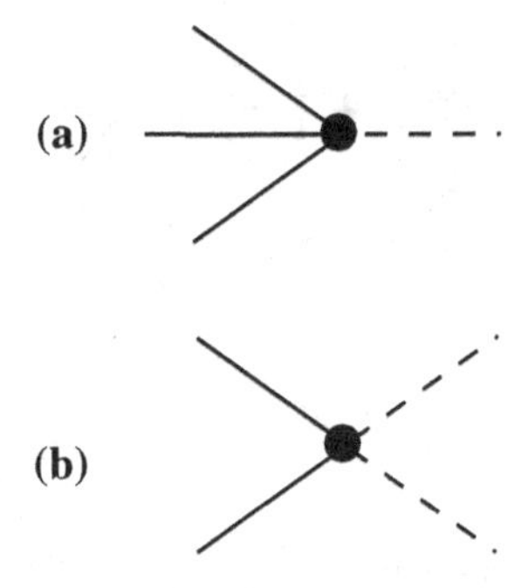

Fig. 12.6. Two vertices.

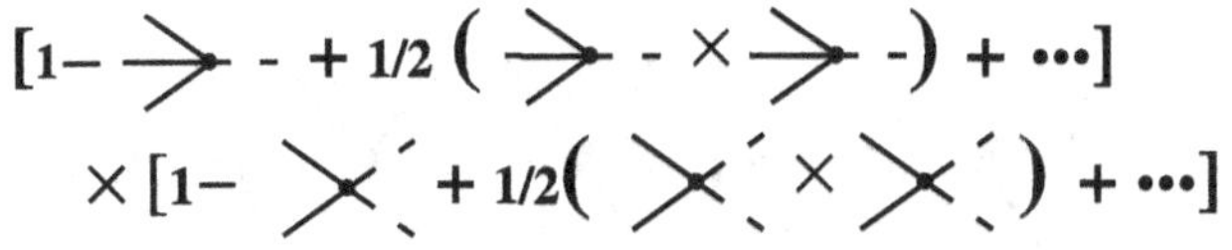

Fig. 12.7. The graphical version of the expansion.

The number of ways of pairing $S(\vec{q}_i)$'s with $\vec{q}_i$ in the shell is a bit complicated to determine. Graphical methods help. Denote a vertex to which an $S(\vec{q})$ with $\vec{q}$ in the shell and with the other $S(\vec{q})$'s in the zone as follows: The vertices depicted in Figure 12.6 are given by[1]

$$(a) = \frac{4u}{N} \sum_{\substack{\vec{q}_1 \text{ in shell} \\ \vec{q}_2,\ldots,\vec{q}_4 \text{ in zone}}} S(\vec{q}_1)\cdots S(\vec{q}_4)\delta_{\vec{q}_1+\cdots+\vec{q}_4} \tag{12.14}$$

$$(b) = \frac{6u}{N} \sum_{\substack{\vec{q}_1,\vec{q}_2 \text{ in shell} \\ \vec{q}_3,\vec{q}_4 \text{ in zone}}} S(\vec{q}_1)\cdots S(\vec{q}_4)\delta_{\vec{q}_1+\cdots+\vec{q}_4} \tag{12.15}$$

The expansion is then as depicted in Figure 12.7 Multiplying the expansion above by $\exp[-\frac{1}{2}\sum_{\vec{q} \text{ in shell}}(r+1)S(\vec{q})S(-\vec{q})]$ and integrating over $S(\vec{q})$'s with $\vec{q}$ in the shell (the integrations required being all of the Gaussian type), we conclude that it is only the dotted lines that are paired off that give a non-zero value to the integral. The types of diagram that are generated are shown in Figure 12.8. A factor of $\prod_{\vec{q} \text{ in shell}}(2\pi/(r+1))^{1/2}$ multiplies the expressions depicted by the diagrams in Figure 12.8. These diagrams can be exponentiated, leaving the expression that is illustrated in Figure 12.9. Note that the terms in the exponential include only connected diagrams; in particular, terms such as those asterisked in Figure 12.8 are

[1] In this chapter the interaction, u, will be represented by a dot, ●, while in Chapter 7 interactions were represented by a dotted line. Here, dotted lines are utilized to represent degrees of freedom whose wave-vectors lie in the inner Brillouin zone.

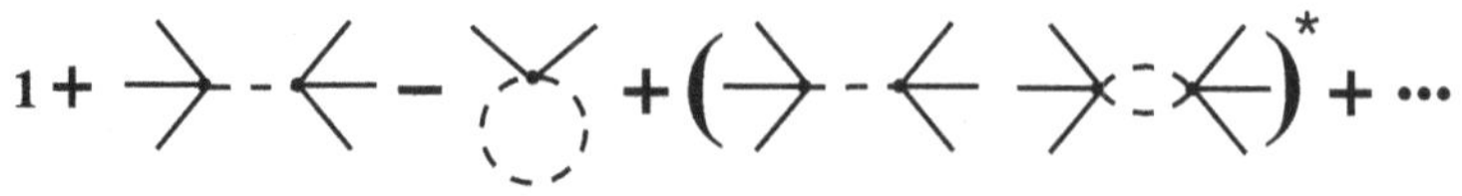

Fig. 12.8. The types of diagram that are generated.

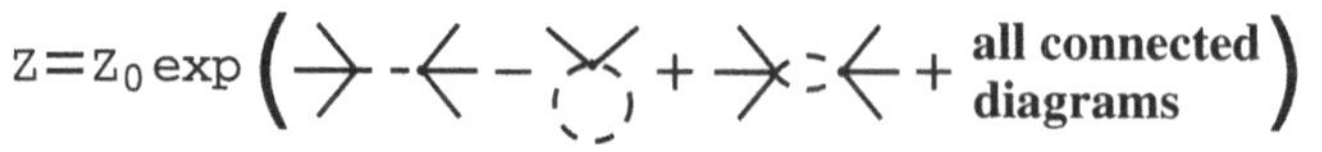

Fig. 12.9. The form of the exponential, in terms of diagrams. The quantity Z_0 in the figure stands for $\exp[-\frac{1}{2}\sum_{\vec{Q}\text{ in shell}} \ln\left(\frac{r+1}{2\pi}\right)]$.

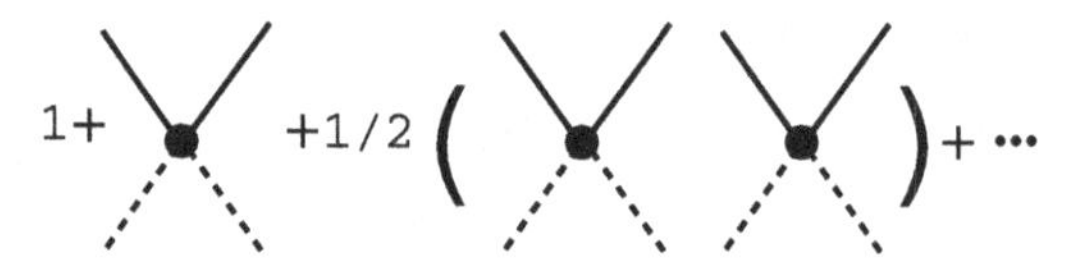

Fig. 12.10. Diagram sum that becomes a single diagram upon exponentiating.

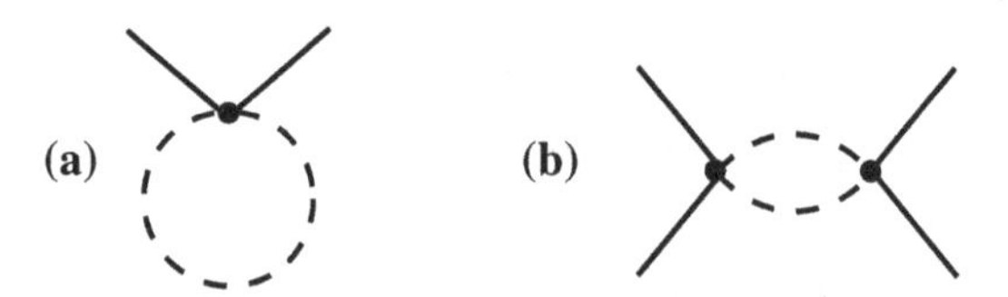

Fig. 12.11. The two diagrams of immediate interest. The first is a one-loop diagram for the renormalized quadratic coefficient and the other is the one-loop diagram for the alteration of the quartic coefficient.

to be excluded. That exponentiating a diagrammatic expansion removes all terms that cannot be represented as linked diagrams is a well-known result in a variety of contexts. In the supplement at the end of this chapter we show in detail how this comes about for the series of diagrams shown in Figure 12.10.

Exercise 12.1

Show, by extending the arguments in the supplement at the end of this chapter, that performing Gaussian averages of the sum of terms in Figure 12.7 leads to the linked cluster result displayed in Figure 12.9.

The two diagrams of immediate interest to us are shown in Figure 12.11. Figure 12.11(a) is quadratic in the $S(\vec{q})$'s, and Figure 12.11(b) is quartic in the spin variables. Diagram (a) contributes to a renormalized effective temperature, r, in the term

$$\frac{1}{2} \sum_{\text{all } \vec{q}\text{'s in zone}} (r + q^2) S(\vec{q}) S(-\vec{q}) \tag{12.16}$$

Figure 12.16(b) contributes to a change in the fourth order term

$$\frac{u}{N} \sum_{\text{all } \vec{q}\text{'s in zone}} S(\vec{q}_1) \cdots S(\vec{q}_4) \delta_{\vec{q}_1 + \cdots + \vec{q}_4} \tag{12.17}$$

Let's see exactly what changes are induced by these two terms. For clarity, denote by $\vec{Q}$ the $\vec{q}$'s that are in the outer shell, and by $\vec{q}$ the $\vec{q}$'s that are in the inner zone. Then, performing the specific Gaussian integrals, we find for the value of the diagrams in Figure 12.11:

$$(\text{a}) = \frac{6u}{N} \frac{1}{r+1} \left(\sum_{\vec{Q} \text{ in shell}} \right) \sum_{\vec{q} \text{ in zone}} S(\vec{q}) S(-\vec{q}) \tag{12.18}$$

$$(\text{b}) = \frac{36u^2}{N^2} \frac{1}{(r+1)^2} \left(\sum_{\vec{Q}, \vec{Q}' \text{ in shell}} \right) \sum_{\vec{q}_1, \dots, \vec{q}_4} S(\vec{q}_1) \cdots S(\vec{q}_4) \delta_{\vec{q}_1 + \cdots + \vec{q}_4} \tag{12.19}$$

The first sum, $\sum_{\vec{Q} \text{ in shell}}$, is proportional to the volume in q-space of the shell. In fact, we can write

$$\sum_{\vec{Q} \text{ in shell}} = N \frac{\text{Volume of shell}}{\text{Volume of BZ}} \tag{12.20}$$

since there are exactly N $\vec{q}$'s in the BZ. Furthermore, we also know that the volume of the BZ is give by

$$V_{\text{BZ}} = K_d \Lambda^d \tag{12.21}$$

where Λ is the radius of the BZ in q-space, and K_d is a factor that depends on the BZ's shape and the system's dimensionality. This factor cancels when the ratio on the right hand side of (12.20) is taken. The volume of the shell in q-space between Λ and $\Lambda - \delta q$, where δq is small compared to Λ, can be approximated by

$$\delta q \frac{\partial}{\partial \Lambda} \left(K_d \Lambda^d \right) + O\left[(\delta q^2) \right] = d \delta q K_d \Lambda^{d-1} \tag{12.22}$$

leading to

$$\begin{aligned} \sum_{\vec{Q} \text{ in shell}} &= N \left(\frac{d}{\Lambda} \delta q \right) \\ &= \frac{N}{\text{e}^{-\ell}} d \delta q \end{aligned} \tag{12.23}$$

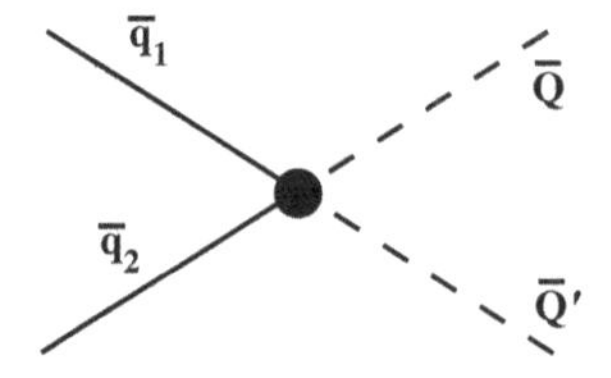

Fig. 12.12. A vertex in a diagram. The lines ending on the vertex are labeled with the wave-vectors that they carry.

Recall, $\Lambda = e^{-\ell}$, which implies

$$dq = e^{-\ell} - e^{-\ell-\delta\ell}$$
$$= e^{-\ell} d\ell \tag{12.24}$$

to lowest order in $\delta\ell$. We are thus left with

$$(a) = \frac{6u}{N(r+1)} N d\delta\ell \sum_{\vec{q} \text{ in zone}} S(\vec{q})S(-\vec{q}) \tag{12.25}$$

Calculation of the term (b)

Because of "momentum conservation," the sum over $\vec{Q}$ and $\vec{Q}'$ in the expression represented by the Figure 12.11(b) is just a sum over one of those wave-vectors. Momentum conservation at a vertex (see Figure 12.12) implies

$$\vec{Q} + \vec{Q}' + \vec{q}_1 + \vec{q}_2 = 0 \tag{12.26}$$

From (12.26), we note that there are restrictions on $\vec{q}_1$ and $\vec{q}_2$ if $\vec{Q}$ and $\vec{Q}'$ are to be in the shell. The sum is actually a function of $\vec{q}_1$ and $\vec{q}_2$, $u(\vec{q}_2, \vec{q}_2)$. But, since the variables $\vec{q}_1$ and $\vec{q}_2$ will eventually be rescaled, the largest contribution turns out to arise from the limits $\vec{q}_1 + \vec{q}_2 = 0$, with $\vec{Q} = -\vec{Q}'$. The argument supporting this approximation comes from expanding $u(\vec{q}_1, \vec{q}_2)$ in powers of $\vec{q}_1$ and $\vec{q}_2$, and then retaining the leading term $u(0, 0)$. The next term, which is quadratic in the $\vec{q}$'s, is reduced by a factor of $1/b^2$, where b is the scale factor, which is, in the case of interest to us, greater than one. We thus arrive at

$$(b) = \frac{36u^2}{N^2} \frac{1}{(r+1)^2} d\delta\ell N \sum_{\vec{q}_1,\ldots,\vec{q}_4 \text{ in zone}} S(\vec{q}_1) \cdots S(\vec{q}_4) \delta_{\vec{q}_1+\cdots+\vec{q}_4} \tag{12.27}$$

Combining these new terms with the original terms in (12.11) that depend only on $S(\vec{q})$'s in which the $\vec{q}$'s lie in the inner zone, we obtain an effective Hamiltonian that has the altered form

$$\sum_{\vec{q}\text{ in zone}} \frac{1}{2}\left(r + \frac{12u}{r+1} d\delta\ell + q^2\right) S(\vec{q})S(-\vec{q})$$
$$+\left[\frac{u}{N} - \frac{36u^2}{(r+1)^2}\frac{d\delta\ell}{N}\right] \sum_{\vec{q}_1,\dots,\vec{q}_4\text{ in zone}} S(\vec{q}_1)\cdots S(\vec{q}_4)\delta_{\vec{q}_1+\cdots+\vec{q}_4} \tag{12.28}$$

Rescaling N, $\vec{q}$ and $S(\vec{q})$

We rescale N first. The number of q-lattice sites in the first BZ is N, but there are fewer sites in the inner zone. The number of sites scales with the volume of the zone, which, in turn, scales as (radius)^d. Thus, the number of sites in the inner zone is reduced by

$$\left(\frac{e^{-\ell-\delta\ell}}{e^{-\ell}}\right)^d = e^{-d\delta\ell} \tag{12.29}$$

To compensate for this reduction, we replace N by $N' e^{d\delta\ell}$.

Next, we rescale the q's. To do this, we observe that in the inner zone the q's range in magnitude from zero to $e^{-\ell-\delta\ell}$, while for the whole zone the upper limit on the magnitude of the wave-vectors is $e^{-\ell}$. To normalize the magnitude of a wavelength to the radius of the BZ, we replace $\vec{q}$ by $\vec{q}\,' e^{-\delta\ell}$. However, implementing this rescaling causes the quadratic term proprtional to q^2 to become

$$S'(\vec{q}\,')S'(-\vec{q}\,')q'^2 e^{-2\delta\ell} \tag{12.30}$$

where

$$\begin{aligned} S'(\vec{q}\,') &= S(q' e^{-\delta\ell}) \\ &= S(\vec{q}) \end{aligned} \tag{12.31}$$

The rescaling of the $S(\vec{q})$'s is performed to undo the effects of the q-rescaling. This can be accomplished by replacing $S'(\vec{q}\,')$ by $S''(\vec{q}\,')e^{\delta\ell}$. This rescaling is allowed because the $S'(\vec{q}\,')$'s are continuous variables. In sum, then, our rescaling is accomplished by the following substitutions.

$$\begin{aligned} N &\to N' e^{d\delta\ell} \\ &\approx N'(1+\delta\ell) \end{aligned} \tag{12.32}$$

$$\begin{aligned} \vec{q} &\to \vec{q}\,' e^{-d\delta\ell} \\ &\approx \vec{q}\,'(1-d\delta\ell) \end{aligned} \tag{12.33}$$

$$\begin{aligned} S(\vec{q}) &\to S''(\vec{q}\,')e^{\delta\ell} \\ &\approx S''(\vec{q}\,')(1+\delta\ell) \end{aligned} \tag{12.34}$$

After rescaling, the Boltzmann factor looks like

$$\exp\left\{-\frac{1}{2}\sum_{q'}\left[\left(r+\frac{12ud\delta\ell}{r+1}\right)(1+2\delta\ell)+q'^2\right]S''(\vec{q}\,')S''(-\vec{q}\,')\right\}$$
$$\times\exp\left\{\left(\frac{u}{N'}-\frac{36u^2}{(r+1)^2}\frac{d\delta\ell}{N'}\right)(1+(4-d)\delta\ell)\sum_{\vec{q}_1',\ldots,\vec{q}_4'}S''(\vec{q}_1')\cdots S''(\vec{q}_4')\delta_{\vec{q}_1'+\cdots+\vec{q}_4'}\right\} \tag{12.35}$$

Because $\delta\ell$ is small, we are allowed to linearize all terms in $\delta\ell$. Also, the fact that the unprimed variables are being summed over or integrated out allow us to eliminate all primes. At the risk of a bit of confusion, we also eliminate the prime on N, in the hope that the reader will keep in mind that the symbol N now stands for the number of remaining degrees of freedom. With all these changes, the free energy takes the form

$$\frac{1}{2}\sum_{\vec{q}}\left[r+\left(2r+\frac{12ud}{r+1}\right)\delta\ell+q^2\right]S(\vec{q})S(-\vec{q})$$
$$+\frac{1}{N}\sum_{\vec{q}_1,\ldots,\vec{q}_4}\left[u+(4-d)\,\delta\ell u-\frac{36u^2}{(1+r)^2}\delta\ell\right]S(\vec{q}_1)\cdots S(\vec{q}_4)\delta_{\vec{q}_1+\cdots+\vec{q}_4} \tag{12.36}$$

12.2.5 Renormalization group recursion relations

Clearly, the process outlined above can be repeated to obtain $r(\ell)$ and $u(\ell)$. If we allow $\delta\ell$ to be an infinitesimal, these two parameters satisfy

$$r(\ell+\delta\ell)=r(\ell)+\left[2r(\ell)+\frac{12u(\ell)d}{r(\ell)+1}\right]\delta\ell \tag{12.37}$$

$$u(\ell+\delta\ell)=u(\ell)+\left[(4-d)\,u(\ell)-\frac{36u(\ell)^2}{(r(\ell)+1)^2}\right]\delta\ell \tag{12.38}$$

Here we have a pair of first order, non-linear differential equations that describe the evolution of the effective Hamiltonian under the renormalization group transformations. They are a specific realization of the general transformation equations discussed in Chapter 10 (see (10.74) and the discussion that follows). Our interest here will not be to obtain the most accurate solution to these equations, but rather to carry out the analysis sufficiently far along for a determination of the critical exponents pertinent to the self-avoiding walk problem. This will be accomplished to lowest order in $\epsilon=4-d$. The basic idea is to solve coupled equations to lowest order in ϵ by iteration.

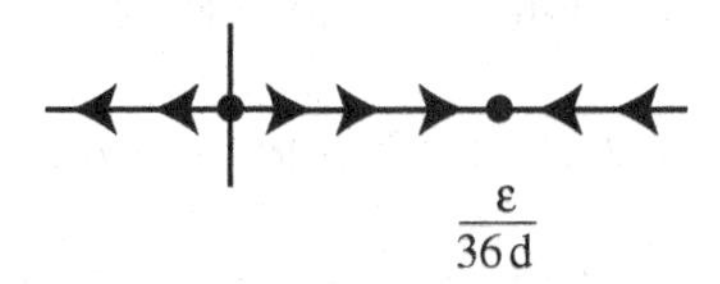

Fig. 12.13. The flow of the renormalized coupling constant, $u(\ell)$, under the renormalization group transformation, through (12.39).

Let's look at the equation for $u(\ell)$ first. We start by setting $r(\ell) = 0$. Then

$$\frac{\mathrm{d}u(\ell)}{\mathrm{d}\ell} = (4-d)\,u(\ell) - 36u(\ell)^2 \tag{12.39}$$

Let $\epsilon = 4 - d$ be greater than zero. Integrating (12.39), we obtain

$$\int_{u_0}^{u(\ell)} \frac{\mathrm{d}u}{\epsilon u - 36du^2} = \int_0^{\ell} \mathrm{d}\ell' = \ell \tag{12.40}$$

After some straightforward algebra we find for $u(\ell)$

$$u(\ell) = \frac{\epsilon u_0 \mathrm{e}^{\epsilon\ell}}{\epsilon + 36u_0 d\left(\mathrm{e}^{\epsilon\ell} - 1\right)} \tag{12.41}$$

Exercise 12.2

Show that integration of (12.40) leads to the result (12.41) for $u(\ell)$.

As $\ell \to \infty$, $u^*(\ell) \to \epsilon/36d$. The coupling constant $u(\ell)$ approaches a fixed point.

Exercise 12.3

Find the other fixed point in the differential equation (12.39). Show that it is unstable, in that if $u(\ell_0)$ is close to that fixed point, but not equal to it, then as ℓ increases from ℓ_0, the difference between $u(\ell)$ and that fixed point grows. Show that the flow (i.e. change) in $u(\ell)$ is as depicted in Figure 12.13.

The flow of the coupling constant under the transformations is shown in Figure 12.13. The renormalization group equation for the effective reduced temperature, $r(\ell)$, is

$$\begin{aligned}\frac{\mathrm{d}r(\ell)}{\mathrm{d}\ell} &= 2r(\ell) + \frac{12u(\ell)d}{r(\ell)+1} \\ &= 2r(\ell) + 12u(\ell)d - 12u(\ell)dr(\ell) + \frac{12u(\ell)dr(\ell)^2}{r(\ell)+1}\end{aligned} \tag{12.42}$$

If we drop the last term on the right hand side of (12.42), we have

$$\frac{\mathrm{d}r(\ell)}{\mathrm{d}\ell} = 2r(\ell) + 12u(\ell)d - 12u(\ell)r(\ell)d \tag{12.43}$$

Writing $r(\ell) = t(\ell) - 6u(\ell)d$, and making use of the fact that in the regime of interest to us $u(\ell)$ is small, while we are taking ϵ to be small by assumption, then, $\mathrm{d}u(\ell)/\mathrm{d}\ell \sim O(\epsilon u, u^2)$, which allows us to write

$$\frac{\mathrm{d}r(\ell)}{\mathrm{d}\ell} \approx \frac{\mathrm{d}t(\ell)}{\mathrm{d}\ell} \tag{12.44}$$

Dropping terms that are of order $u(\ell)^2$, we end up with

$$\frac{\mathrm{d}t(\ell)}{\mathrm{d}\ell} = 2t(\ell) - 12du(\ell)t(\ell) \tag{12.45}$$

Separating variables and integrating, we find

$$t(\ell) = t(0)\mathrm{e}^{2\ell - \int_0^\ell 12du(\ell')\,\mathrm{d}\ell'} \tag{12.46}$$

The integration over ℓ' can be carried out, yielding our solution for $u(\ell)$

$$\int_0^\ell u(\ell')\,\mathrm{d}\ell' = 12d \int_{u_0}^{u(\ell)} u' \left(\frac{\mathrm{d}\ell'}{\mathrm{d}u'}\right) \mathrm{d}u' \tag{12.47}$$

$$= 12d \int_{u_0}^{u} \frac{\mathrm{d}u'}{\epsilon u' - 36\,\mathrm{d}u'^2} \tag{12.48}$$

$$= -\frac{1}{3} \ln\left(\epsilon - 36u'd\right)\Big|_{u_0}^{u} \tag{12.49}$$

which allows us to write for $t(\ell)$

$$t(\ell) = t(0)\mathrm{e}^{2\ell} \left(\frac{\epsilon - 36u(\ell)d}{\epsilon - 36u_0 d}\right)^{1/3} \tag{12.50}$$

Finally, using our previous expression for $u(\ell)$ yields the following result for $t(\ell)$

$$t(\ell) = t(0)\mathrm{e}^{2\ell} \left[\frac{\epsilon}{\epsilon + 36\,du_0\left(\mathrm{e}^{\epsilon\ell} - 1\right)}\right]^{1/3} \tag{12.51}$$

12.2.6 Correlation length and the critical exponent, ν

If $t(0)$ is small, then ℓ must be large in order that $t(\ell)$ becomes sizable. In this case we have

$$\begin{aligned} r(\ell) &\approx t(\ell) \\ &\propto r^*(0)\mathrm{e}^{(2-\epsilon/3)\ell} \end{aligned} \tag{12.52}$$

as $\ell \to \infty$.

Exercise 12.4 Show that (12.53) follows from (12.51) and that $r^*(0)$ is the reduced temperature, while the critical temperature has been shifted. What is the value of this shift, ΔT_c?

For large ℓ, the system is driven from its critical point. The variable ℓ is large but arbitrary. We can choose its value such that the correlation length $\xi(\ell)$ is unity, i.e. $\xi(\ell^*) = 1$. We characterize this property of the system by writing $r(\ell^*) = D$, where D is a constant. Then

$$\mathrm{e}^{(2-\epsilon/3)\ell^*} = \frac{D}{r^*(0)} \tag{12.53}$$

$$\mathrm{e}^{\ell^*} = \left[\frac{D}{r^*(0)}\right]^{\frac{1}{2-\epsilon/3}}$$

$$= \left(\frac{D}{T - T_c^*}\right)^{\frac{1}{2-\epsilon/3}} \tag{12.54}$$

In other words, to order ϵ in the exponent,

$$e^{\ell^*} = \left(T - T_c^*\right)^{-\frac{1}{2}(1+\epsilon/6)} \tag{12.55}$$

We know that the correlation length decreases as $\mathrm{e}^{-\ell}$, Thus

$$1 = \xi(\ell^*)$$

$$= \xi(0)\mathrm{e}^{-\ell^*} \tag{12.56}$$

or

$$\ell^* = -\ln \xi(0) \tag{12.57}$$

which implies

$$\xi(0) \propto \left(T - T_c^*\right)^{-\frac{1}{2}(1+\epsilon/6)}$$

$$= \left(T - T_c^*\right)^{-\nu} \tag{12.58}$$

We finally arrive at the critical exponent for the correlation length, to first order in ϵ,

$$\nu = \frac{1}{2} + \frac{\epsilon}{12}, \qquad \epsilon = 4 - d \tag{12.59}$$

If we set $\epsilon = 4 - 3 = 1$, then

$$\nu = 7/12 = 0.583\,33\ldots \tag{12.60}$$

In the case of the three-dimensional Ising model, $\nu \sim 0.61\ldots$. Thus, (12.60) is

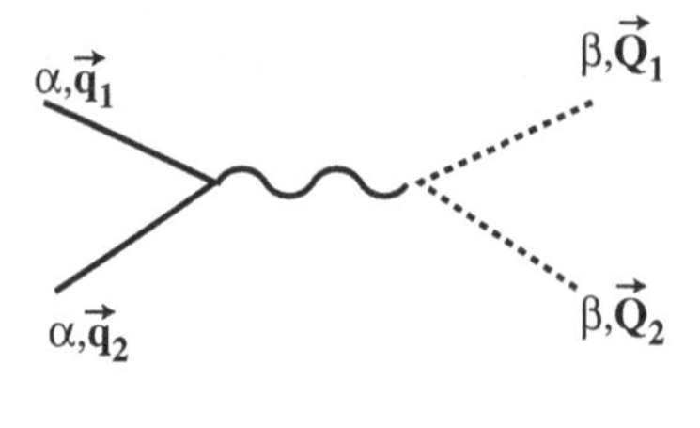

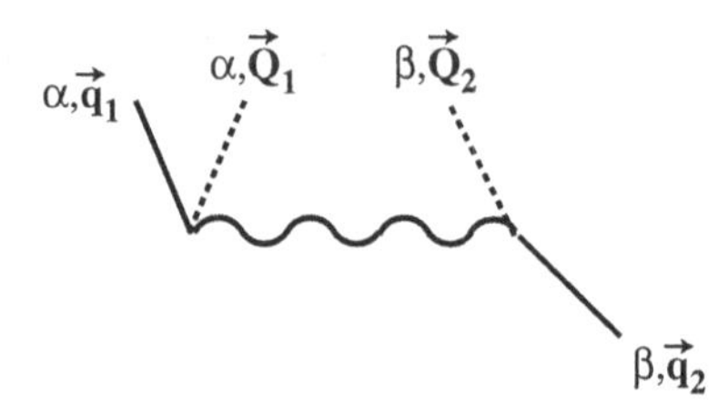

Fig. 12.14. The two terms contained in (12.62).

a considerable improvement over the mean field value, 1/2, for the correlation length exponent ν.

As we have argued in previous chapters, the magnetic system pertinent to the random walk problem is not the one-component Ising system but one in which the spins have n components, in the limit $n \to 0$. We now take up the task of extending the above analysis to the full $O(n)$ Ginzburg–Landau–Wilson effective Hamiltonian.

12.3 The $O(n)$ model: diagrammatics

For the $O(n)$ model, the fourth order term in the effective Hamiltonian has the form

$$\frac{u}{N} \sum_{\vec{q}_1,\ldots,\vec{q}_4} \left(\vec{S}(\vec{q}_1) \cdot \vec{S}(\vec{q}_2)\right) \left(\vec{S}(\vec{q}_3) \cdot \vec{S}(\vec{q}_4)\right) \delta(\vec{q}_1 + \cdots + \vec{q}_4) \tag{12.61}$$

Recall that the momentum-space renormalization prescription separates the sum over wave-vectors into those inside and outside a thin shell in q-space. The latter set of wave-vectors will be denoted as inside the BZ and will be denoted by $\vec{q}$, while the wave-vectors inside the shell carry the label $\vec{Q}$. Applying this procedure, (12.61) is written

$$\frac{u}{N} \sum_{\substack{\vec{q}_1,\vec{q}_2 \\ \text{in BZ}}} \sum_{\substack{\vec{Q}_1,\vec{Q}_2 \\ \text{in shell}}} \sum_{\alpha,\beta=1}^{n} \Big[2S^{\alpha}(\vec{q}_1)\, S^{\alpha}(\vec{q}_2) S^{\beta}(\vec{Q}_1) S^{\beta}(\vec{Q}_2)$$
$$+ 4S^{\alpha}(\vec{q}_1)\, S^{\alpha}(\vec{Q}_1) S^{\beta}(\vec{q}_2)\, S^{\beta}(\vec{Q}_2)\Big]\, \delta\left(\vec{q}_1 + \vec{q}_2 + \vec{Q}_1 + \vec{Q}_2\right) \tag{12.62}$$

The two terms in (12.62) are represented digrammatically in Figure 12.14. The solid lines represent $S^{\alpha}(\vec{q})$'s in the BZ, and dashed lines represent $S^{\beta}(\vec{Q})$'s in the

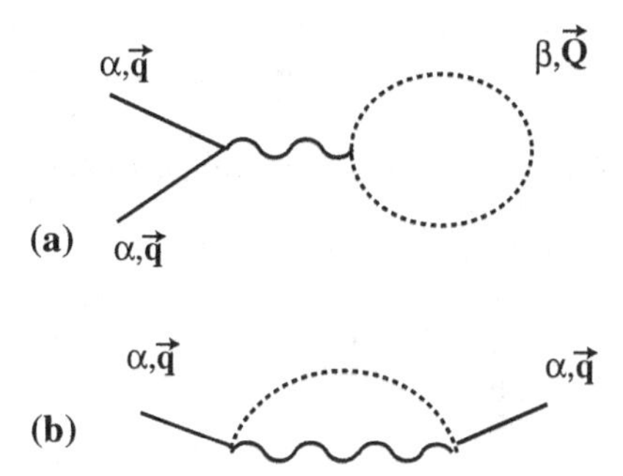

Fig. 12.15. The two contributions to the renormalized reduced temperature, r.

shell. As indicated in the figure, the wavy line is to be associated with the coupling constant, u.[2]

12.3.1 r-renormalization

The renormalized reduced temper-ature parameter, r, is determined simply by averaging (12.62) over the Gaussian distribution of spins in the shell: $\exp[-(1/2)\sum_{\vec{Q}}(r+1)S^{\alpha}(\vec{Q})S^{\alpha}(-\vec{Q})]$. The integrations are straightforward, and we depict the two contributions to r in Figure 12.15.

Exercise 12.5
Show that

$$\langle S^{\alpha}(\vec{Q}_1)S^{\beta}(S(\vec{Q}_2)\rangle = \frac{\delta_{\alpha,\beta}}{1+r}\delta(\vec{Q}_1+\vec{Q}_2) \tag{12.63}$$

where the brackets denote averaging over a Gaussian distribution of the spin variables.

Exercise 12.6
Applying the result of Exercise 12.5, show that the graphical quanitites in Figure 12.15 sum to

$$(\mathrm{a})+(\mathrm{b}) = \frac{2n+4}{1+r}\sum_{\substack{\vec{Q}\\ \text{in shell}}}\left(\sum_{\substack{\vec{Q}\\ \text{in BZ}}}\vec{S}(\vec{q})\cdot\vec{S}(-\vec{q})\right) \tag{12.64}$$

---

[2] While dots are a simpler way of depicting interactions, the nature of the interaction in the $O(n)$ model necessitates the use of some kind of line for the interaction.

Fig. 12.16. The graphs associated with renormalization of the coupling constant, u.

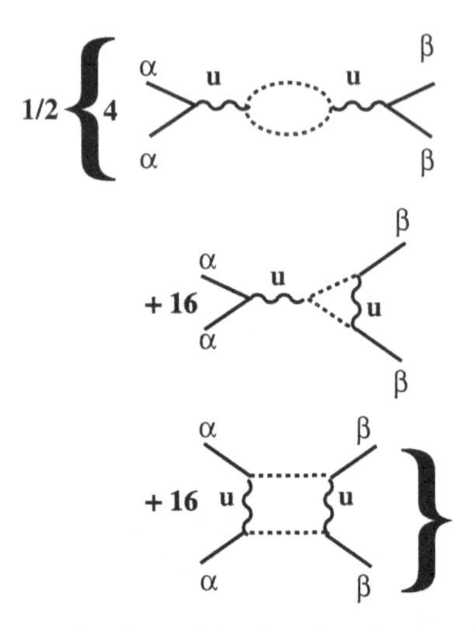

Fig. 12.17. Renormalization of the fourth order coupling constant, u.

12.3.2 *u-renormalization*

As before, the coupling constant u is renormalized by averaging over the graphs shown in Figure 12.16. After performing the Gaussian integrations, the contribution which renormalizes u is depicted in Figure 12.17.

Exercise 12.7

Repeat the averaging process asked for in Exercise 12.5 and show that when applied to the graphs in Figure 12.17 lead, when summed, to

$$\frac{4(n+8)}{(1+r)^2}\frac{u^2}{N^2}\sum_{\substack{\vec{Q}_1,\vec{Q}_2\\ \text{in shell}}}\left(\sum_{\substack{\vec{q}_1,\dots,\vec{q}_4\\ \text{in BZ}}}\left(\vec{S}(\vec{q}_1)\cdot\vec{S}(\vec{q}_2)\right)\left(\vec{S}(\vec{q}_3)\cdot\vec{S}(\vec{q}_4)\right)\delta\left(\vec{q}_1+\cdots+\vec{q}_4\right)\right) \tag{12.65}$$

12.4 $O(n)$ recursion relations

The recipe for deriving the recursion equations in the $O(n)$ case is identical, step-by-step, to the procedure we have already performed for the one-component spin Hamiltonian. Repeating that calculation, we end up with the $O(n)$ recursion relations

$$\frac{dr(\ell)}{d\ell} = 2r(\ell) + \frac{4(n+2)u(\ell)}{1+r(\ell)} \tag{12.66}$$

$$\frac{du(\ell)}{d\ell} = (4-d)u(\ell) - \frac{4(n+8)}{(1+r(\ell))^2}u(\ell)^2 \tag{12.67}$$

The equations are solved, at least to first order in $\epsilon = 4 - d$, following the same analytical steps as before. The correlation length critical exponent, ν, can immediately be discovered by noting that 12 is replaced by $4(n+2)$ in the r-equation, and 36 by $4(n+8)$ in the equation for u. With these substitutions, we find for ν

$$\nu = \frac{1}{2} + \frac{n+2}{4(n+8)}\epsilon + O(\epsilon^2) \tag{12.68}$$

The limit $n = 0$ is pertinent to the case of self-avoiding random walks. We find

$$\nu = \frac{1}{2} + \frac{\epsilon}{16} + O(\epsilon^2) \tag{12.69}$$

Setting $\epsilon = 1$ ($d = 3$), we obtain

$$\nu \approx \frac{1}{2} + \frac{1}{16} = 9/16 = 0.563 \tag{12.70}$$

Exercise 12.8

Perform the steps necessary to arrive at the expression (12.70) for the exponent ν.

It is interesting to compare this result with Flory's formula for ν:

$$\begin{aligned}\nu_F &= \frac{3}{d+2} \\ &= 3/5 \\ &= 0.6\end{aligned} \tag{12.71}$$

For a "fair" comparison with the $n = 0$ prediction, we ought to expand Flory's expression for the exponent ν in powers of ϵ, and retain the first order term in that expansion. If we do this, we find

$$\begin{aligned}\nu_F &= \frac{3}{6-\epsilon} \\ &= \frac{1}{2} + \frac{\epsilon}{12} + O(\epsilon^2) \\ &\approx \frac{7}{12} \quad \text{(to first order in } \epsilon \text{ at } d = 3) \\ &= 0.583\end{aligned} \tag{12.72}$$

Two points are worth making here. First, the ϵ expansion we have just obtained for ν is numerically close to ϵ-expanded version of Flory's ν, carried to the same order. Second, there is a clear difference in the coefficients of the first-order-in-ϵ term of the two expansions. As the ϵ expansion that we have derived here is believed to be exact, we can see that Flory's expression cannot be, at least just below four dimensions. As good as the Flory exponent is, it is not an exact result.

For small ϵ, or, equivalently, for dimensions close to four, we have obtained results for the flow of the effective Hamiltonian from the critical hypersurface and have been able to extract the critical exponent, ν. We are now in a position to determine the scaling properties of the free energy itself. However, a number of questions justifying our procedure have simply been ignored. For example:

(1) The expansion we produce is in the difference between the dimensionality of the space in which the random walk takes place and four. Considering that our primary focus is on three-dimensional walks, the expansion parameter, ϵ, is equal to one. This does not bode well for a series truncated at low order.
(2) In fact, the kind of expansion we generate is well-known to be *asymptotic*, in that, formally, its radius of convergence is equal to zero. That is, at sufficiently high order, coefficients in the series grow faster than exponentially, and standard summation methods fail.
(3) An issue closely related to the one immediately above is that we do not know that our power series in the quantity ϵ is complete. In fact, we have ignored the very real possibility that there are important contributions that are not subsumed in a power series. Such a contribution would be one going as $e^{-A/\epsilon}$.
(4) In fact, it is not proven that the renormalization group is a formally correct way to evaluate the kinds of quantities of interest to us. In the case of statistical mechanical systems, counterexamples have been constructed in which reasonable-looking versions of the renormalization group manifestly fail to correctly predict the behavior of thermodynamic functions.

What this all means is that we cannot blithely assume that the renormalization group provides us with the complete answer to key questions. For the time being, however, the evidence before us indicates that the results it generates are trustworthy, and we will, for the purposes of this book, at least, take that attitude.

12.5 The diagrammatic method

The above discussion of the renormalization group is along the lines of the original work by Wilson (Wilson, 1971a; Wilson, 1971b; Wilson and Kogut, 1974). It is probably the most intuitively appealing version of the method, in that it bears a direct relationship to standard methods for the description and investigation of self-similar systems. However, the momentum-shell method has not proven to be a useful approach to more general field-theoretical systems. Furthermore, it does not lend itself to calculations when the desired results require the evaluation of terms that are high order in the fourth order coupling constant. For those purposes, professional theorists must avail themselves of approaches that arise from the adaptation of perturbation theoretical methods. These approaches are based on differential equations derived from, or related to, the Callan–Symanzik equations (Callan, 1970; Symanzik, 1970) of the Feynman-diagrammatic renormalization group. The advantage of these approaches is that they allow the theorist to avail him or herself of the tricks that have been invented over the course of two thirds of a century of research in quantum field theory. These stratagems have been utilized in the determination of the critical properties of $O(n)$ systems to quite high order in the fourth order coupling constant, or equivalently, $\epsilon = 4 - d$ in the ϵ expansion.

The recursion relation method developed in the last two chapters has the virtue of a direct connection with the intuitive underpinnings of the renormalization group method. The hallmark of systems at a critical point – shared by a long self-avoiding walk – is self-similarity. Both look the same in fundamental respect when viewed at a variety of length scales. Indeed, exactly at the critical point of an $O(n)$ model – and, for the random walk, if it is infinitely long – the system in question will retain, unchanged, essential characteristics no matter what the scale on which it is observed, as long as that length scale is sufficiently large. Recursion relations describe the way in which the properties of a system in one length scale relate to the properties of the same system when viewed at a different length scale. They provide the most natural way of approaching the behavior of a system with scale invariance, in that, when properly framed, they yield fixed points, or unchanging characteristics, for a system whose qualities include self-similarity.

However, the most generally utilized calculational techniques that lead to direct, numerical results for the self-similarity are, in the case of random walks and the $O(n)$ model, derived from the perturbation theoretical methods developed to study quantum field theories. Over the years, a number of powerful techniques have emerged, allowing for the relatively rapid and accurate evaluation of the multiple integrals represented by Feynman diagrams. Indeed, the single indispensable tool of the (pre-string theory) elementary particle theorist,[3] and for just about any theoretical physicist concerned with the interplay between a large number of degrees of freedom, consists of a set of strategems for the analysis of Feynman diagrams.

There is a version of the renormalization group that is formulated for use in conjunction with diagrammatic methods. In fact, the renormalization group first made its appearance in (1953) papers by Stueckelberg and Peterman and by Gell-Mann and Low (1954) in which differential equations of the Renormalization Group type were derived from the diagrammatics of quantum electrodynamics. This

[3] Aside, perhaps, from Lie algebra methods.

development was effected as a central component of an inquiry into the structure of quantum electrodynamics at high energy and momentum. Further refinements of the diagrammatically based renormalization group have led to the Callan–Symanzik equation, which forms the basis of the most widely utilized renormalization group method, in its application to critical phenomena and the random walk.

Because of the computational utility of diagrammatic methods, and because the language of renormalization group is so tied up in these methods, we will in this chapter rederive the renormalization group equations for the $O(n)$ model making use of the diagrammatic approach. The key relations will be a set of differential equations, closely related to the equations introduced in Chapter 11. These equations express the fact that physical quantities are independent of certain details in the method utilized to compute them. The differential equations lead straightforwardly to scaling forms for key quantities, including free energies and correlation functions. We will also see how Feynman diagrams generate expressions containing the values of critical exponents. This approach complements the recursion relation discussion of the previous two chapters. It will also provide the reader with the background and vocabulary necessary to assimilate the methods that are most frequently utilized in the analysis of field theories appropriate to critical point behavior.

There will be no attempt in the following sections to properly indoctrinate the reader in the mysteries of the diagrammatic renormalization group method. The literature is already replete with books offering exhaustive prescriptions for the successful calculation of critical exponents and thermodynamic functions in the vicinity of critical points, and for the analysis of the asymptotic statistical properties of self-avoiding walks.[4] It is, rather, the hope of the authors that the discussion below will prepare the reader to better understand the conceptual underpinnings of the methods described in the pedagogical literature.

12.6 Diagrammatic analysis: the two-point correlation function

A fundamental element of our approach is that we partition the multiple integrals represented by diagrams in terms of the ordering, by magnitude, of wavenumbers or momenta.[5]

For example, recall the two-point correlation function, which has the diagrammatic expansion shown in Figure 12.18 The sum can be written as follows:

$$\frac{1}{r+q^2} + \frac{1}{r+q^2}\Sigma(r,\vec{q})\frac{1}{r+q^2} + \cdots = \frac{1}{r+q^2-\Sigma(r,\vec{q})} \tag{12.73}$$

[4] See, for example, works by Itzykson and Drouffe (1991), Zinn-Justin (2002), Freed (1987), Kleinert and Schulte-Frohlinde (2001), Schäfer (1999), Amit (1984), and Parisi (1998). This collection of references is by no means exhaustive.

[5] We will, from now on, refer to wave numbers as momenta, in keeping with standard usage in physics.

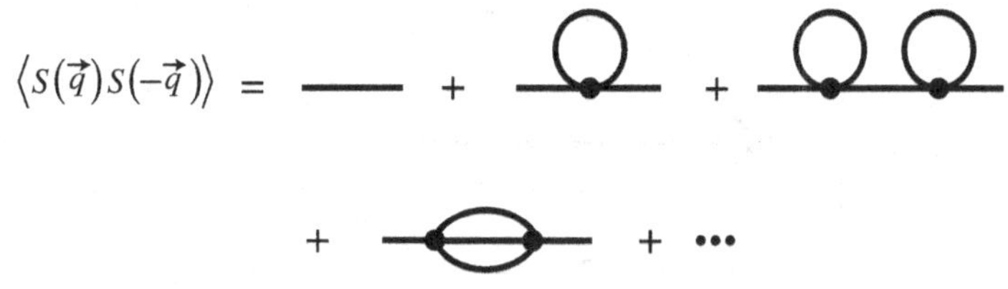

$\langle s(\vec{q})s(-\vec{q})\rangle$ =

Fig. 12.18. The diagrammatic sum for the two-point correlation function.

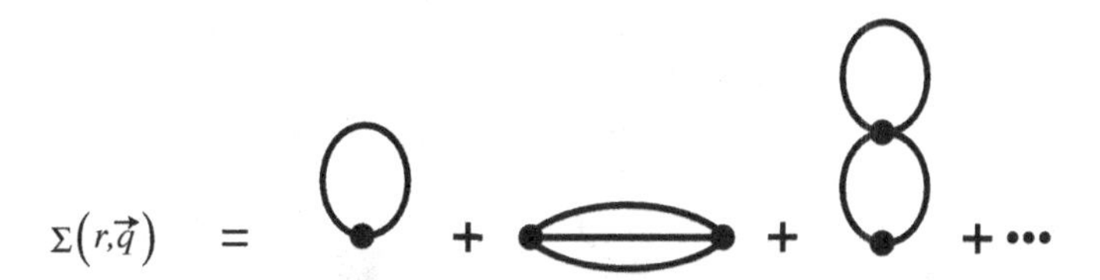

Fig. 12.19. Diagrammatic sum for the quantity $\Sigma(r, \vec{q})$. The sum is over all mass operator insertions.

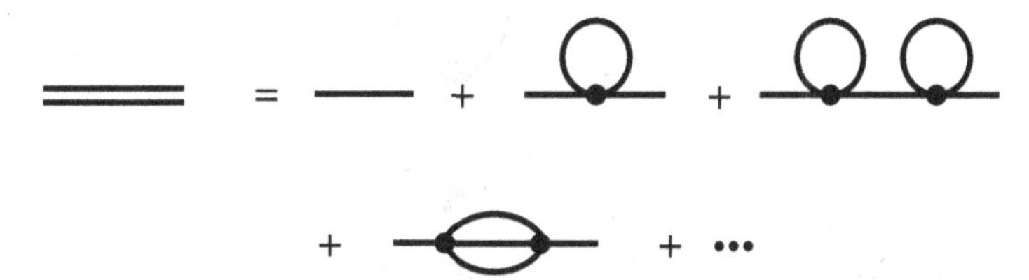

Fig. 12.20. Representation of the renormalized propagator.

where the diagrammatic sum for the quantity $\Sigma(r, \vec{q})$ is as shown in Figure 12.19 The quantity $\Sigma(r, \vec{q})$ is called the mass operator. The sum on the right hand side of the equality in Figure 12.19 is over all mass operator insertions. The quantity $\langle S(\vec{q})S(-\vec{q})\rangle$ is also known as a propagator. When all mass operator insertions are taken into account, it is referred to as a *renormalized* propagator, and is represented as a double line, as indicated in Figure 12.20.

Figure 12.21 represents an integral with the following form:

$$\int \mathrm{d}^d q_1 \int \mathrm{d}^d q_2 \int \mathrm{d}^d q_3 \frac{1}{r+q_1^2} \frac{1}{r+q_2^2} \frac{1}{r+q_3^2} \delta(\vec{q}_1 + \vec{q}_2 + \vec{q}_3 - \vec{q}) \quad (12.74)$$

Here, $\vec{q}$ is the external momentum, carried by the amputated external lines. The delta function in (12.74) effectively reduces the number of integrations from $3d$, where d is the physical dimensionality of the system, to $2d$. In ordering the momenta, we split the multiple integral in (12.74) into a number of different contributions, according to the ordering of the magnitudes of $\vec{q}, \vec{q}_1, \vec{q}_2$, and $\vec{q}_3$. Of particular interest to us will be those in which the external momentum has the smallest magnitude. We will focus on all contributions to the mass operator for which the inequality $q < q_1, q_2, q_3$

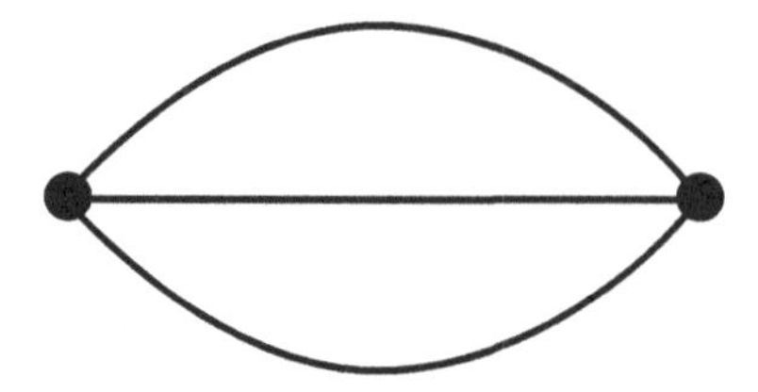

Fig. 12.21. The two-loop diagram for the mass operator, aka the "watermelon diagram."

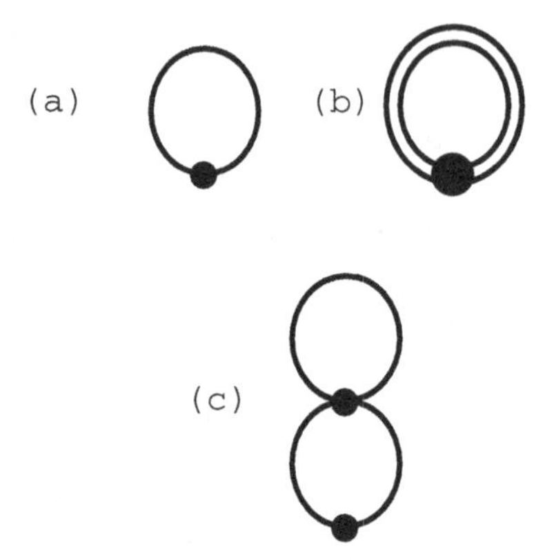

Fig. 12.22. (a) A contribution to the mass operator in which the internal propagator line has the "bare" form, $(r + q^2)^{-1}$. (b) The same contribution, the internal propagator line having been replaced by its fully renormalized version. (c) A diagram that must be eliminated, as it represents a contribution to the diagram (b).

holds. In fact, we will reinforce this inequality and introduce a quantity q_0 that satisfies $q < q_0 < q_1, q_2, q_3$.

The contributions to the two-point correlation function in which this collection of inequalities applies will be denoted by $\Sigma'(q_0; r, \vec{q})$, and the sum $(r + q^2 - \Sigma')^{-1}$ is denoted as a renormalized propagator $G(\vec{q})$. We can go further. The lines in our diagram for Σ stand for the bare propagator $(r + q^2)^{-1} = G_0(\vec{q})$. It can be verified that they may be replaced by the fully renormalized lines, provided we eliminate certain diagrams from the full diagram sum Σ. For example, the diagram in Figure 12.22(a) can be replaced by the diagram in Figure 12.22(b), provided that the diagram in Figure 12.22(c) is eliminated, as it represents a contribution to the renormalization of the doubled propagator line in Figure 12.22(b). This procedure is referred to as propagator renormalization.

Having done this, the contributions to the mass operator will be denoted by the symbol $\Sigma'(q_0; r_R, \vec{q})$. The quantity r_{R} appearing in the argument is an adjusted reduced temperature. This "shifted" version of the bare reduced temperature, r, is equal to zero at the phase transition.

We now assert – without justification as yet – that the combination $r + q^2 + \Sigma'$ has the following form when r_R is equal to zero.

$$q^2 + \Sigma'(q_0; 0, \vec{q}) = q_0^{2-\eta} S(q/q_0) \tag{12.75}$$

We further assert that the function S allows for a power-series expansion in its argument q/q_0, that the series consists only of even-order terms in that argument, and the series begins with at quadratic order $((q/q_0)^2)$. Note that the assertion leads to a correlation function which goes as $q^{2-\eta}$ if q_0 is set equal to q. Now, let's introduce a small but finite r_R. As another assumption, we predict that it is possible to expand the quantity $\Sigma'(q_0; r_R, \vec{q})$ as follows:

$$\begin{aligned} r + q^2 + \Sigma'(q_0; r_R, q) &= q^2 + \Sigma'(q_0; 0, \vec{q}) + r_R \Sigma''(q_0; 0, q) + O\left(r_R^2\right) \\ &= q_0^{2-\eta} S(q/q_0) + r_R q_0^{2-\eta-y} S'(q/q_0) + O\left(r_R^2\right) \end{aligned} \tag{12.76}$$

The right hand side of (12.76) is, thus, an expansion in the adjusted reduced temperature, r_R, and the last line above displays the scaling form of the coefficients in this expansion. The exponent y in the last line of (12.76) is equal to $1/\nu$, ν being the exponent connecting correlation length with reduced temperature ($\xi \propto |r|^{-\nu}$) This correspondence follows from the definition of ξ,

$$\begin{aligned} \xi^2 &= \frac{\int R^2 C(\vec{R})\, \mathrm{d}^d R}{\int C(\vec{R})\, \mathrm{d}^d R} \\ &= -\left(\frac{\partial^2 G(\vec{q})}{\partial q^2}\right)_{q+0} \Big/ G(0) \end{aligned} \tag{12.77}$$

where $C(R)$ is the two-spin correlation function in real space and $G(\vec{q})$ is its Fourier transform. Notice that the expansion in (12.76) is consistent with the following scaling form for the mass operator contribution Σ'

$$q^2 + \Sigma'(q_0; r_R, \vec{q}) = q_0^{2-\eta} \sigma'(q/q_0, r_R q_0^{-y}) \tag{12.78}$$

In fact, we expect that just this form will emerge from further analysis.

Finally, we separate out two of the lowest order terms in the right hand side of (12.78). The first is proportional to q^2, and the second goes as r_R. That is, we write

$$\Sigma'(q_0; r_R, \vec{q}) = q_0^{2-\eta} \left(\left(\frac{q}{q_0}\right)^2 + r_R q_0^{-y} \right) + O\left(q^4, r_R^2, q^2 r_R\right) \tag{12.79}$$

Now, consider a given propagator line, carrying a momentum equal to $\vec{q}$. Suppose we collect all mass operator corrections to this line, the internal momenta of which are larger than $q_0 > q$. Combining those mass operator corrections with the "bare" terms $r + q^2$, we then separate out the first two terms on the right hand side of

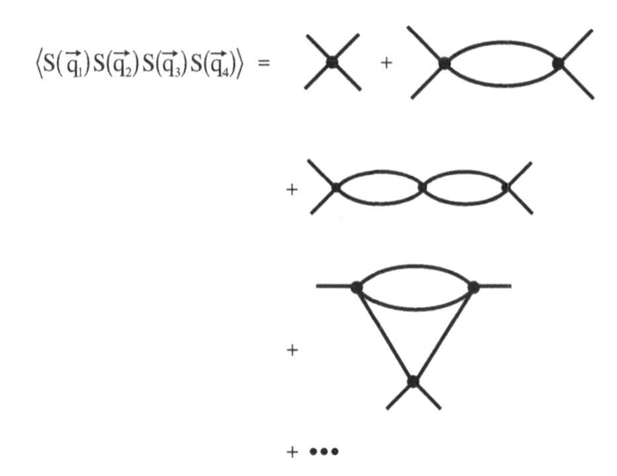

Fig. 12.23. The diagrammatic expansion of the average $\langle S(\vec{q}_1)S(\vec{q}_2)S(\vec{q}_3)S(\vec{q}_4)\rangle$.

(12.79). We then incorporate those terms into a (partially) renormalized propagator line, which takes the form

$$q_0^{-2+\eta} \frac{1}{r_R q_0^{-y} + (q/q_0)^2} \tag{12.80}$$

This new propagator line will be the one used in all diagrammatic expansions. All contributions to the renormalized correlation function will be treated as perturbations.[6]

12.6.1 The renormalized fourth order coupling

The next quantity of interest is the average $\langle S(\vec{q}_1)S(\vec{q}_2)S(\vec{q}_3)S(\vec{q}_4)\rangle$, referred to as a four-point correlation function. The perturbation expansion for this quantity in powers of u is as shown in Figure 12.23. The amputated versions of these diagrams as shown in Figure 12.24. The quantity Γ_4, equal to the sum of the diagrams in Figure 12.24, is of interest to us, in that it amounts to a modification of the fourth order coupling constant, previously denoted as the constant u. As Figure 12.24 makes clear, Γ_4 has the external momentum-carrying lines removed.

Again, there are internal and external momenta. In the case of the fourth order coupling we can identify four external momenta, restricted by the requirement that they sum to zero. If we denote those external momenta by q_i^{ext}, the index i ranging from one to four, then we can also introduce a magnitude q_0 that interposes between the smaller external momenta and larger internal momenta of a class of contributions to the modified fourth order coupling. We then conjecture the following scaling form

[6] A note to the finicky: the result for the propagator in (12.80) is not correct for the fully renormalized propagator. However, it contains the essence of the corrections leading to exact critical point behavior.

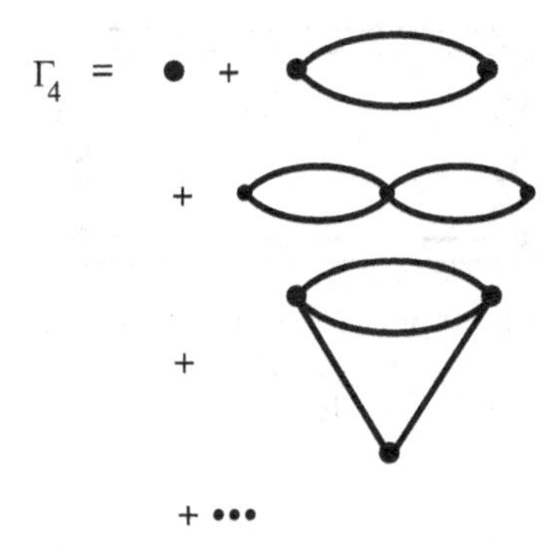

Fig. 12.24. The diagrammatic expansion of the "amputated" modification of the four-point correlation function, $\langle S(\vec{q}_1)S(\vec{q}_2)S(\vec{q}_3)S(\vec{q}_4)\rangle$. The notation for this quantity is Γ_4.

for the sum of the bare fourth order coupling and all corrections to that quantity

$$u_{\rm R}(q_0; r_{\rm R}, \vec{q}_1^{\,\rm ext}, \ldots, \vec{q}_4^{\,\rm ext}) = q_0^{4-d+2\eta} g(q_0, r_{\rm R} q_0^{-y}, \vec{q}_1^{\,\rm ext}/q_0, \ldots, \vec{q}_4^{\,\rm ext}/q_0) \quad (12.81)$$

Note that the quantity g contains an additional explicit q_0-dependence.

In this case, we focus on the contribution to the renormalized fourth order coupling that survives the setting of all external momenta equal to zero. According to (12.81), this term will have the form

$$q_0^{4-d+2\eta} g_0(q_0; r_{\rm R}) \quad (12.82)$$

This is the fourth order term that will be utilized in our diagrammatic expansions.

Relationship of the diagrammatic method to the momentum-shell renormalization group: recursion relations

The idea of taking into account fluctuations greater than a certain infrared cutoff is intrinsic to the momentum shell renormalization group. In fact, this is precisely what occurs as the size of the Brillouin zone shrinks. The modifications to the effective Hamiltonian that result from this process of elimination of degrees of freedom (and rescaling) correspond precisely to the kinds of calculations envisioned in the above discussion. The diagrammatic method refers to the evaluation of effective Hamiltonian modifications according to the rules of perturbation theory, while in the momentum-shell renormalization group those modifications are determined recursively. If both of the above programs are carried out completely, the same expressions would result.

This clearly implies that recursion relations will also arise from the diagrammatic approach. In fact, as we will now show, one is led directly to a differential equation for the quantity $g_0(q_0)$ appearing in (12.82). To see how this comes about, consider a diagram contributing to the renormalized fourth order coupling. One such diagram is

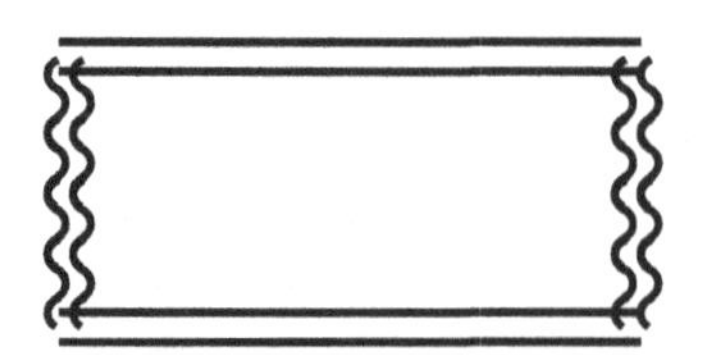

Fig. 12.25. Diagram contributing to the renormalization of the fourth order coupling. The double lines for both propagators and interactions indicate that both quantities are replaced by appropriately renormalized versions, as given in (12.82) and (12.80).

shown in Figure 12.25. Setting all the external momenta equal to zero, the momenta of the internal propagator lines will be $\check{Q}$ and $-\vec{Q}$. This means that the q_0 associated with those lines must be greater than or equal to Q. Let's set the two equal. Then, the integral that this diagram represents goes as

$$\int u_{\mathrm{R}}^2(Q) G^2(Q) \, \mathrm{d}^d Q = \int g_0(Q)^2 Q^{8-2d+4\eta} \frac{Q^{-4+2\eta}}{(r_{\mathrm{R}} Q^{-y} + 1)^2} Q^{d-1} \, \mathrm{d}Q \tag{12.83}$$

In keeping with the philosophy of this approach, we will set the lower limit of the integration over Q equal to q_0, and we take the limit $r_{\mathrm{R}} = 0$. The self-consistent contribution to the renormalized coupling strength $g_0(q_0)$ is, then, proportional to

$$\int_{q_0} g_0(Q)^2 Q^{3-d+2\eta} \, \mathrm{d}Q \tag{12.84}$$

Now, this integral is a contribution to the renormalized coupling strength u_{R}. The quantity g is related to u_{R} by

$$g_0(q_0) = q_0^{-4+d-2\eta} u_{\mathrm{R}}(q_0; r_{\mathrm{R}} = 0, \vec{q}_1 = 0, \ldots, \vec{q}_4 = 0) \tag{12.85}$$

If we take the derivative of this integral with respect to the lower limit, q_0, we end up with a contribution to the renormalized coupling strength that goes as $g_0(q_0)^2/q_0$ and the contribution $(-4 + d - 2\eta) g_0(q_0)/q_0$. Both these terms combine to generate the following differential equation for $g_0(q_0)$:

$$q_0 \frac{\mathrm{d}g_0(q_0)}{\mathrm{d}q_0} = (-4 + d - 2\eta) g_0(q_0) + \mathcal{B} g_0(q_0)^2 \tag{12.86}$$

This is precisely the form of the equation that describes the development of the fourth order coupling in the momentum-shell renormalization group, as embodied in (12.39) and (12.67). Given the connection between the diagrammatic method outlined here and that version of the renormalization group, the connection between the two equations should be no surprise.

A corresponding recursion relation can be derived for the renormalized propagator. In this case, we consider the diagrams contributing to the renormalization

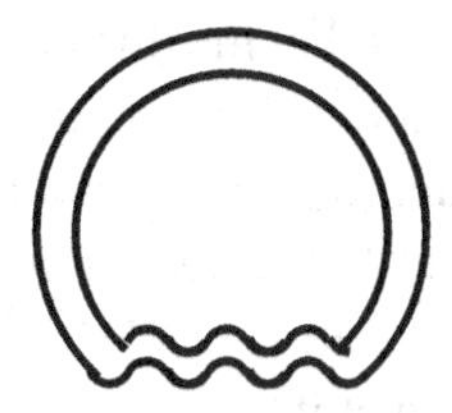

Fig. 12.26. The one-loop diagram leading to contributions to the renormalized propagator shown in (12.87) and (12.88).

of the new reduced temperature, r_{R}. There are two principal ways in which these contributions arise. The first is through an alteration in the temperature at which $r_{\mathrm{R}} = 0$. This alteration arises from contributions to the mass operator for which the external momentum is equal to zero, and for which the effective reduced temperature appearing in internal propagator lines is also zero. At lowest level in perturbation theory, the expression for this contribution, depicted diagrammatically in Figure 12.26, is proportional to

$$\begin{aligned}\int_{Q>q_0} u_{\mathrm{R}}(Q)G(Q)\,\mathrm{d}^dQ &= \int_{Q>q_0} (q_0(Q))^{4-d+2\eta}\, g_0(q_0(Q))\frac{(q_0(Q))^{-2+\eta}}{(Q/q_0(Q))^2}Q^{d-1}\,\mathrm{d}Q \\ &= \int_{Q>q_0} Q^{1-\eta}g(Q)\,\mathrm{d}Q \end{aligned} \tag{12.87}$$

In this expresion, we have taken the appropriate value for the "internal" cutoff, $q_0(Q)$ to be equal to Q, in the case of the effective fourth order coupling and the mass operator that modifies the propagator line inside the integration. The result of this integration will be a contribution to the renormalized propagator going as $q_0^{2-\eta}$, due to the behavior of the integrand in (12.87) near its lower limit of $Q = q_0$, and another term that is essentially independent of q_0, arising from all other contributions to the integration. These latter contributions are what control the actual shift in the transition temperature.

The second type of contribution to the renormalized propagator is linear in the new reduced temperature. It follows from an expansion of the propagator line in the contribution shown in Figure 12.26 to first order in r_{R}. The integral form of this contribution is

$$\begin{aligned}&\int_{Q>q_0} g(q_0(Q))\,(q_0(Q))^{4-d+2\eta}\,r_{\mathrm{R}}\,(q_0(Q))\,(q_0(Q))^{2-\eta-y}\,Q^{d-1}\,\mathrm{d}q \\ &\quad= \int_{Q>q_0} g(Q)Q^{-y+1-\eta}r_{\mathrm{R}}\,\mathrm{d}Q \\ &\quad\to \frac{r_{\mathrm{R}}g(q_0)q_0^{2-\eta-y}}{y-2+\eta}\end{aligned} \tag{12.88}$$

To achieve consistency, this implies that the difference between the exponent y and $2-\eta$ is proportional to the asymptotic fixed point value of the modified coupling strength $g(q_0)$. In fact, if the appropriate combinatorial weights and prefactors are inserted into the integrals in (12.87), we end up with the correct value for y by requiring that the last line of that equation be identical to $r_R q_0^{2-\eta-y}$. The relationship (12.88) can also be processed so that it takes the form of a differential equation. In that case, we have an equation equivalent to (12.66).

The calculation of the critical exponent η proceeds along similar lines. In this case, one extracts the term proportional to q^2 in the correlation function diagram shown in Figure 12.21, in which the fourth order couplings are partially renormalized according to the prescription laid out in the previous pages and the propagators are modified by the mass operator Σ'. The calculations are a bit more involved than those outlined immediately above, in that the enforcement of the requirement that internal momenta are greater than the cutoff q_0 is a bit more involved. However, one finds that this can be done, and a value for the exponent η appears as if by magic.

12.6.2 Construction of a complete, self-consistent diagrammatic method based on the above results

The calculation of the exponents y and η, and of the behavior of the coupling strength, g, provides the basis of a self-consistent and convergent calculation of quantities in the vicinity of the critical point. In particular, their incorporation into a modified perturbation theory suffices to "tame" the diagrammatic evaluation of correlation functions and thermodynamic quantities in the vicinity of the critical point of an $O(n)$ model. This means that we can also utilize a suitably "renormalized" perturbation theory to obtain the statistical properties of a long self-avoiding walk.

The key to the convergence of the diagrammatic method lies in the scaling properties of the fourth order coupling constant and of propagators. Recall the assumed scaling form of the modified fourth order coupling constant, u_R, as embodied in (12.81). Assuming a fixed point for the amplitude g, we have $u_R(q_0; r_R = 0, \vec{q}_i = 0) \propto q_0^{4-d+2\eta}$. In fewer than four dimensions, the (partially) renormalized fourth order coupling constant at the critical temperature, and in the limit of arbitrarily small external momenta, disappears as the lower limit on internal momenta in the diagrammatic contributions is reduced to zero. This implies that a fully renormalized version of the fourth order coupling at the critical point – and for which all the external momenta have been set equal to zero – has vanishingly small amplitude. It is this evaporation of the fourth coupling at the critical point that "rescues" perturbation theory.

To see in broad outline how this comes about, we'll revisit the power counting argument in Chapter 7. Naive power counting tells us that the contribution to a

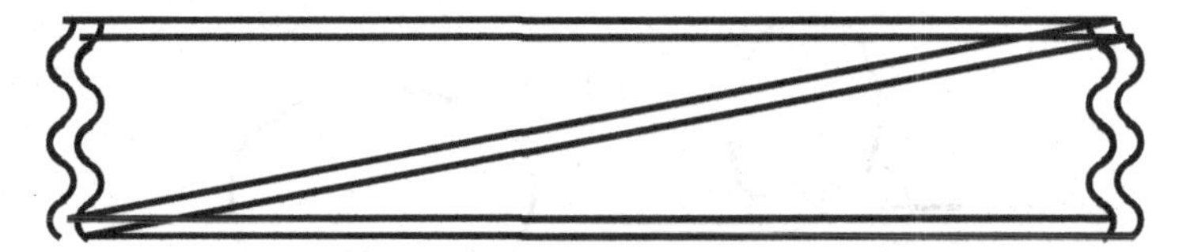

Fig. 12.27. Diagram contributing to the renormalization of the mass operator. Both the propagator lines and the interactions are renormalized.

given quantity that is nth order in the "bare" coupling strength, u, contains a term going $(ur^{-(4-d)/2})^n$. This results in a divergent perturbation series as $r \to 0$ when the spatial dimensionality of the system, d, is less than four. However, perturbation theory is rescued through renormalization. The divergent factor $r^{-(4-d)/2}$ is cancelled in all orders of the perturbation expansion by a renormalized fourth order coupling constant that goes to zero as $r^{(4-d)/2}$, neglecting the critical exponent η. As a simplified illustration consider the second order contribution to the mass operator shown in Fig. 12.27. This diagram is proportional to the following multiple integral.

$$\int u^2 G(Q_1)G(Q_2)G(Q_3)\,\mathrm{d}^dQ_1\,\mathrm{d}^dQ_2\,\mathrm{d}^dQ_3\,\delta(\vec{Q}_1 + \vec{Q}_2 + \vec{Q}_3 - \vec{q}) \quad (12.89)$$

If the interaction strength, u and the propagator lines, $G(Q_i)$, were to take on their "bare" forms, so that u is constant and $G(Q_i) = 1/(r + Q_i^2)$, then, scaling the reduced temperature, r, out of the integrand we are left with an expression going as $u^2 r^{2d-3} = r^{-1} \times (ur^{-(4-d)/2})^2$. On the other hand, if $u \to g q_0^{4-d+2\eta}$ and $G \propto q_0^{-2+\eta} f(rq_0^{-y}, Q_i/q_0)$, then, scaling the cutoff q_0 out of the integrand, we end up with a contribution to the mass operator going as $q^{2-\eta}u^2\mathcal{F}(rq_0^{-y})$. This is consistent with the assumed form of the mass operator contributions to the correlation function, and, hence, the propagator.

This argument can be made more rigorous, and the results above can be shown to follow at all orders in perturbation theory.

Counterterms

The introduction of renormalized vertices and propagators into diagrammatic calculations carries with it a redundancy for which one must compensate. As an example, consider Figure 12.28 in which (a), (b), and (c) correspond to three different depictions of a single diagram for a mass operator insertion on a propagator line. There are two fourth order couplings in each diagram. The portions of the diagrams that are surrounded by the dashed oval can be associated with a modification of the fourth order coupling in diagram (a). This fourth order coupling is depicted as a doubled wavy line, to indicate that it has been modified by the incorporation of corrections. Now, replacing the fourth order coupling by the modified version calculated as described in the sections above amounts to the inclusion of diagrams

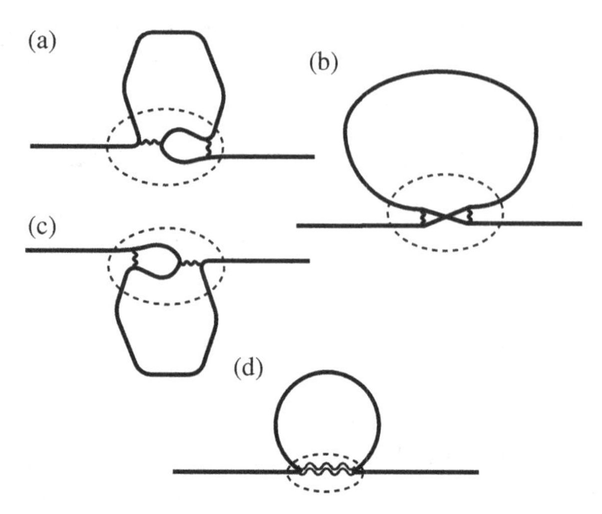

Fig. 12.28. Illustrating how portions of a diagram ((a), (b), and (c)) can be associated with a renormalization of the fourth order coupling, which is then incorporated into a lower order diagram, (d).

such as (a) in the putative low order diagram (d). The separate evaluation of the higher order diagram then introduces a "double counting" of its influence on the properties of the system under consideration. It is necessary to perform subtractions to remove this double counting. This is done by the introduction of counterterms, which compensate for the incorporation of higher order diagrams into lower order ones through the modifications of fourth order couplings generated by the higher order diagrams.

This can be done in more than one way. The most straightforward is to order the multiple integral represented by a diagram according to the magnitudes of the internal momenta. Then, when the internal momenta in a portion of the diagram that can be associated with a modification of a fourth order coupling are all greater than the momenta of the propagator lines attached to that portion, one subtracts from the multiple integral a "diagram" in which that portion is replaced by one in which all external momenta have been set equal to zero, and for which (in the portion) the reduced temperature is equal to zero. This approach can be shown to work. However, as a practical matter, it is extremely unwieldy. The jumble of multiple integrals that emerge at higher order quickly becomes intractable. A simpler calculational trick involves the introduction of what are called counterterms in the original effective Hamiltonian, corresponding to the corrections to the fourth order coupling. Those counterterms are then inserted strategically into diagrams. If properly exploited, they account for all corrections and cancel the overcounting alluded to above.

Similar stratagems allow the theorist to avoid redundancies associated with the modifications of the mass operator leading to the exponents η and $y = 1/\nu$.

12.6.3 Differential equations for correlation functions

Once the renormalized fourth order coupling has been determined with the use of recursion relations and the exponents y and η have been obtained via calculations of the properties of the partially modified mass operator, and after one has reconstituted correlation or thermodynamic functions by performing diagrammatic calculations with the new quantities, one is left with expressions for physically meaningful quantities, such as correlation functions, or thermodynamic potentials, or susceptibilities, or numbers of random walks. In the course of the calculations, the quantity q_0 has been introduced, and, as a practical matter, it will have been kept equal to a non-zero value. However, this cutoff is associated with a calculational method. It is not a physical quantity. In fact, one generally has the freedom to adjust it at will, within certain limits. This means that the result must be independent of the precise value of q_0. One can formalize this independence by taking a derivative of the quantity of interest with respect to q_0 and checking that this derivative is equal to zero. In fact, one can turn q_0-independence into a requirement on all calculated quantities. Such an approach leads to a differential equation for key functions, known generically as a Callan–Symanzik equation (Callan, 1970; Symanzik, 1970). This equation plays a central role in field-theoretical approaches to the renormalization group. The reader is directed to any one of the many pedagogical books that have been written on the subject, mentioned earlier in a footnote in Section 12.5.

12.7 Supplement: linked cluster expansion

Here, we will demonstrate that averaging an infinite series of diagrams leads to the exponential of a linked cluster expansion. This equality is depicted in Figure 12:S-1. The quantity $\langle \ldots \rangle$ signifies a Gaussian average of all $S(\vec{q})$'s in the shell (see (12.13) in Section 12.2.4).

We start with the basic form of the effective Hamiltonian, as given in (12.3). Then, we divide the spin degrees of freedom, $S(\vec{q})$, into those having argument $\vec{q}$ in the inner Brillouin zone and those whose argument lies in the shell surrounding that zone. To keep things clear, we replace the argument $\vec{q}$ by the argument $\vec{Q}$ when it lies in the shell. The quadratic portion of the Hamiltonian containing $S(\vec{Q})$'s is

$$\sum_{\vec{Q}} \frac{1}{2} \left(r + Q^2\right) S(\vec{Q}) S(-\vec{Q}) \qquad (12\text{:S-1})$$

The quartic term splits up as indicated in (12.11). The quartic term with two $S(\vec{Q})$'s and two $s(\vec{q})$'s is shown in Figure 12.6b, or Figure 12.12. The exponential of this vertex is expressed pictorially as an expansion in Figure 12.10. If we integrate over the $S(\vec{Q})$'s we are performing a Gaussian average of this infinite sum of diagrams.

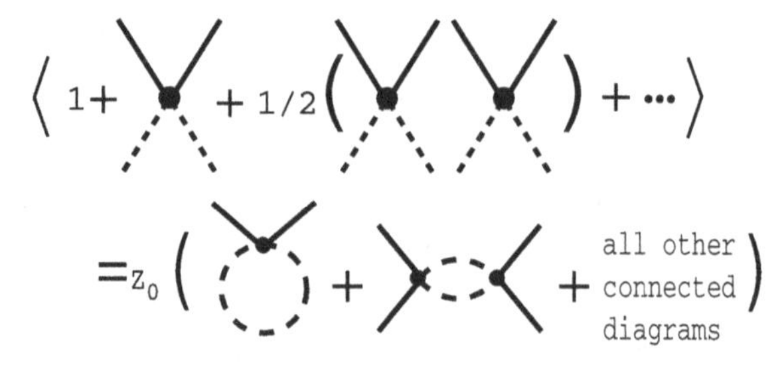

Fig. 12:S-1. The equality that is encapsulated in the linked cluster expansion. The left hand side of the equality is a sum of all diagrams generated by splitting the $S(\vec{q})$'s into spin degrees of freedom for which the argument lies in the inner Brillouin zone and degrees of freedom whose arguments lie in the shell surrounding that Brillouin zone, as depicted in Figure 12.5.

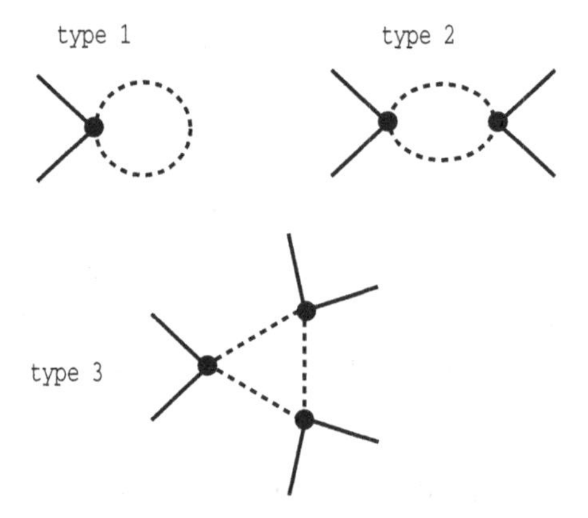

Fig. 12:S-2. Ways in which the four-point diagrammatic fragments in the top line of Figure 12:S-1 can be joined by the Gaussian pairing of lines corresponding to $S(\vec{q})$'s for which $\vec{q}$ is in the spherical shell surrounding the reduced Brillouin zone.

Each term in the sum consists of a product of the quartic vertex, which we will call a *quartic fragment* from now on. In averaging over a quartic fragment, we pair $S(\vec{Q})$'s up in all possible ways, performing the Gaussian average over each pair. This Gaussian average leads to the result for each pair

$$\langle S(\vec{Q}_1)S(\vec{Q}_2)\rangle = \delta\left(\vec{Q}_1 + \vec{Q}_2\right)\frac{1}{r + Q_1^2} \tag{12:S-2}$$

Some of the expressions that result are illustrated in Figure 12:S-2. The dashed lines in the figure now correspond to averages of pairs of $S(\vec{Q})$)'s, and are equal to $1/(r + Q^2)$. Because the shell surrounding the BZ is thin, the Q in this factor can be set equal to a constant in all the diagrams. Note that all the diagrams shown in Figure 12:S-2 are *connected*, in that they cannot be separated into two different parts, except by severing one or more dashed lines. In fact, in the case at hand,

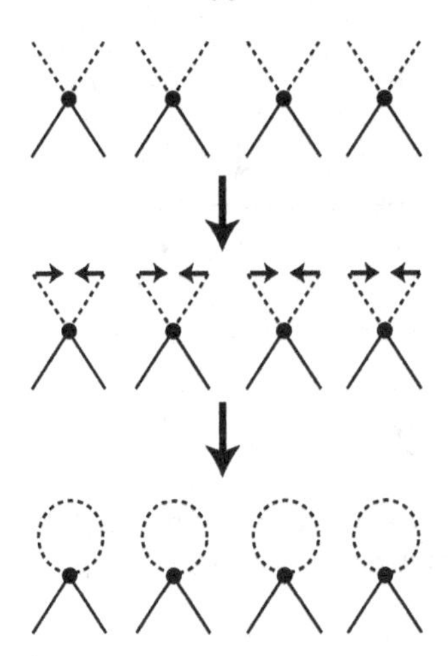

Fig. 12:S-3. The generation of n type 1 diagrams from a product of n quartic fragments. The top line is the product of n fragments. The second line indicates how the $S(\vec{Q})$'s are paired. The bottom line displays the resulting product of diagrams.

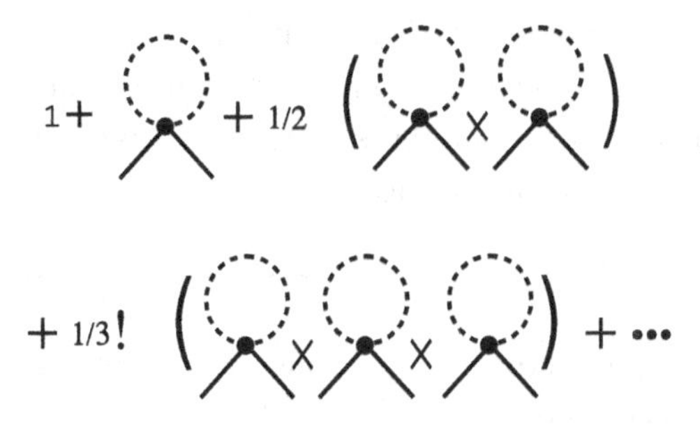

Fig. 12:S-4. The sum of diagrams when spins are paired to produce type 1 diagrams.

separating a connected diagram into two component parts requires the severing of at least two dashed lines.

Let's start by considering what happens if we count only the "type 1" connected diagram in Figure 12:S-2. The process by which a set of n diagrams of this type is generated upon averaging over $S(\vec{Q})$'s associated with the product of n quartic fragments is shown in Figure 12:S-3. There is clearly only one way to achieve the pairings indicated in Figure 12:S-3. This means that the averaged sum shown in the top line of Figure 12:S-1 generates the sum shown in Figure 12:S-4. If we denote by $\mathcal{T}_1$ the expression corresponding to the type 1 diagram, then the result of the operations depicted in Figure 12:S-4 is

$$1 + \mathcal{T}_1 + \frac{1}{2!}\mathcal{T}_1^2 + \frac{1}{3!}\mathcal{T}_1^3 + \cdots = e^{\mathcal{T}_1} \qquad (12:\text{S-3})$$

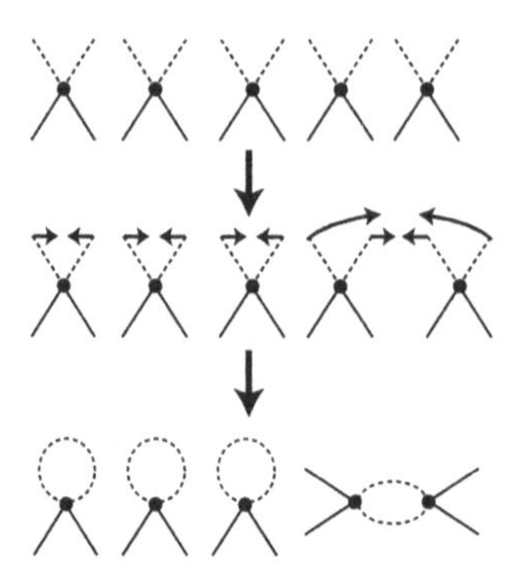

Fig. 12:S-5. The pairing of the lines in a product of five quartic fragments producing a product of three type 1 diagrams and 1 type 2 diagram.

As the next step in this demonstration, we consider what happens when the $S(\vec{Q})$'s in a product of quartic fragments are paired so as to produce both type 1 and type 2 diagrams. This process is illustrated in Figure 12:S-5. There is a combinatorial factor associated with the collection of diagrams on the bottom line in Figure 12:S-5. This factor is equal to the number of ways in which the $S(\vec{Q})$'s can be paired so as to form those diagrams. The middle line in the figure depicts one of the ways in which this can be done. There are others, however. A bit of investigation, and we find that there are $5!/(3!\ 2!)$ ways of doing this.

Exercise 12.9

Perform the investigation leading to the above result for the number of ways of forming the diagrams at the bottom of Figure 12:S-5.

It is possible to argue for the factor as follows. Start by ordering the five fragments in all possible ways. Then, take the first three and form type 1 diagrams from them, and form a type two diagram from the last two. This yields the factor $5!$, the number of ways of ordering five objects. However, we have overcounted the distinct number of ways of forming the diagrams using this method. If we were to interchange any two of the first three fragments, we would end up with the same collection of diagrams. This also holds true if we were to interchange the last two fragments. In order to compensate for this overcounting, we divide 5! by $3!$, the number of ways of permuting the first three fragments, and also by $2!$, the number of ways of permuting the last two. In this way, we end up with the overall combinatorial factor of $5!/(2!\ 1!)$.

To make things a bit more complicated, imagine that we had $n = n_1 + 2n_2$ fragments, and that we paired $S(\vec{Q})$'s so as to construct n_1 type 1 diagrams and n_2 type 2 diagrams, in a generalization of the process depicted in Fig. 12:S-6. The

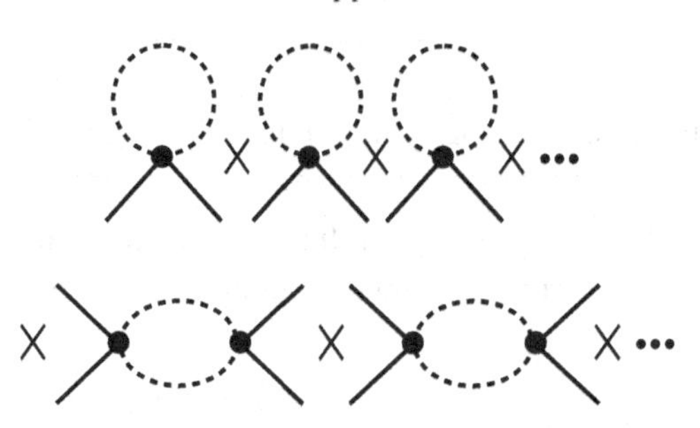

Fig. 12:S-6. An overall diagram consisting of n_1 type 1 diagrams and n_2 type 2 diagrams.

combinatorial factor that tells us how many ways there are to do this

$$\frac{n!}{n_1!n_2!(2!)^{n_2}} \tag{12:S-4}$$

Exercise 12.10

Show, by a generalization of the argument just given, that the combinatorial factor (12:S-4) correctly counts the number of ways of combining $n = n_1 + 2n_2$ quartic fragments so as to produce n_1 type 1 diagrams and n_2 type 2 diagrams.

As there is a factor of $1/n!$ associated with the original set of fragments in this example, we see that the product of n_1 type 1 diagrams and n_2 type 2 diagrams that have been formed as a result of the pairing of $S(\vec{Q})$'s will have the overall combinatorial factor $1/(n_1!n_2!(2!)^{n_2})$ associated with it. If we denote by $2!\mathcal{T}_2$ the expression depicted by the type 2 diagram and sum over all possible values of n_1 and n_2, we obtain

$$\sum_{n_1=0}^{\infty}\sum_{n_2=0}^{\infty} \frac{\mathcal{T}_1^{n_1}}{n_1!}\frac{\mathcal{T}_2^{n_2}}{n_2!} = \exp\left[\mathcal{T}_1 + \mathcal{T}_2\right] \tag{12:S-5}$$

We can now generalize. Suppose we have a set of n fragments, which, as a result of the Gaussian averaging, leads to a product of n_1 type 1 diagrams, n_2 type 2 diagrams, n_3 type 3 diagrams, ... The combinatorial factor that tells us how many ways this can be done will be equal to

$$\frac{n!}{n_1!\,(n_2!(2!)^{n_2})\,(n_3!(f_3)^{n_3})\ldots} \tag{12:S-6}$$

The factor f_3 is the number of ways of permuting vertices in a type 3 diagram without disrupting the pairing of $S(\vec{Q})$'s that leads to that diagram.

Exercise 12.11

Show that $f_3 = 6$ in the case of the type 3 diagram shown in Figure 12:S-2.

If, now, we denote by $f_n \mathcal{T}_n$ the type n diagram, we then discover that summing over all possible ways of forming connected diagrams yields the following:

$$\sum_{n_1=1}^{\infty}\sum_{n_2=0}^{\infty}\sum_{n_3=0}^{\infty}\cdots\frac{\mathcal{T}_1^{n_1}}{n_1!}\frac{\mathcal{T}_2^{n_3}}{n_2!}\frac{\mathcal{T}_3^{n_3}}{n_3!}\cdots = \exp\left[\mathcal{T}_1 + \mathcal{T}_2 + \mathcal{T}_3 + \ldots\right] \qquad (12\text{:S-}7)$$

The right hand side of Figure 12:S-1 is the graphical representation of the right hand side of (12:S-7). We now have our desired result.

References

Abramowitz, M. and Stegun, I. A. (1970). *Handbook of Mathematical Functions: with Formulas, Graphs, and Mathematical Tables.* Dover Publications, New York.

Amit, D. J. (1984). *Field Theory, The Renormalization Group, and Critical Phenomena.* World Scientific, Singapore, revised 2nd edn.

Aronovitz, J. A. and Nelson, D. R. (1986). Universal features of polymer shapes. *Journal de Physique*, **47**(9):1445–56.

Ashcroft, N. W. and Mermin, N. D. (1976). *Solid State Physics.* Holt, Rinehart and Winston, New York.

Barber, M. N. and Ninham, B. W. (1970). *Random and Restricted Walks; Theory and Applications.* Gordon and Breach, New York.

Berg, H. C. (1993). *Random Walks in Biology.* Princeton University Press, Princeton, NJ, expanded edn.

Berg, H. C. and Purcell, E. M. (1977). Physics of chemoreception. *Biophysical Journal*, **20**(2):193–219.

Bishop, M. and Michels, J. P. J. (1986). Polymer shapes in three dimensions. *Journal of Chemical Physics*, **85**(10):5961–2.

Boas, M. L. (1983). *Mathematical Methods in the Physical Sciences.* Wiley, New York, 2nd edn. (A good introduction to combinatorial counting and elementary concepts in probability theory.)

Bookstein, F. L. (1978). *The Measurement of Biological Shape and Shape Change.* Lecture notes in biomathematics; 24. Springer-Verlag, Berlin, New York.

Callan, C. G., J. (1970). Broken scale invariance in scalar field theory. *Physical Review D*, **2**(8):1541–7.

Costa, L. d. F. and Cesar, R. M. (2001). *Shape Analysis and Classification: Theory and Practice.* Image processing series. CRC Press, Boca Raton, FL.

de Gennes, P. G. (1972). Exponents for the excluded volume problem as derived by the wilson method. *Physics Letters A*, **38**(5):339–340.

—(1979). *Scaling Concepts in Polymer Physics.* Cornell University Press, Ithaca, NY.

des Closeaux, J. and Jannink, G. (1990). *Polymers in Solution: Their Modeling and Structure.* Clarendon Press, Oxford.

Diehl, H. W. and Eisenriegler, E. (1989). Universal shape ratios for open and closed random walks: exact results for all *d*. *Journal of Physics A (Mathematical and General)*, **22**(3):L87–91.

Emery, V. J. (1975). Critical properties of many-component systems. *Physical Review B (Condensed Matter)*, **11**(1):239.

Feller, W. (1968). *An Introduction to Probability Theory and its Applications.* Wiley series in probability and mathematical statistics. Wiley, New York.
Feynman, R. P. and Hibbs, A. R. (1965). *Quantum Mechanics and Path Integrals.* International series in pure and applied physics. McGraw-Hill, New York.
Fixman, M. (1962). Radius of gyration of polymer chains. *Journal of Chemical Physics,* **36**:306–10.
Flory, P. J. (1953). *Principles of Polymer Chemistry.* Cornell University Press, Ithaca, NY.
—(1969). *Statistical Mechanics of Chain Molecules.* Interscience, New York.
Freed, K. F. (1987). *Renormalization Group Theory of Macromolecules.* Wiley, New York.
Gaspari, G., Rudnick, J., and Beldjenna, A. (1987). The shapes of open and closed random walks: a $1/d$ expansion. *Journal of Physics A (Mathematical and General),* **20**(11):3393–414.
Gell-Mann, M. and Low, F. E. (1954). Quantum electrodynamics at small distances. *Physical Review,* **95**:1300–1312.
Gradshteyn, I. S., Ryzhik, I. M., and Jeffrey, A. (2000). *Table of Integrals, Series, and Products.* Academic Press, San Diego, 6th edn.
Hughes, B. D. (1995). *Random Walks and Random Environments.* Clarendon Press, Oxford; Oxford University Press, New York.
Itzykson, C. and Drouffe, J.-M. (1991). *Statistical Field Theory.* Cambridge monographs on mathematical physics. Cambridge University Press, Cambridge, 1st paperback edn.
Jackson, J. D. (1999). *Classical Electrodynamics.* Wiley, New York, 3rd edn.
Jeffreys, H. (1972). *Methods of Mathematical Physics.* Cambridge University Press, Cambridge, 1st paperback of 3rd edn. (A good introduction to the theory of asymptotic expansions and the method of steepest descents.)
Kleinert, H. (1995). *Path Integrals in Quantum Mechanics, Statistics, and Polymer Physics.* World Scientific, Singapore.
Kleinert, H. and Schulte-Frohlinde, V. (2001). *Critical Properties of ϕ^4-theories.* World Scientific, River Edge, NJ.
Kosmas, M. K. and Freed, K. F. (1978). On scaling theories of polymer solutions. *Journal of Chemical Physics,* **69**(8):3647–59.
Kramers, H. A. (1946). The behavior of macromolecules in inhomogeneous flow. *Journal of Chemical Physics,* 14:415–24.
Mandelbrot, B. B. (1982). *The Fractal Geometry of Nature.* Freeman, San Francisco.
Maris, H. J. and Kadanoff, L. P. (1978). Teaching the renormalization group. *American Journal of Physics,* **46**(6):652–7.
Montroll, E. W. (1956). Random walks on multidimensional spaces, especially on periodic lattices. *Journal of the Society for Applied and Industrial Mathematics,* **4**:241–60.
Montroll, D. and Shlesinger, M. F. (1983). The wonderful world of random walks. In Falk, H., ed., *CCNY Physics Symposium in Celebration of Melvin Lax's Sixtieth Birthday,* page 364, City College of New York Physics Dept., New York.
Montroll, E. W. and Weiss, G. H. (1965). Random walks on lattices ii. *Journal of Mathematical Physics,* **6**:364.
Morse, P. M. and Feshbach, H. (1953). *Methods of Theoretical Physics.* McGraw-Hill, New York.
Nauenberg, M. (1975). Renormalization group solution of the one-dimensional Ising model. *Journal of Mathematical Physics,* **16**(3):703–5.
Nelson, D. R. and Fisher, M. E. (1973). Exact renormalisation groups for one-dimensional spin systems. *19th Annual Conference on Magnetism and Magnetic Materials,* Boston, MA, USA, 13-16 Nov. 1973. AIP Conf. Proc. (USA), pages 888–90.

Parisi, G. (1998). *Statistical Field Theory*. Perseus Books, Reading, MA.
Pólya, G. (1919). Quelques problèmes de probabilité se rapportant à la 'promenade au hasard' (Some problems of probability associated with the 'random walk'). *L'Enseignement Mathématique*, **20**:444–445.
—(1921). Über eine aufgabe der wahrscheinlichkeitsrechnung betreffen die irrfahrt im straßennetz (On a theorem of probability calculus concerning wandering in a network of streets). *Mathematische Annalen*, **83**:149–160.
Redner, S. and Reynolds, P. J. (1981). Single-scaling-field approach for an isolated polymer chain. *Journal of Physics A (Mathematical and General)*, **14**(3):L55–61.
Rudnick, J. and Gaspari, G. (1986a). The asphericity of random walks. *Journal of Physics A (Mathematical and General)*, **19**(4):L191–3.
—(1986b). Bond percolation on a finite lattice: the one-state Potts model reconsidered. *Journal of Statistical Physics*, **42**(5–6):833–60.
Rudnick, J., Beldjenna, A., and Gaspari, G. (1987). The shapes of high-dimensional random walks. *Journal of Physics A (Mathematical and General)*, **20**(4):971–84.
Schäfer, L. (1999). *Excluded Volume Effects in Polyer Solutions, As Explained by the Renormalization Group*. Springer-Verlag, Berlin.
Solc, K. and Stockmayer, W. H. (1971). Shape of a random flight chain. *Journal of Chemical Physics*, **54**:2756–57.
Stueckelberg, E. C. G. and Peterman, A. (1953). *Helvetia Physica Acta*, **26**:499.
Symanzik, K. (1970). Small distance behaviour in field theory and power counting. *Communications in Mathematical Physics*, **18**(3):227–46.
Theodorou, D. N. and Suter, U. (1985). Shape of unperturbed linear polymers: polypropylene. *Macromolecules*, **18**:1206–14.
Watson, G. N. (1939). Three triple integrals. *Quarterly Journal of Mathematics, Oxford Series (1)*, **10**:266.
Weiss, G. H. (1994). *Aspects and Applications of the Random Walk*. Random materials and processes. North-Holland, Amsterdam.
Weiss, G. H. and Rubin, R. J. (1976). The theory of ordered spans of unrestricted random walks. *Journal of Statistical Physics*, **14**(4):333–50.
Widom, B. (1965). Surface tension and molecular correlations near the critical point. *Journal of Chemical Physics*, **43**:3892–905.
Wilf, H. S. (1994). *Generatingfunctionology*. Academic Press, Boston, 2nd edn. (A very nice, informal but comprehensive discussion of generating functions and their applications in a variety of contexts.)
Wilson, K. G. (1971a). Renormalization group and critical phenomena. I. Renormalization group and the Kadanoff scaling picture. *Physical Review B (Solid State)*, **4**(9):3174–83.
—(1971b). Renormalization group and critical phenomena. II. Phase-space cell analysis of critical behavior. *Physical Review B (Solid State)*, **4**(9):3184–205.
Wilson, K. G. and Kogut, J. (1974). The renormalization group and the epsilon expansion. *Physics Reports. Physics Letters Section C*, **12**(2):75–200.
Ziman, J. M. (1979). *Principles of the Theory of Solids*. Cambridge University Press, Cambridge, 1st paperback edn.
Zinn-Justin, J. (2002). *Quantum Field Theory and Critical Phenomena*. Clarendon Press, Oxford; Oxford University Press, New York, 4th edn.

Index

Printed in the United States
By Bookmasters